Introductory Biology

Introductory Biology

Edited by
Jerry Thorpe

Larsen & Keller
www.larsen-keller.com

Introductory Biology
Edited by Jerry Thorpe
ISBN: 978-1-63549-300-9 (Hardback)

© 2017 Larsen & Keller

⊟ Larsen & Keller

Published by Larsen and Keller Education,
5 Penn Plaza,
19th Floor,
New York, NY 10001, USA

Cataloging-in-Publication Data

Introductory biology / edited by Jerry Thorpe.
    p. cm.
Includes bibliographical references and index.
ISBN 978-1-63549-300-9
1. Biology. 2. Botany. 3. Zoology. I. Thorpe, Jerry.
QH303 .I58 2017
570--dc23

The publisher's policy is to use permanent paper from mills that operate a sustainable forestry policy. Furthermore, the publisher ensures that the text paper and cover boards used have met acceptable environmental accreditation standards.

Printed and bound in the United States of America.

For more information regarding Larsen and Keller Education and its products, please visit the publisher's website www.larsen-keller.com

Table of Contents

Permissions

Index

Preface

Biology is the branch of science that studies the structure, habitat, surroundings, evolution, identification, taxonomy, and functions of living organisms and the biological components of life on earth. It is a vast subject, which covers subfields like botany, molecular biology, evolutionary biology, biochemistry, etc. The aim of this textbook is to familiarize students with the basic and most fundamental concepts of this field. This book is a valuable compilation of topics, ranging from the theories to the practices in the field of biology. Some of the diverse topics covered in it address the varied branches that fall under this category. This textbook is a complete source of knowledge on the present status of this important field.

To facilitate a deeper understanding of the contents of this book a short introduction of every chapter is written below:

Chapter 1- Biology is the study of life and of living organisms. It studies the functions, the life span, growth and the evolution of these organisms. It has a number of branches and subdisciplines. This chapter will provide an integrated understanding of biology.

Chapter 2- The main branches of biology are anatomy, biomechanics, evolutionary biology, astrobiology, cryobiology, marine biology and mycology. The study of the form of animals or plants or of any living organism is known as anatomy whereas astrobiology is the study of evolution and of the future of the universe. This text is a compilation of the various branches of biology that forms an integral part of the broader subject matter.

Chapter 3- The process by which organisms/humans are born is known as reproduction. In asexual reproduction the offspring is reproduced without the participation of another organism. The types of reproduction explained are budding, vegetative reproduction, fission and apomixes. This section on reproduction offers an insightful focus, keeping in mind the theme of the chapter.

Chapter 4- Cells have different structures and different functions, the study of these is known as cell biology. Some of the topics explained in this chapter are cell membrane, cytoskeleton, organelle, cell growth, cell cycle, mitosis and meiosis. This section is an overview of the subject matter incorporating all the major aspects of cell biology.

Chapter 5- Botany is the study of plants. Plant anatomy is the study of the structure of plants. The themes covered in this chapter are plant breeding, plant propagation, plant morphology, plant ecology, plant pathology etc. The chapter serves as a source to understand the major categories related to plant biology.

Chapter 6- Zoology is the branch of biology that studies animals, their structures, evolution and habits. Zoology includes both living animals and the extinct ones as well. Animal science, embryology, ethology and behavioral ecology are the topics explained in the following section. The text helps the reader in developing an in-depth understanding of the subject.

I owe the completion of this book to the never-ending support of my family, who supported me throughout the project.

Editor

Introduction to Biology

Biology is the study of life and of living organisms. It studies the functions, the life span, growth and the evolution of these organisms. It has a number of branches and subdisciplines. This chapter will provide an integrated understanding of biology.

Biology

Biology is a natural science concerned with the study of life and living organisms, including their structure, function, growth, evolution, distribution, identification and taxonomy. Modern biology is a vast and eclectic field, composed of many branches and subdisciplines. However, despite the broad scope of biology, there are certain general and unifying concepts within it that govern all study and research, consolidating it into single, coherent field. In general, biology recognizes the cell as the basic unit of life, genes as the basic unit of heredity, and evolution as the engine that propels the synthesis and creation of new species. It is also understood today that all the organisms survive by consuming and transforming energy and by regulating their internal environment to maintain a stable and vital condition known as homeostasis.

Sub-disciplines of biology are defined by the scale at which organisms are studied, the kinds of organisms studied, and the methods used to study them: biochemistry examines the rudimentary chemistry of life; molecular biology studies the complex interactions among biological molecules; botany studies the biology of plants; cellular biology examines the basic building-block of all life, the cell; physiology examines the physical and chemical functions of tissues, organs, and organ systems of an organism; evolutionary biology examines the processes that produced the diversity of life; and ecology examines how organisms interact in their environment.

History

A Diagram of a fly from Robert Hooke's innovative Micrographia, 1665

The Latin-language form of the term first appeared in 1736 when Swedish scientist Carl Linnaeus (Carl von Linné) used *biologi* in his *Bibliotheca botanica*. It was used again in 1766 in a work entitled *Philosophiae naturalis sive physicae: tomus III, continens geologian, biologian, phytologian generalis*, by Michael Christoph Hanov, a disciple of Christian Wolff. The first German use, *Biologie*, was in a 1771 translation of Linnaeus' work. In 1797, Theodor Georg August Roose used the term in the preface of a book, *Grundzüge der Lehre van der Lebenskraft*. Karl Friedrich Burdach used the term in 1800 in a more restricted sense of the study of human beings from a morphological, physiological and psychological perspective (*Propädeutik zum Studien der gesammten Heilkunst*). The term came into its modern usage with the six-volume treatise *Biologie, oder Philosophie der lebenden Natur* (1802–22) by Gottfried Reinhold Treviranus, who announced:

Ernst Haeckel's Tree of Life (1879)

> The objects of our research will be the different forms and manifestations of life, the conditions and laws under which these phenomena occur, and the causes through which they have been effected. The science that concerns itself with these objects we will indicate by the name biology [Biologie] or the doctrine of life [Lebenslehre].

Although modern biology is a relatively recent development, sciences related to and included within it have been studied since ancient times. Natural philosophy was studied as early as the ancient civilizations of Mesopotamia, Egypt, the Indian subcontinent, and China. However, the origins of modern biology and its approach to the study of nature are most often traced back to ancient Greece. While the formal study of medicine dates back to Hippocrates (ca. 460 BC – ca. 370 BC), it was Aristotle (384 BC – 322 BC) who contributed most extensively to the development of biology. Especially important are his History of Animals and other works where he showed naturalist leanings, and later more empirical works that focused on biological causation and the diversity of life. Aristotle's successor at the Lyceum, Theophrastus, wrote a series of books on botany that survived as the most important contribution of antiquity to the plant sciences, even into the Middle Ages.

Scholars of the medieval Islamic world who wrote on biology included al-Jahiz (781–869), Al-Dīnawarī (828–896), who wrote on botany, and Rhazes (865–925) who wrote on anatomy and physiology. Medicine was especially well studied by Islamic scholars working in Greek philosopher traditions, while natural history drew heavily on Aristotelian thought, especially in upholding a fixed hierarchy of life.

Biology began to quickly develop and grow with Anton van Leeuwenhoek's dramatic improvement of the microscope. It was then that scholars discovered spermatozoa, bacteria, infusoria and the diversity of microscopic life. Investigations by Jan Swammerdam led to new interest in entomology and helped to develop the basic techniques of microscopic dissection and staining.

Advances in microscopy also had a profound impact on biological thinking. In the early 19th century, a number of biologists pointed to the central importance of the cell. Then, in 1838, Schleiden and Schwann began promoting the now universal ideas that (1) the basic unit of organisms is the cell and (2) that individual cells have all the characteristics of life, although they opposed the idea that (3) all cells come from the division of other cells. Thanks to the work of Robert Remak and Rudolf Virchow, however, by the 1860s most biologists accepted all three tenets of what came to be known as cell theory.

Meanwhile, taxonomy and classification became the focus of natural historians. Carl Linnaeus published a basic taxonomy for the natural world in 1735 (variations of which have been in use ever since), and in the 1750s introduced scientific names for all his species. Georges-Louis Leclerc, Comte de Buffon, treated species as artificial categories and living forms as malleable—even suggesting the possibility of common descent. Though he was opposed to evolution, Buffon is a key figure in the history of evolutionary thought; his work influenced the evolutionary theories of both Lamarck and Darwin.

Serious evolutionary thinking originated with the works of Jean-Baptiste Lamarck, who was the first to present a coherent theory of evolution. He posited that evolution was the result of environmental stress on properties of animals, meaning that the more frequently and rigorously an organ was used, the more complex and efficient it would become, thus adapting the animal to its environment. Lamarck believed that these acquired traits could then be passed on to the animal's offspring, who would further develop and perfect them. However, it was the British naturalist Charles Darwin, combining the biogeographical approach of Humboldt, the uniformitarian geology of Lyell, Malthus's writings on population growth, and his own morphological expertise and extensive natural observations, who forged a more successful evolutionary theory based on natural selection; similar reasoning and evidence led Alfred Russel Wallace to independently reach the same conclusions. Although it was the subject of controversy (which continues to this day), Darwin's theory quickly spread through the scientific community and soon became a central axiom of the rapidly developing science of biology.

The discovery of the physical representation of heredity came along with evolutionary principles and population genetics. In the 1940s and early 1950s, experiments pointed to DNA as the component of chromosomes that held the trait-carrying units that had become known as genes. A focus on new kinds of model organisms such as viruses and bacteria, along with the discovery of the double helical structure of DNA in 1953, marked the transition to the era of molecular genetics. From the 1950s to present times, biology has been vastly extended in the molecular domain. The genetic code was cracked by Har Gobind Khorana, Robert W. Holley and Marshall Warren Nirenberg after DNA was understood to contain codons. Finally, the Human Genome Project was launched in 1990 with the goal of mapping the general human genome. This project was essentially completed in 2003, with further analysis still being published. The Human Genome Project was the first step in a globalized effort to incorporate accumulated knowledge of biology into a functional, molecular definition of the human body and the bodies of other organisms.

Foundations of Modern Biology

Cell Theory

Human cancer cells with nuclei (specifically the DNA) stained blue. The central and rightmost cell are in interphase, so the entire nuclei are labeled. The cell on the left is going through mitosis and its DNA has condensed.

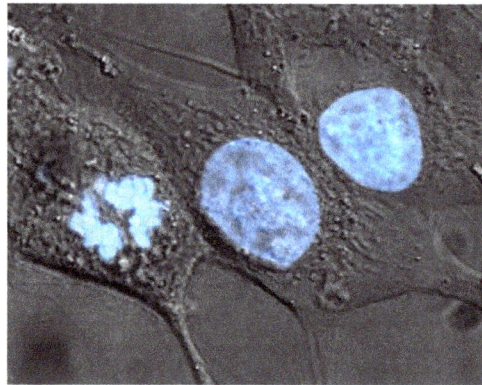

Cell theory states that the cell is the fundamental unit of life, and that all living things are composed of one or more cells or the secreted products of those cells (e.g. shells, hairs and nails etc.). All cells arise from other cells through cell division. In multicellular organisms, every cell in the organism's body derives ultimately from a single cell in a fertilized egg. The cell is also considered to be the basic unit in many pathological processes. In addition, the phenomenon of energy flow occurs in cells in processes that are part of the function known as metabolism. Finally, cells contain hereditary information (DNA), which is passed from cell to cell during cell division.

Evolution

A central organizing concept in biology is that life changes and develops through evolution, and that all life-forms known have a common origin. The theory of evolution postulates that all organisms on the Earth, both living and extinct, have descended from a common ancestor or an ancestral gene pool. This last universal common ancestor of all organisms is believed to have appeared about 3.5 billion years ago. Biologists generally regard the universality and ubiquity of the genetic code as definitive evidence in favor of the theory of universal common descent for all bacteria, archaea, and eukaryotes.

Introduced into the scientific lexicon by Jean-Baptiste de Lamarck in 1809, evolution was established by Charles Darwin fifty years later as a viable scientific model when he articulated its driving force: natural selection. (Alfred Russel Wallace is recognized as the co-discoverer of this concept as he helped research and experiment with the concept of evolution.) Evolution is now used to explain the great variations of life found on Earth.

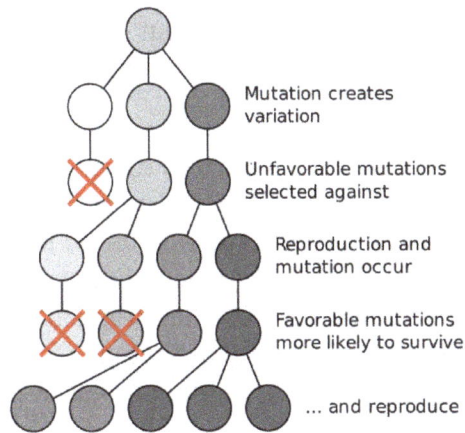

Natural selection of a population for dark coloration.

Darwin theorized that species and breeds developed through the processes of natural selection and artificial selection or selective breeding. Genetic drift was embraced as an additional mechanism of evolutionary development in the modern synthesis of the theory.

The evolutionary history of the species—which describes the characteristics of the various species from which it descended—together with its genealogical relationship to every other species is known as its phylogeny. Widely varied approaches to biology generate information about phylogeny. These include the comparisons of DNA sequences conducted within molecular biology or genomics, and comparisons of fossils or other records of ancient organisms in paleontology. Biologists organize and analyze evolutionary relationships through various methods, including phylogenetics, phenetics, and cladistics.

Genetics

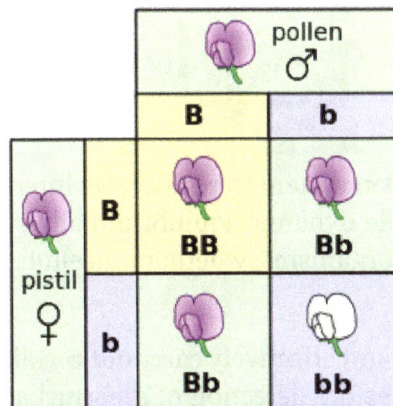

A Punnett square depicting a cross between two pea plants heterozygous for purple (B) and white (b) blossoms

Genes are the primary units of inheritance in all organisms. A gene is a unit of heredity and corresponds to a region of DNA that influences the form or function of an organism in specific ways. All organisms, from bacteria to animals, share the same basic machinery that copies and translates DNA into proteins. Cells transcribe a DNA gene into an RNA version of the gene, and a ribosome then translates the RNA into a protein, a sequence of amino acids. The translation code from RNA codon to amino acid is the same for most organisms, but slightly different for some. For example, a sequence of DNA that codes for insulin in humans also codes for insulin when inserted into other organisms, such as plants.

DNA usually occurs as linear chromosomes in eukaryotes, and circular chromosomes in prokaryotes. A chromosome is an organized structure consisting of DNA and histones. The set of chromosomes in a cell and any other hereditary information found in the mitochondria, chloroplasts, or other locations is collectively known as its genome. In eukaryotes, genomic DNA is located in the cell nucleus, along with small amounts in mitochondria and chloroplasts. In prokaryotes, the DNA is held within an irregularly shaped body in the cytoplasm called the nucleoid. The genetic information in a genome is held within genes, and the complete assemblage of this information in an organism is called its genotype.

Homeostasis

The hypothalamus secretes CRH, which directs the pituitary gland to secrete ACTH. In turn, ACTH directs the adrenal cortex to secrete glucocorticoids, such as cortisol. The GCs then reduce the rate of secretion by the hypothalamus and the pituitary gland once a sufficient amount of GCs has been released.

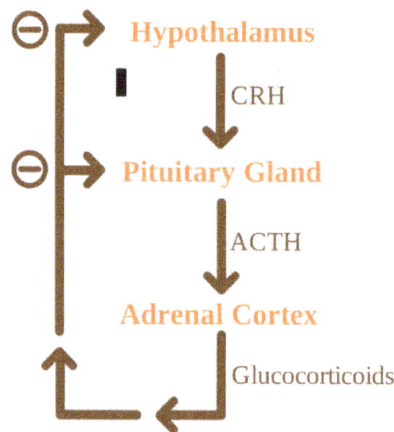

Homeostasis is the ability of an open system to regulate its internal environment to maintain stable conditions by means of multiple dynamic equilibrium adjustments controlled by interrelated regulation mechanisms. All living organisms, whether unicellular or multicellular, exhibit homeostasis.

To maintain dynamic equilibrium and effectively carry out certain functions, a system must detect and respond to perturbations. After the detection of a perturbation, a biological system normally responds through negative feedback. This means stabilizing conditions by either reducing or increasing the activity of an organ or system. One example is the release of glucagon when sugar levels are too low.

Energy and human life

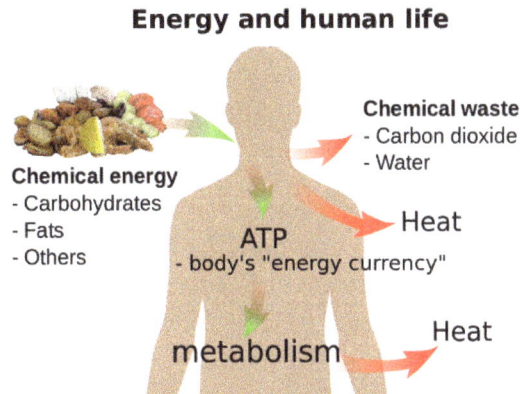

Basic overview of energy and human life.

Energy

The survival of a living organism depends on the continuous input of energy. Chemical reactions that are responsible for its structure and function are tuned to extract energy from substances that act as its food and transform them to help form new cells and sustain them. In this process, molecules of chemical substances that constitute food play two roles; first, they contain energy that can be transformed for biological chemical reactions; second, they develop new molecular structures made up of biomolecules.

The organisms responsible for the introduction of energy into an ecosystem are known as producers or autotrophs. Nearly all of these organisms originally draw energy from the sun. Plants and other phototrophs use solar energy via a process known as photosynthesis to convert raw materials into organic molecules, such as ATP, whose bonds can be broken to release energy. A few ecosystems, however, depend entirely on energy extracted by chemotrophs from methane, sulfides, or other non-luminal energy sources.

Some of the captured energy is used to produce biomass to sustain life and provide energy for growth and development. The majority of the rest of this energy is lost as heat and waste molecules. The most important processes for converting the energy trapped in chemical substances into energy useful to sustain life are metabolism and cellular respiration.

Study and Research

Structural

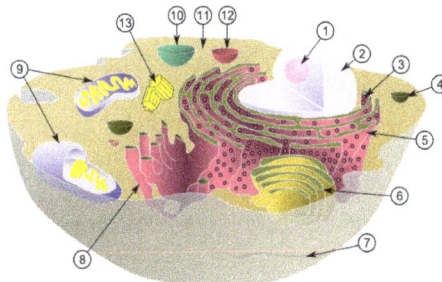

Schematic of typical animal cell depicting the various organelles and structures.

Molecular biology is the study of biology at a molecular level. This field overlaps with other areas of biology, particularly with genetics and biochemistry. Molecular biology chiefly concerns itself with understanding the interactions between the various systems of a cell, including the interrelationship of DNA, RNA, and protein synthesis and learning how these interactions are regulated.

Cell biology studies the structural and physiological properties of cells, including their behaviors, interactions, and environment. This is done on both the microscopic and molecular levels, for unicellular organisms such as bacteria, as well as the specialized cells in multicellular organisms such as humans. Understanding the structure and function of cells is fundamental to all of the biological sciences. The similarities and differences between cell types are particularly relevant to molecular biology.

Anatomy Considers The Forms of Macroscopic Structures Such as Organs and Organ Systems.

Genetics is the science of genes, heredity, and the variation of organisms. Genes encode the information necessary for synthesizing proteins, which in turn play a central role in influencing the final phenotype of the organism. In modern research, genetics provides important tools in the investigation of the function of a particular gene, or the analysis of genetic interactions. Within organisms, genetic information generally is carried in chromosomes, where it is represented in the chemical structure of particular DNA molecules.

Developmental biology studies the process by which organisms grow and develop. Originating in embryology, modern developmental biology studies the genetic control of cell growth, differentiation, and "morphogenesis," which is the process that progressively gives rise to tissues, organs, and anatomy. Model organisms for developmental biology include the round worm *Caenorhabditis elegans*, the fruit fly *Drosophila melanogaster*, the zebrafish *Danio rerio*, the mouse *Mus musculus*, and the weed *Arabidopsis thaliana*. (A model organism is a species that is extensively studied to understand particular biological phenomena, with the expectation that discoveries made in that organism provide insight into the workings of other organisms.)

Physiological

Physiology studies the mechanical, physical, and biochemical processes of living organisms by attempting to understand how all of the structures function as a whole. The theme of "structure to function" is central to biology. Physiological studies have traditionally been divided into plant physiology and animal physiology, but some principles of physiology are universal, no matter what particular organism is being studied. For example, what is learned about the physiology of yeast cells can also apply to human cells. The field of animal physiology extends the tools and methods of human physiology to non-human species. Plant physiology borrows techniques from both research fields.

Physiology studies how for example nervous, immune, endocrine, respiratory, and circulatory systems, function and interact. The study of these systems is shared with medically oriented disciplines such as neurology and immunology.

Evolutionary

Evolutionary research is concerned with the origin and descent of species, as well as their change over time, and includes scientists from many taxonomically oriented disciplines. For example, it generally involves scientists who have special training in particular organisms such as mammalogy, ornithology, botany, or herpetology, but use those organisms as systems to answer general questions about evolution.

Evolutionary biology is partly based on paleontology, which uses the fossil record to answer questions about the mode and tempo of evolution, and partly on the developments in areas such as population genetics. In the 1980s, developmental biology re-entered evolutionary biology from its initial exclusion from the modern synthesis through the study of evolutionary developmental biology. Related fields often considered part of evolutionary biology are phylogenetics, systematics, and taxonomy.

Systematic

A phylogenetic tree of all living things, based on rRNA gene data, showing the separation of the three domains bacteria, archaea, and eukaryotes as described initially by Carl Woese. Trees constructed with other genes are generally similar, although they may place some early-branching groups very differently, presumably owing to rapid rRNA evolution. The exact relationships of the three domains are still being debated.

Phylogenetic Tree of Life

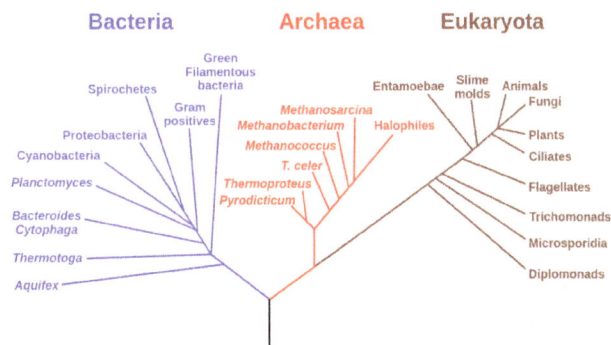

Multiple speciation events create a tree structured system of relationships between species. The role of systematics is to study these relationships and thus the differences and similarities between species and groups of species. However, systematics was an active field of research long before evolutionary thinking was common.

Traditionally, living things have been divided into five kingdoms: Monera; Protista; Fungi; Plantae; Animalia. However, many scientists now consider this five-kingdom system outdated. Modern alternative classification systems generally begin with the three-domain system: Archaea (originally Archaebacteria); Bacteria (originally Eubacteria) and Eukaryota (including protists, fungi, plants, and animals) These domains reflect whether the cells have nuclei or not, as well as differences in the chemical composition of key biomolecules such as ribosomes.

Further, each kingdom is broken down recursively until each species is separately classified. The order is: Domain; Kingdom; Phylum; Class; Order; Family; Genus; Species.

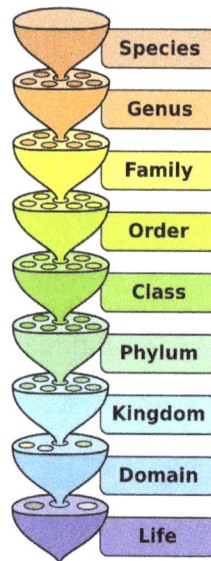

The hierarchy of biological classification's eight major taxonomic ranks. Intermediate minor rankings are not shown. This diagram uses a 3 Domains / 6 Kingdoms format

Outside of these categories, there are obligate intracellular parasites that are "on the edge of life" in terms of metabolic activity, meaning that many scientists do not actually classify these structures as alive, due to their lack of at least one or more of the fundamental functions or characteristics that define life. They are classified as viruses, viroids, prions, or satellites.

The scientific name of an organism is generated from its genus and species. For example, humans are listed as *Homo sapiens*. *Homo* is the genus, and *sapiens* the species. When writing the scientific name of an organism, it is proper to capitalize the first letter in the genus and put all of the species in lowercase. Additionally, the entire term may be italicized or underlined.

The dominant classification system is called the Linnaean taxonomy. It includes ranks and binomial nomenclature. How organisms are named is governed by international agreements such as the International Code of Nomenclature for algae, fungi, and plants (ICN), the International Code of Zoological Nomenclature (ICZN), and the International Code of Nomenclature of Bacteria (ICNB). The classification of viruses, viroids, prions, and all other sub-viral agents that demonstrate biological characteristics is conducted by the International Committee on Taxonomy of Viruses (ICTV) and is known as the International Code of Viral Classification and Nomenclature (ICVCN). However, several other viral classification systems do exist.

A merging draft, BioCode, was published in 1997 in an attempt to standardize nomenclature in these three areas, but has yet to be formally adopted. The BioCode draft has received little attention since 1997; its originally planned implementation date of January 1, 2000, has passed unnoticed. A revised BioCode that, instead of replacing the existing codes, would provide a unified context for them, was proposed in 2011. However, the International Botanical Congress of 2011 declined to consider the BioCode proposal. The ICVCN remains outside the BioCode, which does not include viral classification.

Kingdoms

Animalia – Bos primigenius taurus

Fungi – Morchella esculenta

Planta – Triticum

Ecological and Environmental

Mutual symbiosis between clownfish of the genus Amphiprion that dwell among the tentacles of tropical sea anemones. The territorial fish protects the anemone from anemone-eating fish, and in turn the stinging tentacles of the anemone protects the clown fish from its predators.

Ecology studies the distribution and abundance of living organisms, and the interactions between organisms and their environment. The habitat of an organism can be described as the local abiotic factors such as climate and ecology, in addition to the other organisms and biotic factors that share its environment. One reason that biological systems can be difficult to study is that so many different interactions with other organisms and the environment are possible, even on small scales. A microscopic bacterium in a local sugar gradient is responding to its environment as much as a lion searching for food in the African savanna. For any species, behaviors can be co-operative, competitive, parasitic, or symbiotic. Matters become more complex when two or more species interact in an ecosystem.

Ecological systems are studied at several different levels, from individuals and populations to eco-systems and the biosphere. The term population biology is often used interchangeably with pop-ulation ecology, although *population biology* is more frequently used when studying diseases, viruses, and microbes, while population ecology is more commonly used when studying plants and animals. Ecology draws on many subdisciplines.

Ethology studies animal behavior (particularly that of social animals such as primates and canids), and is sometimes considered a branch of zoology. Ethologists have been particularly concerned with the evolution of behavior and the understanding of behavior in terms of the theory of natural selection. In one sense, the first modern ethologist was Charles Darwin, whose book, *The Expression of the Emotions in Man and Animals,* influenced many etholo-gists to come.

Biogeography studies the spatial distribution of organisms on the Earth, focusing on topics like plate tectonics, climate change, dispersal and migration, and cladistics.

Basic Unresolved Problems in Biology

Despite the profound advances made over recent decades in our understanding of life's funda-mental processes, some basic problems have remained unresolved. For example, one of the major unresolved problems in biology is the primary adaptive function of sex, and particularly its key processes in eukaryotes, meiosis and homologous recombination. One view is that sex evolved pri-marily as an adaptation for increasing genetic diversity. An alternative view is that sex is an adap-tation for promoting accurate DNA repair in germ-line DNA, and that increased genetic diversity is primarily a byproduct that may be useful in the long run.

Another basic unresolved problem in biology is the biologic basis of aging. At present, there is no consensus view on the underlying cause of aging. Various competing theories are outlined in Age-ing Theories.

Life

Life is a characteristic distinguishing physical entities having biological processes, such as sig-naling and self-sustaining processes, from those that do not, either because such functions have ceased, or because they never had such functions and are classified as inanimate. Various forms of life exist such as plants, animals, fungi, protists, archaea, and bacteria. The criteria can at times be ambiguous and may or may not define viruses, viroids or potential artificial life as living. Biology is the primary science concerned with the study of life, although many other sciences are involved.

The definition of life is controversial. The current definition is that organisms maintain homeo-stasis, are composed of cells, undergo metabolism, can grow, adapt to their environment, respond to stimuli, and reproduce. However, many other biological definitions have been proposed, and there are also some borderline cases, such as viruses. Biophysicists have also proposed some defi-nitions, many being based on chemical systems. There are also some living systems theories, such as the Gaia hypothesis, the idea that the Earth is alive; the former first developed by James Grier

Miller. Another one is that life is the property of ecological systems, and yet another is the complex systems biology, a branch or subfield of mathematical biology. Some other systemic definitions includes the theory involving the darwinian dynamic, and the operator theory. However, throughout history, there have been many other theories and definitions about life such as materialism, the belief that everything is made out of matter and that life is merely a complex form of it; hylomorphism, the belief that all things are a combination of matter and form, and the form of a living thing is its soul; spontaneous generation, the belief that life repeatedly emerge from non-life; and vitalism, a discredited scientific hypothesis that living organisms possess a "life force" or "vital spark". Abiogenesis is the natural process of life arising from non-living matter, such as simple organic compounds. Life on Earth arose 3.8–4.1 billion years ago. It is widely accepted that current life on Earth descended from an RNA world, but RNA based life may not have been the first. The mechanism by which life began on Earth is unknown, although many hypotheses have been formulated, most based on the Miller–Urey experiment. In July 2016, scientists reported identifying a set of 355 genes from the Last Universal Common Ancestor (LUCA) of all organisms living on Earth.

Since appearing, life on Earth has changed its environment on a geologic time scale. To survive in most ecosystems, life can adapt and thrive in a wide range of conditions. Some organisms, called extremophiles, can thrive in physically or geochemically extreme conditions that are detrimental to most other life on Earth. Properties common to all organisms are the need for certain core chemical elements needed for biochemical functioning. Aristotle was the first person to classify organisms. Later, Carl Linnaeus introduced his system of binomial nomenclature for the classification of species. Fungi was later classified as its own kingdom. Eventually new groups of life were revealed, such as cells and microorganisms, and even non-cellular reproducing agents, such as viruses and viroids. Cells are the smallest units of life, often called the "building blocks of life". There are two kind of cells, prokaryotic and eukaryotic. Cells consist of cytoplasm enclosed within a membrane, which contains many biomolecules such as proteins and nucleic acids. Cells reproduce through a process of cell division in which the parent cell divides into two or more daughter cells.

Though only known on Earth, many believe in the existence of extraterrestrial life. Artificial life is a computer simulation of any aspect of life, which is used to examine systems related to life. Death is the permanent termination of all biological functions which sustain an organism, and as such, is the end of its life. Extinction is the process by which a group of taxa, normally a species, dies out. Fossils are the preserved remains or traces of organisms.

Definitions

It is a challenge for scientists and philosophers to define life. This is partially because life is a process, not a substance. Any definition must be general enough to both encompass all known life and any unknown life that may be different from life on Earth.

Biology

Since there is no unequivocal definition of life, most current definitions in biology are descriptive. Life is considered a characteristic of something that exhibits all or most of the following traits:

1.   Homeostasis: regulation of the internal environment to maintain a constant state; for ex-

ample, sweating to reduce temperature

2. Organization: being structurally composed of one or more cells — the basic units of life

3. Metabolism: transformation of energy by converting chemicals and energy into cellular components (anabolism) and decomposing organic matter (catabolism). Living things require energy to maintain internal organization (homeostasis) and to produce the other phenomena associated with life.

4. Growth: maintenance of a higher rate of anabolism than catabolism. A growing organism increases in size in all of its parts, rather than simply accumulating matter.

5. Adaptation: the ability to change over time in response to the environment. This ability is fundamental to the process of evolution and is determined by the organism's heredity, diet, and external factors.

6. Response to stimuli: a response can take many forms, from the contraction of a unicellular organism to external chemicals, to complex reactions involving all the senses of multicellular organisms. A response is often expressed by motion; for example, the leaves of a plant turning toward the sun (phototropism), and chemotaxis.

7. Reproduction: the ability to produce new individual organisms, either asexually from a single parent organism, or sexually from two parent organisms.

These complex processes, called physiological functions, have underlying physical and chemical bases, as well as signaling and control mechanisms that are essential to maintaining life.

Alternative Definitions

From physics perspective, living beings are thermodynamic systems with an organized molecular structure that can reproduce itself and evolve as survival dictates. Thermodynamically, life has been described as an open system, which makes use of gradients in its surroundings to create imperfect copies of itself. Hence, life is a self-sustained chemical system capable of undergoing Darwinian evolution. A major strength of this definition is that it distinguishes life by the evolutionary process rather than its chemical composition.

Others take a systemic viewpoint that does not necessarily depend on molecular chemistry. One systemic definition of life is that living things are self-organizing and autopoietic (self-producing). Variations of this definition include Stuart Kauffman's definition as an autonomous agent or a multi-agent system capable of reproducing itself or themselves, and of completing at least one thermodynamic work cycle.

Viruses

Whether or not viruses should be considered as alive is controversial. They are most often considered as just replicators rather than forms of life. They have been described as "organisms at the edge of life," since they possess genes, evolve by natural selection, and replicate by creating multiple copies of themselves through self-assembly. However, viruses do not metabolize and they require a host cell to make new products. Virus self-assembly within host cells has implications for the study of the origin of life, as it may support the hypothesis that life could have started as self-assembling organic molecules.

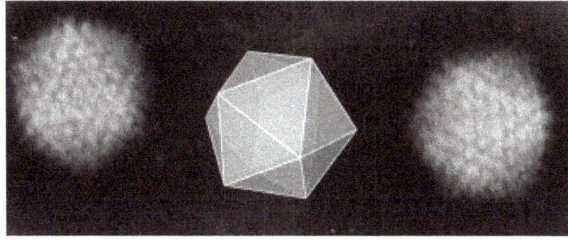
Adenovirus with icosahedral diagram

Biophysics

To reflect the minimum phenomena required, other biological definitions of life have been proposed, with many of these being based upon chemical systems. Biophysicists have commented that living things function on negative entropy. In other words, living processes can be viewed as a delay of the spontaneous diffusion or dispersion of the internal energy of biological molecules towards more potential microstates. In more detail, according to physicists such as John Bernal, Erwin Schrödinger, Eugene Wigner, and John Avery, life is a member of the class of phenomena that are open or continuous systems able to decrease their internal entropy at the expense of substances or free energy taken in from the environment and subsequently rejected in a degraded form.

Living Systems Theories

Living systems are open self-organizing living things that interact with their environment. These systems are maintained by flows of information, energy and matter.

Some scientists have proposed in the last few decades that a general living systems theory is required to explain the nature of life. Such a general theory, arising out of the ecological and biological sciences, attempts to map general principles for how all living systems work. Instead of examining phenomena by attempting to break things down into components, a general living systems theory explores phenomena in terms of dynamic patterns of the relationships of organisms with their environment.

Gaia Hypothesis

The idea that the Earth is alive is found in philosophy and religion, but the first scientific discussion of it was by the Scottish scientist James Hutton. In 1785, he stated that the Earth was a super-organism and that its proper study should be physiology. Hutton is considered the father of geology, but his idea of a living Earth was forgotten in the intense reductionism of the 19th century. The Gaia hypothesis, proposed in the 1960s by scientist James Lovelock, suggests that life on Earth functions as a single organism that defines and maintains environmental conditions necessary for its survival. This hypothesis served as one of the foundations of the modern Earth system science.

Nonfractionability

The first attempt at a general living systems theory for explaining the nature of life was in 1978, by American biologist James Grier Miller. Such a general theory, arising out of the ecological and biological sciences, attempts to map general principles for how all living systems work. Instead of examining phenomena by attempting to break things down into component parts, a general living systems theory explores phenomena in terms of dynamic patterns of the relationships of organisms with their environment. Robert Rosen (1991) built on this by defining a system component as "a unit of organization; a part with a function, i.e., a definite relation between part and whole." From this and other starting concepts, he developed a "relational theory of systems" that attempts to explain the special properties of life. Specifically, he identified the "nonfractionability of components in an organism" as the fundamental difference between living systems and "biological machines."

Life as A Property of Ecosystems

A systems view of life treats environmental fluxes and biological fluxes together as a "reciprocity of influence", and a reciprocal relation with environment is arguably as important for understanding life as it is for understanding ecosystems. As Harold J. Morowitz (1992) explains it, life is a property of an ecological system rather than a single organism or species. He argues that an ecosystemic definition of life is preferable to a strictly biochemical or physical one. Robert Ulanowicz (2009) highlights mutualism as the key to understand the systemic, order-generating behavior of life and ecosystems.

Complex Systems Biology

Complex systems biology (CSB) is a field of science that studies the emergence of complexity in functional organisms from the viewpoint of dynamic systems theory. The latter is often called also systems biology and aims to understand the most fundamental aspects of life. A closely related approach to CSB and systems biology, called relational biology, is concerned mainly with understanding life processes in terms of the most important relations, and categories of such relations among the essential functional components of organisms; for multicellular organisms, this has been defined as "categorical biology", or a model representation of organisms as a category theory of biological relations, and also an algebraic topology of the functional organization of living organisms in terms of their dynamic, complex networks of metabolic, genetic, epigenetic processes and signaling pathways.

Darwinian Dynamic

It has also been argued that the evolution of order in living systems and certain physical systems obey a common fundamental principle termed the Darwinian dynamic. The Darwinian dynamic was formulated by first considering how macroscopic order is generated in a simple non-biological system far from thermodynamic equilibrium, and then extending consideration to short, replicating RNA molecules. The underlying order generating process for both types of system was concluded to be basically similar.

Operator Theory

Another systemic definition, called the Operator theory, proposes that 'life is a general term for the presence of the typical closures found in organisms; the typical closures are a membrane and an autocatalytic set in the cell', and also proposes that an organism is 'any system with an organisation that complies with an operator type that is at least as complex as the cell. Life can also be modeled as a network of inferior negative feedbacks of regulatory mechanisms subordinated to a superior positive feedback formed by the potential of expansion and reproduction.

History of Study

Materialism

Some of the earliest theories of life were materialist, holding that all that exists is matter, and that life is merely a complex form or arrangement of matter. Empedocles (430 BC) argued that everything in the universe is made up of a combination of four eternal "elements" or "roots of all": earth, water, air, and fire. All change is explained by the arrangement and rearrangement of these four elements. The various forms of life are caused by an appropriate mixture of elements.

Herds of zebra and impala gathering on the Maasai Mara plain

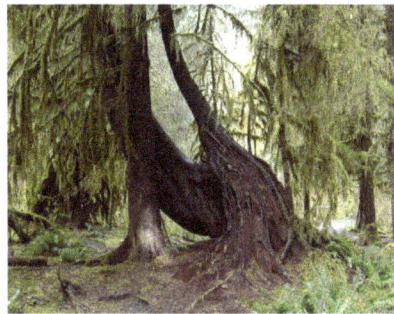
Plant growth in the Hoh Rainforest

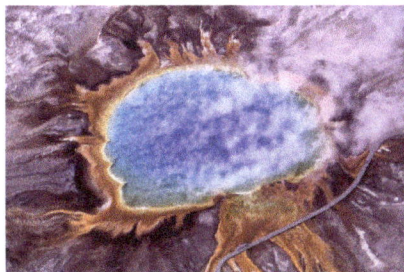
An aerial photo of microbial mats around the Grand Prismatic Spring of Yellowstone National Park

Democritus (460 BC) thought that the essential characteristic of life is having a soul (*psyche*). Like other ancient writers, he was attempting to explain what makes something a *living* thing. His explanation was that fiery atoms make a soul in exactly the same way atoms and void account for any other thing. He elaborates on fire because of the apparent connection between life and heat, and because fire moves.

Plato's world of eternal and unchanging Forms, imperfectly represented in matter by a divine Artisan, contrasts sharply with the various mechanistic Weltanschauungen, of which atomism was, by the fourth century at least, the most prominent ... This debate persisted throughout the ancient world. Atomistic mechanism got a shot in the arm from Epicurus ... while the Stoics adopted a divine teleology ... The choice seems simple: either show how a structured, regular world could arise out of undirected processes, or inject intelligence into the system.

— R. J. Hankinson, Cause and Explanation in Ancient Greek Thought

The mechanistic materialism that originated in ancient Greece was revived and revised by the French philosopher René Descartes, who held that animals and humans were assemblages of parts that together functioned as a machine. In the 19th century, the advances in cell theory in biological science encouraged this view. The evolutionary theory of Charles Darwin (1859) is a mechanistic explanation for the origin of species by means of natural selection.

Hylomorphism

Hylomorphism is a theory first expressed by Greek philosopher Aristotle (322 BC). The application of Hylomorphism to biology was important to Aristotle, and biology is extensively covered in his extant writings. In this view, everything in the material universe has both matter and form, and the form of a living thing is its soul. There are three kinds of souls: the *vegetative soul* of plants, which causes them to grow and decay and nourish themselves, but does not cause motion and sensation; the *animal soul*, which causes animals to move and feel; and the *rational soul*, which is the source of consciousness and reasoning, which (Aristotle believed) is found only in man. Each higher soul has all of the attributes of the lower ones. Aristotle believed that while matter can exist without form, form cannot exist without matter, and that therefore the soul cannot exist without the body.

This account is consistent with teleological explanations of life, which account for phenomena in terms of purpose or goal-directedness. Thus, the whiteness of the polar bear's coat is explained by its purpose of camouflage. The direction of causality (from the future to the past) is in contradiction with the scientific evidence for natural selection, which explains the consequence in terms of a prior cause. Biological features are explained not by looking at future optimal results, but by looking at the past evolutionary history of a species, which led to the natural selection of the features in question.

Spontaneous Generation

Spontaneous generation was the belief on the ordinary formation of living organisms without descent from similar organisms. Typically, the idea was that certain forms such as fleas could arise from inanimate matter such as dust or the supposed seasonal generation of mice and insects from mud or garbage.

The theory of spontaneous generation was proposed by Aristotle, who compiled and expanded the work of prior natural philosophers and the various ancient explanations of the appearance of organisms; it held sway for two millennia. It was decisively dispelled by the experiments of Louis Pasteur in 1859, who expanded upon the investigations of predecessors (such as Francesco Redi. Disproof of the traditional ideas of spontaneous generation is no longer controversial among biologists.

Vitalism

Vitalism is the belief that the life-principle is non-material. This originated with Georg Ernst Stahl (17th century), and remained popular until the middle of the 19th century. It appealed to philosophers such as Henri Bergson, Friedrich Nietzsche, Wilhelm Dilthey, anatomists like Marie François Xavier Bichat, and chemists like Justus von Liebig. Vitalism included the idea that there was a fundamental difference between organic and inorganic material, and the belief that organic material can only be derived from living things. This was disproved in 1828, when Friedrich Wöhler prepared urea from inorganic materials. This Wöhler synthesis is considered the starting point of modern organic chemistry. It is of historical significance because for the first time an organic compound was produced in inorganic reactions.

During the 1850s, Hermann von Helmholtz, anticipated by Julius Robert von Mayer, demonstrated that no energy is lost in muscle movement, suggesting that there were no "vital forces" necessary to move a muscle. These results led to the abandonment of scientific interest in vitalistic theories, although the belief lingered on in pseudoscientific theories such as homeopathy, which interprets diseases and sickness as caused by disturbances in a hypothetical vital force or life force.

Origin

The age of the Earth is about 4.54 billion years. Evidence suggests that life on Earth has existed for at least 3.5 billion years, with the oldest physical traces of life dating back 3.7 billion years; however, some theories, such as the Late Heavy Bombardment theory, suggest that life on Earth may have started even earlier, as early as 4.1–4.4 billion years ago, but the chemistry leading to life may have begun shortly after the Big Bang, 13.8 billion years ago, during an epoch when the universe was only 10–17 million years old. All known life forms share fundamental molecular mechanisms, reflecting their common descent; based on these observations, hypotheses on the origin of life attempt to find a mechanism explaining the formation of a universal common ancestor, from simple organic molecules via pre-cellular life to protocells and metabolism. Models have been divided into "genes-first" and "metabolism-first" categories, but a recent trend is the emergence of hybrid models that combine both categories.

There is no current scientific consensus as to how life originated. However, most accepted scientific models build on the Miller–Urey experiment, and the work of Sidney Fox, which shows that conditions on the primitive Earth favored chemical reactions that synthesize amino acids and other organic compounds from inorganic precursors, and phospholipids spontaneously forming lipid bilayers, the basic structure of a cell membrane.

Living organisms synthesize proteins, which are polymers of amino acids using instructions encoded by deoxyribonucleic acid (DNA). Protein synthesis entails intermediary ribonucleic acid (RNA) polymers. One possibility for how life began is that genes originated first, followed by proteins; the alternative being that proteins came first and then genes.

However, since genes and proteins are both required to produce the other, the problem of considering which came first is like that of the chicken or the egg. Most scientists have adopted the hypothesis that because of this, it is unlikely that genes and proteins arose independently.

Therefore, a possibility, first suggested by Francis Crick, is that the first life was based on RNA, which has the DNA-like properties of information storage and the catalytic properties of some proteins. This is called the RNA world hypothesis, and it is supported by the observation that many of the most critical components of cells (those that evolve the slowest) are composed mostly or entirely of RNA. Also, many critical cofactors (ATP, Acetyl-CoA, NADH, etc.) are either nucleotides or substances clearly related to them. The catalytic properties of RNA had not yet been demonstrated when the hypothesis was first proposed, but they were confirmed by Thomas Cech in 1986.

One issue with the RNA world hypothesis is that synthesis of RNA from simple inorganic precursors is more difficult than for other organic molecules. One reason for this is that RNA precursors are very stable and react with each other very slowly under ambient conditions, and it has also been proposed that living organisms consisted of other molecules before RNA. However, the successful synthesis of certain RNA molecules under the conditions that existed prior to life on Earth has been achieved by adding alternative precursors in a specified order with the precursor phosphate present throughout the reaction. This study makes the RNA world hypothesis more plausible.

Geological findings in 2013 showed that reactive phosphorus species (like phosphite) were in abundance in the ocean before 3.5 Ga, and that Schreibersite easily reacts with aqueous glycerol to generate phosphite and glycerol 3-phosphate. It is hypothesized that Schreibersite-containing meteorites from the Late Heavy Bombardment could have provided early reduced phosphorus, which could react with prebiotic organic molecules to form phosphorylated biomolecules, like RNA.

In 2009, experiments demonstrated Darwinian evolution of a two-component system of RNA enzymes (ribozymes) *in vitro*. The work was performed in the laboratory of Gerald Joyce, who stated, "This is the first example, outside of biology, of evolutionary adaptation in a molecular genetic system."

Prebiotic compounds may have extraterrestrial origin. NASA findings in 2011, based on studies with meteorites found on Earth, suggest DNA and RNA components (adenine, guanine and related organic molecules) may be formed in outer space.

In March 2015, NASA scientists reported that, for the first time, complex DNA and RNA organic compounds of life, including uracil, cytosine and thymine, have been formed in the laboratory under outer space conditions, using starting chemicals, such as pyrimidine, found in meteorites. Pyrimidine, like polycyclic aromatic hydrocarbons (PAHs), the most carbon-rich chemical found in the Universe, may have been formed in red giants or in interstellar dust and gas clouds, according to the scientists.

According to the panspermia hypothesis, microscopic life—distributed by meteoroids, asteroids and other small Solar System bodies—may exist throughout the universe.

Environmental Conditions

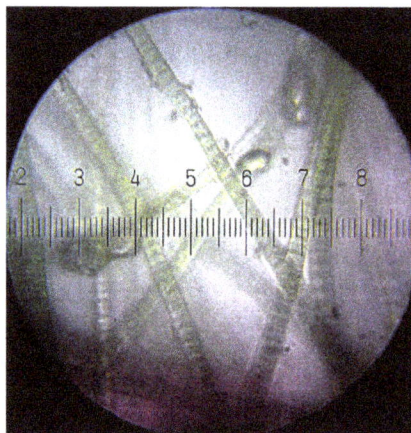

Cyanobacteria dramatically changed the composition of life forms on Earth by leading to the near-extinction of oxygen-intolerant organisms.

The diversity of life on Earth is a result of the dynamic interplay between genetic opportunity, metabolic capability, environmental challenges, and symbiosis. For most of its existence, Earth's habitable environment has been dominated by microorganisms and subjected to their metabolism and evolution. As a consequence of these microbial activities, the physical-chemical environment on Earth has been changing on a geologic time scale, thereby affecting the path of evolution of subsequent life. For example, the release of molecular oxygen by cyanobacteria as a by-product of photosynthesis induced global changes in the Earth's environment. Since oxygen was toxic to most life on Earth at the time, this posed novel evolutionary challenges, and ultimately resulted in the formation of Earth's major animal and plant species. This interplay between organisms and their environment is an inherent feature of living systems.

Biosphere

The biosphere is the global sum of all ecosystems. It can also be termed as the zone of life on Earth, a closed system (apart from solar and cosmic radiation and heat from the interior of the Earth), and largely self-regulating. By the most general biophysiological definition, the biosphere is the global ecological system integrating all living beings and their relationships, including their interaction with the elements of the lithosphere, geosphere, hydrosphere, and atmosphere. The biosphere is postulated to have evolved, beginning with a process of biopoesis (life created naturally from non-living matter, such as simple organic compounds) or biogenesis (life created from living matter), at least some 3.5 billion years ago. The earliest evidence for life on Earth includes biogenic graphite found in 3.7 billion-year-old metasedimentary rocks from Western Greenland and microbial mat fossils found in 3.48 billion-year-old sandstone from Western Australia. More recently, in 2015, "remains of biotic life" were found in 4.1 billion-year-old rocks in Western Australia. According to one of the researchers, "If life arose relatively quickly on Earth ... then it could be common in the universe."

In a general sense, biospheres are any closed, self-regulating systems containing ecosystems. This includes artificial biospheres such as Biosphere 2 and BIOS-3, and potentially ones on other planets or moons.

Range of Tolerance

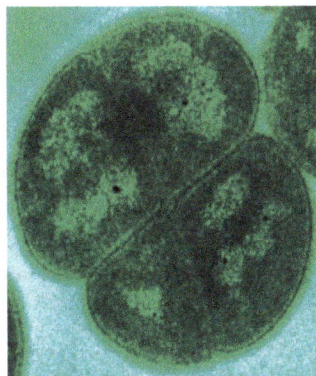

Deinococcus radiodurans is an extremophile that can resist extremes of cold, dehydration, vacuum, acid, and radiation exposure.

The inert components of an ecosystem are the physical and chemical factors necessary for life — energy (sunlight or chemical energy), water, temperature, atmosphere, gravity, nutrients, and ultraviolet solar radiation protection. In most ecosystems, the conditions vary during the day and from one season to the next. To live in most ecosystems, then, organisms must be able to survive a range of conditions, called the "range of tolerance". Outside that are the "zones of physiological stress", where the survival and reproduction are possible but not optimal. Beyond these zones are the "zones of intolerance", where survival and reproduction of that organism is unlikely or impossible. Organisms that have a wide range of tolerance are more widely distributed than organisms with a narrow range of tolerance.

Extremophiles

To survive, selected microorganisms can assume forms that enable them to withstand freezing, complete desiccation, starvation, high levels of radiation exposure, and other physical or chemical challenges. These microorganisms may survive exposure to such conditions for weeks, months, years, or even centuries. Extremophiles are microbial life forms that thrive outside the ranges where life is commonly found. They excel at exploiting uncommon sources of energy. While all organisms are composed of nearly identical molecules, evolution has enabled such microbes to cope with this wide range of physical and chemical conditions. Characterization of the structure and metabolic diversity of microbial communities in such extreme environments is ongoing.

Microbial life forms thrive even in the Mariana Trench, the deepest spot on the Earth. Microbes also thrive inside rocks up to 1900 feet below the sea floor under 8500 feet of ocean.

Investigation of the tenacity and versatility of life on Earth, as well as an understanding of the molecular systems that some organisms utilize to survive such extremes, is important for the search for life beyond Earth. For example, lichen could survive for a month in a simulated Martian environment.

Chemical Elements

All life forms require certain core chemical elements needed for biochemical functioning. These include carbon, hydrogen, nitrogen, oxygen, phosphorus, and sulfur—the elemental macronutrients for all organisms—often represented by the acronym CHNOPS. Together these make up nucleic acids, proteins and lipids, the bulk of living matter. Five of these six elements comprise the chemical components of DNA, the exception being sulfur. The latter is a component of the amino acids cysteine and methionine. The most biologically abundant of these elements is carbon, which has the desirable attribute of forming multiple, stable covalent bonds. This allows carbon-based (organic) molecules to form an immense variety of chemical arrangements. Alternative hypothetical types of biochemistry have been proposed that eliminate one or more of these elements, swap out an element for one not on the list, or change required chiralities or other chemical properties.

DNA

Deoxyribonucleic acid is a molecule that carries most of the genetic instructions used in the growth, development, functioning and reproduction of all known living organisms and many viruses. DNA and RNA are nucleic acids; alongside proteins and complex carbohydrates, they are one of the three major types of macromolecule that are essential for all known forms of life. Most DNA molecules consist of two biopolymer strands coiled around each other to form a double helix. The two DNA strands are known as polynucleotides since they are composed of simpler units called nucleotides. Each nucleotide is composed of a nitrogen-containing nucleobase—either cytosine (C), guanine (G), adenine (A), or thymine (T)—as well as a sugar called deoxyribose and a phosphate group. The nucleotides are joined to one another in a chain by covalent bonds between the sugar of one nucleotide and the phosphate of the next, resulting in an alternating sugar-phosphate backbone. According to base pairing rules (A with T, and C with G), hydrogen bonds bind the nitrogenous bases of the two separate polynucleotide strands to make double-stranded DNA. The total amount of related DNA base pairs on Earth is estimated at 5.0×10^{37}, and weighs 50 billion tonnes. In comparison, the total mass of the biosphere has been estimated to be as much as 4 TtC (trillion tons of carbon).

DNA stores biological information. The DNA backbone is resistant to cleavage, and both strands of the double-stranded structure store the same biological information. Biological information is replicated as the two strands are separated. A significant portion of DNA (more than 98% for humans) is non-coding, meaning that these sections do not serve as patterns for protein sequences.

The two strands of DNA run in opposite directions to each other and are therefore anti-parallel. Attached to each sugar is one of four types of nucleobases (informally, *bases*). It is the sequence of these four nucleobases along the backbone that encodes biological information. Under the genetic code, RNA strands are translated to specify the sequence of amino acids within proteins. These RNA strands are initially created using DNA strands as a template in a process called transcription.

Within cells, DNA is organized into long structures called chromosomes. During cell division these chromosomes are duplicated in the process of DNA replication, providing each cell its own complete set of chromosomes. Eukaryotic organisms (animals, plants, fungi, and protists) store most of their DNA inside the cell nucleus and some of their DNA in organelles, such as mitochondria or chloroplasts. In contrast, prokaryotes (bacteria and archaea) store their DNA only in the cytoplasm. Within the chromosomes, chromatin proteins such as histones compact and organize DNA. These compact structures guide the interactions between DNA and other proteins, helping control which parts of the DNA are transcribed.

DNA was first isolated by Friedrich Miescher in 1869. Its molecular structure was identified by James Watson and Francis Crick in 1953, whose model-building efforts were guided by X-ray diffraction data acquired by Rosalind Franklin.

Classification

Life is usually classified by eight levels of taxa—domains, kingdoms, phyla, class, order, family, genus, and species. In May 2016, scientists reported that 1 trillion species are estimated to be on Earth currently with only one-thousandth of one percent described.

The first known attempt to classify organisms was conducted by the Greek philosopher Aristotle (384–322 BC), who classified all living organisms known at that time as either a plant or an animal, based mainly on their ability to move. He also distinguished animals with blood from animals without blood (or at least without red blood), which can be compared with the concepts of vertebrates and invertebrates respectively, and divided the blooded animals into five groups: viviparous quadrupeds (mammals), oviparous quadrupeds (reptiles and amphibians), birds, fishes and whales. The bloodless animals were also divided into five groups: cephalopods, crustaceans, insects (which included the spiders, scorpions, and centipedes, in addition to what we define as insects today), shelled animals (such as most molluscs and echinoderms) and "zoophytes". Though Aristotle's work in zoology was not without errors, it was the grandest biological synthesis of the time and remained the ultimate authority for many centuries after his death.

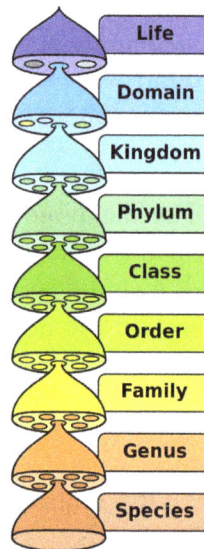

The hierarchy of biological classification's eight major taxonomic ranks. Life is divided into domains, which are subdivided into further groups. Intermediate minor rankings are not shown.

The exploration of the American continent revealed large numbers of new plants and animals that needed descriptions and classification. In the latter part of the 16th century and the beginning of the 17th, careful study of animals commenced and was gradually extended until it formed a sufficient body of knowledge to serve as an anatomical basis for classification. In the late 1740s, Carl Linnaeus introduced his system of binomial nomenclature for the classification of species. Linnaeus attempted to improve the composition and reduce the length of the previously used many-worded names by abolishing unnecessary rhetoric, introducing new descriptive terms and precisely defining their meaning.

The fungi were originally treated as plants. For a short period Linnaeus had classified them in the taxon Vermes in Animalia, but later placed them back in Plantae. Copeland classified the Fungi in his Protoctista, thus partially avoiding the problem but acknowledging their special status. The problem was eventually solved by Whittaker, when he gave them their own kingdom in his five-kingdom system. Evolutionary history shows that the fungi are more closely related to animals than to plants.

As new discoveries enabled detailed study of cells and microorganisms, new groups of life were revealed, and the fields of cell biology and microbiology were created. These new organisms were originally described separately in protozoa as animals and protophyta/thallophyta as plants, but were united by Haeckel in the kingdom Protista; later, the prokaryotes were split off in the kingdom Monera, which would eventually be divided into two separate groups, the Bacteria and the Archaea. This led to the six-kingdom system and eventually to the current three-domain system, which is based on evolutionary relationships. However, the classification of eukaryotes, especially of protists, is still controversial.

As microbiology, molecular biology and virology developed, non-cellular reproducing agents were discovered, such as viruses and viroids. Whether these are considered alive has been a matter of debate; viruses lack characteristics of life such as cell membranes, metabolism and the ability to grow or respond to their environments. Viruses can still be classed into "species" based on their biology and genetics, but many aspects of such a classification remain controversial.

In the 1960s a trend called cladistics emerged, arranging taxa based on clades in an evolutionary or phylogenetic tree.

Linnaeus 1735	Haeckel 1866	Chatton 1925	Copeland 1938	Whittaker 1969	Woese et al. 1990	Cavalier-Smith 1998
2 kingdoms	3 kingdoms	2 empires	4 kingdoms	5 kingdoms	3 domains	6 kingdoms
(not treated)	Protista	Prokaryota	Monera	Monera	Bacteria / Archaea	Bacteria
		Eukaryota	Protoctista	Protista	Eucarya	Protozoa / Chromista
Vegetabilia	Plantae		Plantae	Plantae		Plantae
				Fungi		Fungi
Animalia	Animalia		Animalia	Animalia		Animalia

Cells

Cells are the basic unit of structure in every living thing, and all cells arise from pre-existing cells by division. Cell theory was formulated by Henri Dutrochet, Theodor Schwann, Rudolf Virchow and others during the early nineteenth century, and subsequently became widely accepted. The activity of an organism depends on the total activity of its cells, with energy flow occurring within and between them. Cells contain hereditary information that is carried forward as a genetic code during cell division.

There are two primary types of cells. Prokaryotes lack a nucleus and other membrane-bound organelles, although they have circular DNA and ribosomes. Bacteria and Archaea are two domains of prokaryotes. The other primary type of cells are the eukaryotes, which have distinct nuclei bound by a nuclear membrane and membrane-bound organelles, including mitochondria, chloroplasts, lysosomes, rough and smooth endoplasmic reticulum, and vacuoles. In addition, they possess organized chromosomes that store genetic material. All species of large complex organisms are eukaryotes, including animals, plants and fungi, though most species of eukaryote are protist microorganisms. The conventional model is that eukaryotes evolved from prokaryotes, with the main organelles of the eukaryotes forming through endosymbiosis between bacteria and the progenitor eukaryotic cell.

The molecular mechanisms of cell biology are based on proteins. Most of these are synthesized by the ribosomes through an enzyme-catalyzed process called protein biosynthesis. A sequence of amino acids is assembled and joined together based upon gene expression of the cell's nucleic acid. In eukaryotic cells, these proteins may then be transported and processed through the Golgi apparatus in preparation for dispatch to their destination.

Cells reproduce through a process of cell division in which the parent cell divides into two or more daughter cells. For prokaryotes, cell division occurs through a process of fission in which the DNA is replicated, then the two copies are attached to parts of the cell membrane. In eukaryotes, a more complex process of mitosis is followed. However, the end result is the same; the resulting cell copies are identical to each other and to the original cell (except for mutations), and both are capable of further division following an interphase period.

Multicellular organisms may have first evolved through the formation of colonies like cells. These cells can form group organisms through cell adhesion. The individual members of a colony are capable of surviving on their own, whereas the members of a true multi-cellular organism have developed specializations, making them dependent on the remainder of the organism for survival. Such organisms are formed clonally or from a single germ cell that is capable of forming the various specialized cells that form the adult organism. This specialization allows multicellular organisms to exploit resources more efficiently than single cells. In January 2016, scientists reported that, about 800 million years ago, a minor genetic change in a single molecule, called GK-PID, may have allowed organisms to go from a single cell organism to one of many cells.

Cells have evolved methods to perceive and respond to their microenvironment, thereby enhancing their adaptability. Cell signaling coordinates cellular activities, and hence governs the basic functions of multicellular organisms. Signaling between cells can occur through direct cell contact using juxtacrine signalling, or indirectly through the exchange of agents as in the endocrine system. In more complex organisms, coordination of activities can occur through a dedicated nervous system.

Extraterrestrial

Though life is confirmed only on Earth, many think that extraterrestrial life is not only plausible, but probable or inevitable. Other planets and moons in the Solar System and other planetary systems are being examined for evidence of having once supported simple life, and projects such as SETI are trying to detect radio transmissions from possible alien civilizations. Other locations within the Solar System that may host microbial life include the subsurface of Mars, the upper atmosphere of Venus, and subsurface oceans on some of the moons of the giant planets. Beyond the Solar System, the region around another main-sequence star that could support Earth-like life on an Earth-like planet is known as the habitable zone. The inner and outer radii of this zone vary with the luminosity of the star, as does the time interval during which the zone survives. Stars more massive than the Sun have a larger habitable zone, but remain on the main sequence for a shorter time interval. Small red dwarfs have the opposite problem, with a smaller habitable zone that is subject to higher levels of magnetic activity and the effects of tidal locking from close orbits. Hence, stars in the intermediate mass range such as the Sun may have a greater likelihood for Earth-like life to develop. The location of the star within a galaxy may also affect the likelihood of life forming. Stars in regions with a greater abundance of heavier elements that can form planets, in combination with a low rate of potentially habitat-damaging supernova events, are predicted to have a higher probability of hosting planets with complex life. The variables of the Drake equation are used to discuss the conditions in planetary systems where civilization is most likely to exist. This suggests that life could also form on other planets.

Artificial

Artificial life is a field of study that examines systems related to life, its processes, and its evolution through simulations using computer models, robotics, and biochemistry. The study of artificial life imitates traditional biology by recreating some aspects of biological phenomena. Scientists study the logic of living systems by creating artificial environments—seeking to understand the complex information processing that defines such systems. While life is, by definition, alive, artificial life is generally referred to as data confined to a digital environment and existence.

Synthetic biology is a new area of biotechnology that combines science and biological engineering. The common goal is the design and construction of new biological functions and systems not found in nature. Synthetic biology includes the broad redefinition and expansion of biotechnology, with the ultimate goals of being able to design and build engineered biological systems that process information, manipulate chemicals, fabricate materials and structures, produce energy, provide food, and maintain and enhance human health and the environment.

Death

Death is the permanent termination of all vital functions or life processes in an organism or cell. It can occur as a result of an accident, medical conditions, biological interaction, malnutrition, poisoning, senescence, or suicide. After death, the remains of an organism re-enter the biogeochemical cycle. Organisms may be consumed by a predator or a scavenger and leftover organic material may then be further decomposed by detritivores, organisms that recycle detritus, returning it to the environment for reuse in the food chain.

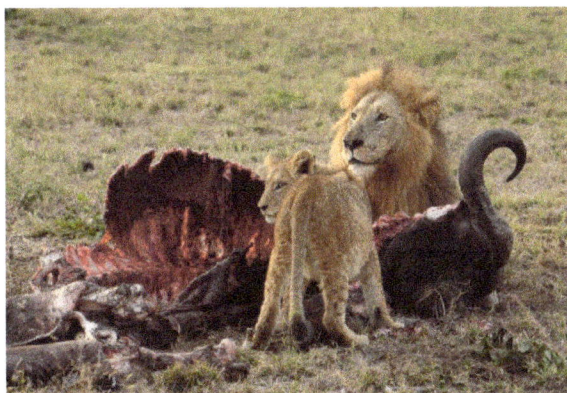

Animal corpses, like this African buffalo, are recycled by the ecosystem, providing energy and nutrients for living creatures

One of the challenges in defining death is in distinguishing it from life. Death would seem to refer to either the moment life ends, or when the state that follows life begins. However, determining when death has occurred requires drawing precise conceptual boundaries between life and death. This is problematic, however, because there is little consensus over how to define life. The nature of death has for millennia been a central concern of the world's religious traditions and of philosophical inquiry. Many religions maintain faith in either a kind of afterlife or reincarnation for the soul, or resurrection of the body at a later date.

Extinction

Extinction is the process by which a group of taxa or species dies out, reducing biodiversity. The moment of extinction is generally considered the death of the last individual of that species. Because a species' potential range may be very large, determining this moment is difficult, and is usually done retrospectively after a period of apparent absence. Species become extinct when they are no longer able to survive in changing habitat or against superior competition. In Earth's history, over 99% of all the species that have ever lived are extinct; however, mass extinctions may have accelerated evolution by providing opportunities for new groups of organisms to diversify.

Fossils

Fossils are the preserved remains or traces of animals, plants, and other organisms from the remote past. The totality of fossils, both discovered and undiscovered, and their placement in fossil-containing rock formations and sedimentary layers (strata) is known as the *fossil record*. A preserved specimen is called a fossil if it is older than the arbitrary date of 10,000 years ago. Hence, fossils range in age from the youngest at the start of the Holocene Epoch to the oldest from the Archaean Eon, up to 3.4 billion years old.

References

- Richards, Robert J. (2002). The Romantic Conception of Life: Science and Philosophy in the Age of Goethe. University of Chicago Press. ISBN 0-226-71210-9.

- Fahd, Toufic (1996). "Botany and agriculture". In Morelon, Régis; Rashed, Roshdi. Encyclopedia of the History of Arabic Science. 3. Routledge. p. 815. ISBN 0-415-12410-7.

- Magner, Lois N. (2002). A History of the Life Sciences, Revised and Expanded. CRC Press. pp. 133–144. ISBN 978-0-203-91100-6.

- Coleman, William (1977) Biology in the Nineteenth Century: Problems of Form, Function, and Transformation, Ch. 2. Cambridge University Press: New York. ISBN 0-521-29293-X

- Gould, Stephen Jay. The Structure of Evolutionary Theory. The Belknap Press of Harvard University Press: Cambridge, 2002. ISBN 0-674-00613-5. p. 187.

- Larson, Edward J. (2006). "Ch. 3". Evolution: The Remarkable History of a Scientific Theory. Random House Publishing Group. ISBN 978-1-58836-538-5.

- De Duve, Christian (2002). Life Evolving: Molecules, Mind, and Meaning. New York: Oxford University Press. p. 44. ISBN 0-19-515605-6.

- Packard, Alpheus Spring (1901). Lamarck, the founder of Evolution: his life and work with translations of his writings on organic evolution. New York: Longmans, Green. ISBN 0-405-12562-3.

- Simpson, George Gaylord (1967). The Meaning of Evolution (Second ed.). Yale University Press. ISBN 0-300-00952-6.

- Sang, James H. (2001). "Drosophila melanogaster: The Fruit Fly". In Eric C. R. Reeve. Encyclopedia of genetics. USA: Fitzroy Dearborn Publishers, I. p. 157. ISBN 978-1-884964-34-3.

- Vassiliki Betta Smocovitis (1996) Unifying Biology: the evolutionary synthesis and evolutionary biology. Princeton University Press. ISBN 0-691-03343-9.

- Neill, Campbell (1996). Biology; Fourth edition. The Benjamin/Cummings Publishing Company. p. G-21 (Glossary). ISBN 0-8053-1940-9.

- Margulis, L; Schwartz, KV (1997). Five Kingdoms: An Illustrated Guide to the Phyla of Life on Earth (3rd ed.). WH Freeman & Co. ISBN 978-0-7167-3183-2. OCLC 223623098.

- Silyn-Roberts, Heather (2000). Writing for Science and Engineering: Papers, Presentation. Oxford: Butterworth-Heinemann. p. 198. ISBN 0-7506-4636-5.

- Begon, M.; Townsend, C. R.; Harper, J. L. (2006). Ecology: From individuals to ecosystems. (4th ed.). Blackwell. ISBN 1-4051-1117-8.

- Hörandl, Elvira (2013). Meiosis and the Paradox of Sex in Nature, Meiosis, Dr. Carol Bernstein (Ed.), ISBN 978-953-51-1197-9, InTech, doi:10.5772/56542.

- Mautner, Michael N. (2000). Seeding the Universe with Life: Securing Our Cosmological Future (PDF). Washington D. C.: Legacy Books (www.amazon.com). ISBN 978-0-476-00330-9.

- Robert, Rosen (November 1991). Life Itself: A Comprehensive Inquiry into the Nature, Origin, and Fabrication of Life. ISBN 978-0-231-07565-7.

- Morowitz, Harold J. (1992). Beginnings of cellular life: metabolism recapitulates biogenesis. Yale University Press. ISBN 978-0-300-05483-5.

- Ulanowicz, Robert W.; Ulanowicz, Robert E. (2009). A third window: natural life beyond Newton and Darwin. Templeton Foundation Press. ISBN 978-1-59947-154-9.

- Michod RE. (1999) Darwinian Dynamics: Evolutionary Transitions in Fitness and Individuality. Princeton University Press, Princeton, New Jersey ISBN 978-0-691-05011-9, 9780691050119

- Hankinson, R. J. (1997). Cause and Explanation in Ancient Greek Thought. Oxford University Press. p. 125. ISBN 978-0-19-924656-4.

- Thagard, Paul (2012). The Cognitive Science of Science: Explanation, Discovery, and Conceptual Change. MIT Press. pp. 204–205. ISBN 978-0-262-01728-2.

- Stewart-Williams, Steve (2010). Darwin, God and the meaning of life: how evolutionary theory undermines everything you thought you knew of life. Cambridge University Press. pp. 193–194. ISBN 978-0-521-76278-6.

- André Brack (1998). "Introduction" (PDF). In André Brack. The Molecular Origins of Life. Cambridge University Press. p. 1. ISBN 978-0-521-56475-5. Retrieved 2009-01-07.

- Zubay, Geoffrey. Origins of Life, Second Edition: On Earth and in the Cosmos. Academic Press 2000. ISBN 978-0-12-781910-5

- Smith, John Maynard; Szathmary, Eors (1997). The Major Transitions in Evolution. Oxford Oxfordshire: Oxford University Press. ISBN 978-0-19-850294-4.

- Schwartz, Sanford (2009). C. S. Lewis on the Final Frontier: Science and the Supernatural in the Space Trilogy. Oxford University Press. p. 56. ISBN 978-0-19-988839-9.

- Rabinbach, Anson (1992). The Human Motor: Energy, Fatigue, and the Origins of Modernity. University of California Press. pp. 124–125. ISBN 978-0-520-07827-7.

Branches of Biology

The main branches of biology are anatomy, biomechanics, evolutionary biology, astrobiology, cryobiology, marine biology and mycology. The study of the form of animals or plants or of any living organism is known as anatomy whereas astrobiology is the study of evolution and of the future of the universe. This text is a compilation of the various branches of biology that forms an integral part of the broader subject matter.

Anatomy

One of the large, detailed illustrations in Andreas Vesalius's *De humani corporis fabrica* 16th century, marking the rebirth of anatomy

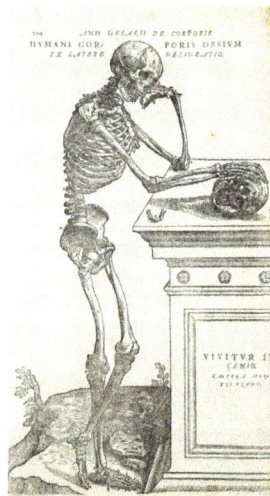

Anatomy is the branch of biology concerned with the study of the structure of organisms and their parts. Anatomy is inherently tied to embryology, comparative anatomy, evolutionary biology, and phylogeny, as these are the processes by which anatomy is generated over immediate (embryology) and long (evolution) timescales. Human anatomy is one of the basic essential sciences of medicine.

The discipline of anatomy is divided into macroscopic and microscopic anatomy. Macroscopic anatomy, or gross anatomy, is the examination of an animal's body parts using unaided eyesight. Gross anatomy also includes the branch of superficial anatomy. Microscopic anatomy involves the use of optical instruments in the study of the tissues of various structures, known as histology, and also in the study of cells.

The history of anatomy is characterized by a progressive understanding of the functions of the organs and structures of the human body. Methods have also improved dramatically, advancing from the examination of animals by dissection of carcasses and cadavers (corpses) to 20th century medical imaging techniques including X-ray, ultrasound, and magnetic resonance imaging.

Anatomy and physiology, which study (respectively) the structure and function of organisms and their parts, make a natural pair of related disciplines, and they are often studied together.

Definition

Anatomy is the scientific study of the structure of organisms including their systems, organs and tissues. It includes the appearance and position of the various parts, the materials from which they are composed, their locations and their relationships with other parts. Anatomy is quite distinct from physiology and biochemistry, which deal respectively with the functions of those parts and the chemical processes involved. For example, an anatomist is concerned with the shape, size, position, structure, blood supply and innervation of an organ such as the liver; while a physiologist is interested in the production of bile, the role of the liver in nutrition and the regulation of bodily functions.

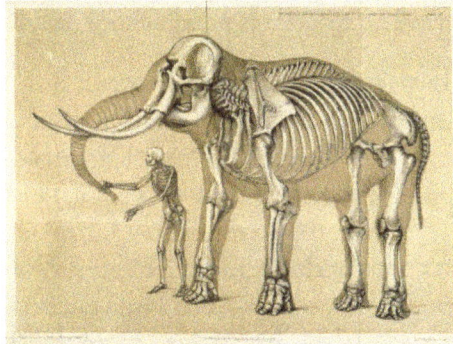

Human compared to elephant frame. Benjamin Waterhouse Hawkins, 1860

The discipline of anatomy can be subdivided into a number of branches including gross or macroscopic anatomy and microscopic anatomy. Gross anatomy is the study of structures large enough to be seen with the naked eye, and also includes superficial anatomy or surface anatomy, the study by sight of the external body features. Microscopic anatomy is the study of structures on a microscopic scale, including histology (the study of tissues), and embryology (the study of an organism in its immature condition).

Anatomy can be studied using both invasive and non-invasive methods with the goal of obtaining information about the structure and organization of organs and systems. Methods used include dissection, in which a body is opened and its organs studied, and endoscopy, in which a video camera-equipped instrument is inserted through a small incision in the body wall and used to explore the internal organs and other structures. Angiography using X-rays or magnetic resonance angiography are methods to visualize blood vessels.

The term "anatomy" is commonly taken to refer to human anatomy. However, substantially the same structures and tissues are found throughout the rest of the animal kingdom and the term also includes the anatomy of other animals. The term *zootomy* is also sometimes used to specifically refer to animals. The structure and tissues of plants are of a dissimilar nature and they are studied in plant anatomy.

Animal Tissues

The kingdom Animalia or metazoa, contains multicellular organisms that are heterotrophic and motile (although some have secondarily adopted a sessile lifestyle). Most animals have bodies differentiated into separate tissues and these animals are also known as eumetazoans. They have an internal digestive chamber, with one or two openings; the gametes are produced in multicellular sex organs, and the zygotes include a blastula stage in their embryonic development. Metazoans do not include the sponges, which have undifferentiated cells.

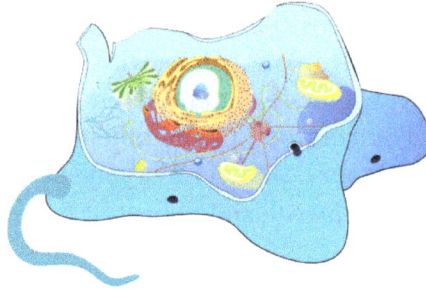

Stylized cutaway diagram of an animal cell (with flagella)

Unlike plant cells, animal cells have neither a cell wall nor chloroplasts. Vacuoles, when present, are more in number and much smaller than those in the plant cell. The body tissues are composed of numerous types of cell, including those found in muscles, nerves and skin. Each typically has a cell membrane formed of phospholipids, cytoplasm and a nucleus. All of the different cells of an animal are derived from the embryonic germ layers. Those simpler invertebrates which are formed from two germ layers of ectoderm and endoderm are called diploblastic and the more developed animals whose structures and organs are formed from three germ layers are called triploblastic. All of a triploblastic animal's tissues and organs are derived from the three germ layers of the embryo, the ectoderm, mesoderm and endoderm.

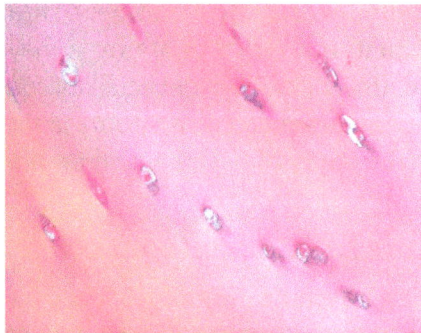

Hyaline cartilage at high magnification (H&E stain)

Animal tissues can be grouped into four basic types: connective, epithelial, muscle and nervous tissue.

Connective Tissue

Connective tissues are fibrous and made up of cells scattered among inorganic material called the extracellular matrix. Connective tissue gives shape to organs and holds them in place. The main types are loose connective tissue, adipose tissue, fibrous connective tissue, cartilage and bone. The extracellular matrix contains proteins, the chief and most abundant of which is collagen. Collagen plays a major part in organizing and maintaining tissues. The matrix can be modified to form a skeleton to support or protect the body. An exoskeleton is a thickened, rigid cuticle which is stiffened by mineralization, as in crustaceans or by the cross-linking of its proteins as in insects. An endoskeleton is internal and present in all developed animals, as well as in many of those less developed.

Epithelium

Epithelial tissue is composed of closely packed cells, bound to each other by cell adhesion molecules, with little intercellular space. Epithelial cells can be squamous (flat), cuboidal or columnar and rest on a basal lamina, the upper layer of the basement membrane, the lower layer is the reticular lamina lying next to the connective tissue in the extracellular matrix secreted by the epithelial cells. There are many different types of epithelium, modified to suit a particular function. In the respiratory tract there is a type of ciliated epithelial lining; in the small intestine there are microvilli on the epithelial lining and in the large intestine there are intestinal villi. Skin consists of an outer layer of keratinized stratified squamous epithelium that covers the exterior of the vertebrate body. Keratinocytes make up to 95% of the cells in the skin. The epithelial cells on the external surface of the body typically secrete an extracellular matrix in the form of a cuticle. In simple animals this may just be a coat of glycoproteins. In more advanced animals, many glands are formed of epithelial cells.

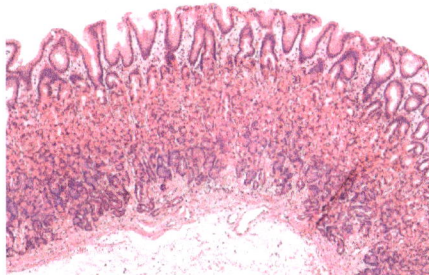

Gastric mucosa at low magnification (H&E stain)

Muscle Tissue

Muscle cells (myocytes) form the active contractile tissue of the body. Muscle tissue functions to produce force and cause motion, either locomotion or movement within internal organs. Muscle is formed of contractile filaments and is separated into three main types; smooth muscle, skeletal muscle and cardiac muscle. Smooth muscle has no striations when examined microscopically. It contracts slowly but maintains contractibility over a wide range of stretch lengths. It is found in such organs as sea anemone tentacles and the body wall of sea cucumbers. Skeletal muscle contracts rapidly but has a limited range of extension. It is found in the movement of appendages and jaws. Obliquely striated muscle is intermediate between the other two. The filaments are staggered and this is the type of muscle found in earthworms that can extend slowly or make rapid contractions. In higher animals striated muscles occur in bundles attached to bone to provide movement and are often arranged in antagonistic sets. Smooth muscle is found in the walls of the uterus, bladder, intestines, stomach, oesophagus, respiratory airways, and blood vessels. Cardiac muscle is found only in the heart, allowing it to contract and pump blood round the body.

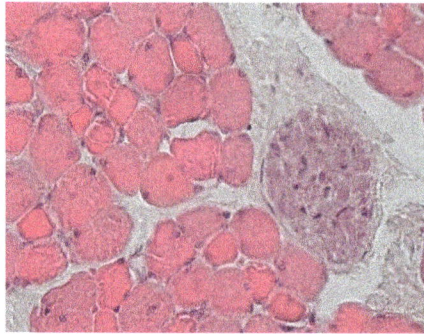

Cross section through skeletal muscle and a small nerve at high magnification (H&E stain)

Nervous Tissue

Nervous tissue is composed of many nerve cells known as neurons which transmit information. In some slow-moving radially symmetrical marine animals such as ctenophores and cnidarians (including sea anemones and jellyfish), the nerves form a nerve net, but in most animals they are organized longitudinally into bundles. In simple animals, receptor neurons in the body wall cause a local reaction to a stimulus. In more complex animals, specialized receptor cells such as chemoreceptors and photoreceptors are found in groups and send messages along neural networks to other parts of the organism. Neurons can be connected together in ganglia. In higher animals, specialized receptors are the basis of sense organs and there is a central nervous system (brain and spinal cord) and a peripheral nervous system. The latter consists of sensory nerves that transmit information from sense organs and motor nerves that influence target organs. The peripheral nervous system is divided into the somatic nervous system which conveys sensation and controls voluntary muscle, and the autonomic nervous system which involuntarily controls smooth muscle, certain glands and internal organs, including the stomach.

Vertebrate Anatomy

All vertebrates have a similar basic body plan and at some point in their lives, (mostly in the embryonic stage), share the major chordate characteristics; a stiffening rod, the notochord; a dorsal hollow tube of nervous material, the neural tube; pharyngeal arches; and a tail posterior to the anus. The spinal cord is protected by the vertebral column and is above the notochord and the gastrointestinal tract is below it. Nervous tissue is derived from the ectoderm, connective tissues are derived from mesoderm, and gut is derived from the endoderm. At the posterior end is a tail which continues the spinal cord and vertebrae but not the gut. The mouth is found at the anterior end of the animal, and the anus at the base of the tail. The defining characteristic of a vertebrate is the vertebral column, formed in the development of the segmented series of vertebrae. In most vertebrates the notochord becomes the nucleus pulposus of the intervertebral discs. However, a few vertebrates, such as the sturgeon and the coelacanth retain the notochord into adulthood. Jawed vertebrates are typified by paired appendages, fins or legs, which may be secondarily lost. The limbs of vertebrates are considered to be homologous because the same underlying skeletal structure was inherited from their last common ancestor. This is one of the arguments put forward by Charles Darwin to support his theory of evolution.

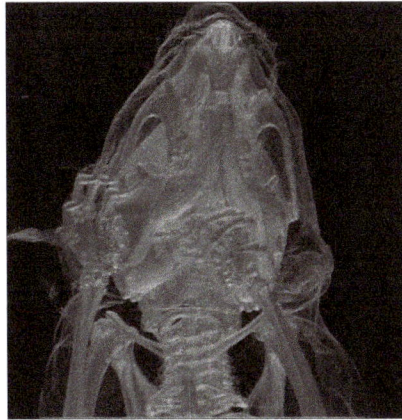
Mouse skull

Fish Anatomy

The body of a fish is divided into a head, trunk and tail, although the divisions between the three are not always externally visible. The skeleton, which forms the support structure inside the fish, is either made of cartilage, in cartilaginous fish, or bone in bony fish. The main skeletal element is the vertebral column, composed of articulating vertebrae which are lightweight yet strong. The ribs attach to the spine and there are no limbs or limb girdles. The main external features of the fish, the fins, are composed of either bony or soft spines called rays, which with the exception of the caudal fins, have no direct connection with the spine. They are supported by the muscles which compose the main part of the trunk. The heart has two chambers and pumps the blood through the respiratory surfaces of the gills and on round the body in a single circulatory loop. The eyes are adapted for seeing underwater and have only local vision. There is an inner ear but no external or middle ear. Low frequency vibrations are detected by the lateral line system of sense organs that run along the length of the sides of fish, and these respond to nearby movements and to changes in water pressure.

Cutaway diagram showing various organs of a fish

Sharks and rays are basal fish with numerous primitive anatomical features similar to those of ancient fish, including skeletons composed of cartilage. Their bodies tend to be dorso-ventrally flattened, they usually have five pairs of gill slits and a large mouth set on the underside of the head. The dermis is covered with separate dermal placoid scales. They have a cloaca into which the urinary and genital passages open, but not a swim bladder. Cartilaginous fish produce a small number of large, yolky eggs. Some species are ovoviviparous and the young develop internally but others are oviparous and the larvae develop externally in egg cases.

The bony fish lineage shows more derived anatomical traits, often with major evolutionary changes from the features of ancient fish. They have a bony skeleton, are generally laterally flattened, have five pairs of gills protected by an operculum, and a mouth at or near the tip of the snout. The dermis is covered with overlapping scales. Bony fish have a swim bladder which helps them maintain a constant depth in the water column, but not a cloaca. They mostly spawn a large number of small eggs with little yolk which they broadcast into the water column.

Amphibian Anatomy

Amphibians are a class of animals comprising frogs, salamanders and caecilians. They are tetrapods, but the caecilians and a few species of salamander have either no limbs or their limbs are much reduced in size. Their main bones are hollow and lightweight and are fully ossified and the vertebrae interlock with each other and have articular processes. Their ribs are usually short and may be fused to the vertebrae. Their skulls are mostly broad and short, and are often incompletely ossified. Their skin contains little keratin and lacks scales, but contains many mucous glands and in some species, poison glands. The hearts of amphibians have three chambers, two atria and one ventricle. They have a urinary bladder and nitrogenous waste products are excreted primarily as urea. Amphibians breathe by means of buccal pumping, a pump action in which air is first drawn into the buccopharyngeal region through the nostrils. These are then closed and the air is forced into the lungs by contraction of the throat. They supplement this with gas exchange through the skin which needs to be kept moist.

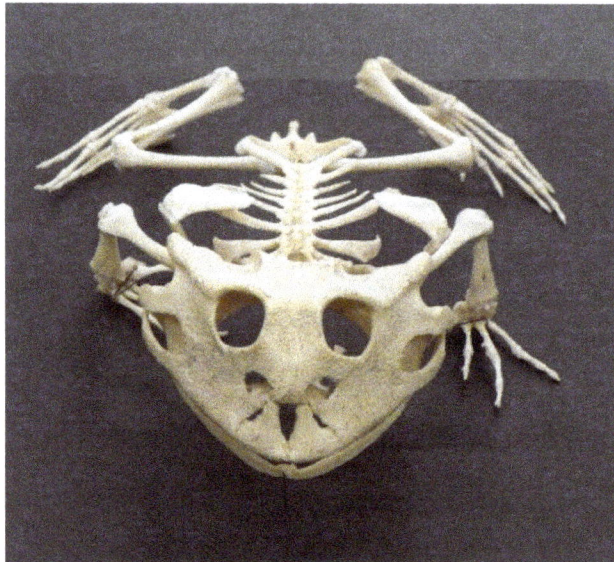

Skeleton of Surinam horned frog (*Ceratophrys cornuta*)

In frogs the pelvic girdle is robust and the hind legs are much longer and stronger than the forelimbs. The feet have four or five digits and the toes are often webbed for swimming or have suction pads for climbing. Frogs have large eyes and no tail. Salamanders resemble lizards in appearance; their short legs project sideways, the belly is close to or in contact with the ground and they have a long tail. Caecilians superficially resemble earthworms and are limbless. They burrow by means of zones of muscle contractions which move along the body and they swim by undulating their body from side to side.

Plastic model of a frog

Reptile Anatomy

Reptiles are a class of animals comprising turtles, tuataras, lizards, snakes and crocodiles. They are tetrapods, but the snakes and a few species of lizard either have no limbs or their limbs are much reduced in size. Their bones are better ossified and their skeletons stronger than those of amphibians. The teeth are conical and mostly uniform in size. The surface cells of the epidermis are modified into horny scales which create a waterproof layer. Reptiles are unable to use their skin for respiration as do amphibians and have a more efficient respiratory system drawing air into their lungs by expanding their chest walls. The heart resembles that of the amphibian but there is a septum which more completely separates the oxygenated and deoxygenated bloodstreams. The reproductive system is designed for internal fertilization, with a copulatory organ present in most species. The eggs are surrounded by amniotic membranes which prevents them from drying out and are laid on land, or develop internally in some species. The bladder is small as nitrogenous waste is excreted as uric acid.

Skeleton of a diamondback rattlesnake

Turtles are notable for their protective shells. They have an inflexible trunk encased in a horny carapace above and a plastron below. These are formed from bony plates embedded in the dermis which are overlain by horny ones and are partially fused with the ribs and spine. The neck is long and flexible and the head and the legs can be drawn back inside the shell. Turtles are vegetarians and the typical reptile teeth have been replaced by sharp, horny plates. In aquatic species, the front legs are modified into flippers.

Tuataras superficially resemble lizards but the lineages diverged in the Triassic period. There is one living species, *Sphenodon punctatus*. The skull has two openings (fenestrae) on either side and the jaw is rigidly attached to the skull. There is one row of teeth in the lower jaw and this fits between the two rows in the upper jaw when the animal chews. The teeth are merely projections of bony material from the jaw and eventually wear down. The brain and heart are more primitive than those of other reptiles, and the lungs have a single chamber and lack bronchi. The tuatara has a well-developed parietal eye on its forehead.

Lizards have skulls with only one fenestra on each side, the lower bar of bone below the second fenestra having been lost. This results in the jaws being less rigidly attached which allows the mouth to open wider. Lizards are mostly quadrupeds, with the trunk held off the ground by short, sideways-facing legs, but a few species have no limbs and resemble snakes. Lizards have moveable eyelids, eardrums are present and some species have a central parietal eye.

Snakes are closely related to lizards, having branched off from a common ancestral lineage during the Cretaceous period, and they share many of the same features. The skeleton consists of a skull, a hyoid bone, spine and ribs though a few species retain a vestige of the pelvis and rear limbs in the form of pelvic spurs. The bar under the second fenestra has also been lost and the jaws have extreme flexibility allowing the snake to swallow its prey whole. Snakes lack moveable eyelids, the eyes being covered by transparent "spectacle" scales. They do not have eardrums but can detect ground vibrations through the bones of their skull. Their forked tongues are used as organs of taste and smell and some species have sensory pits on their heads enabling them to locate warm-blooded prey.

Crocodilians are large, low-slung aquatic reptiles with long snouts and large numbers of teeth. The head and trunk are dorso-ventrally flattened and the tail is laterally compressed. It undulates from side to side to force the animal through the water when swimming. The tough keratinized scales provide body armour and some are fused to the skull. The nostrils, eyes and ears are elevated above the top of the flat head enabling them to remain above the surface of the water when the animal is floating. Valves seal the nostrils and ears when it is submerged. Unlike other reptiles, crocodilians have hearts with four chambers allowing complete separation of oxygenated and deoxygenated blood.

Bird Anatomy

Part of a wing. Albrecht Dürer, c. 1500–1512

Birds are tetrapods but though their hind limbs are used for walking or hopping, their front limbs are wings covered with feathers and adapted for flight. Birds are endothermic, have a high metabolic rate, a light skeletal system and powerful muscles. The long bones are thin, hollow and very light. Air sac extensions from the lungs occupy the centre of some bones. The sternum is wide and usually has a keel and the caudal vertebrae are fused. There are no teeth and the narrow jaws are adapted into a horn-covered beak. The eyes are relatively large, particularly in nocturnal species such as owls. They face forwards in predators and sideways in ducks.

The feathers are outgrowths of the epidermis and are found in localized bands from where they fan out over the skin. Large flight feathers are found on the wings and tail, contour feathers cover the bird's surface and fine down occurs on young birds and under the contour feathers of water birds. The only cutaneous gland is the single uropygial gland near the base of the tail. This produces an oily secretion that waterproofs the feathers when the bird preens. There are scales on the legs, feet and claws on the tips of the toes.

Mammal Anatomy

Mammals are a diverse class of animals, mostly terrestrial but some are aquatic and others have evolved flapping or gliding flight. They mostly have four limbs but some aquatic mammals have no limbs or limbs modified into fins and the forelimbs of bats are modified into wings. The legs of most mammals are situated below the trunk, which is held well clear of the ground. The bones of mammals are well ossified and their teeth, which are usually differentiated, are coated in a layer of prismatic enamel. The teeth are shed once (milk teeth) during the animal's lifetime or not at all, as is the case in cetaceans. Mammals have three bones in the middle ear and a cochlea in the inner ear. They are clothed in hair and their skin contains glands which secrete sweat. Some of these glands are specialized as mammary glands, producing milk to feed the young. Mammals breathe with lungs and have a muscular diaphragm separating the thorax from the abdomen which helps them draw air into the lungs. The mammalian heart has four chambers and oxygenated and deoxygenated blood are kept entirely separate. Nitrogenous waste is excreted primarily as urea.

Mammals are amniotes, and most are viviparous, giving birth to live young. The exception to this are the egg-laying monotremes, the platypus and the echidnas of Australia. Most other mammals have a placenta through which the developing foetus obtains nourishment, but in marsupials, the foetal stage is very short and the immature young is born and finds its way to its mother's pouch where it latches on to a nipple and completes its development.

Human Anatomy

Modern anatomic technique showing sagittal sections of the head as seen by a MRI scan

Humans have the overall body plan of a mammal. Humans have a head, neck, trunk (which includes the thorax and abdomen), two arms and hands and two legs and feet.

Generally, students of certain biological sciences, paramedics, prosthetists and orthotists, physiotherapists, occupational therapists, nurses, and medical students learn gross anatomy and microscopic anatomy from anatomical models, skeletons, textbooks, diagrams, photographs, lectures and tutorials, and in addition, medical students generally also learn gross anatomy through practical experience of dissection and inspection of cadavers. The study of microscopic anatomy (or histology) can be aided by practical experience examining histological preparations (or slides) under a microscope.

In the human, the development of skilled hand movements and increased brain size is likely to have evolved simultaneously.

Human anatomy, physiology and biochemistry are complementary basic medical sciences, which are generally taught to medical students in their first year at medical school. Human anatomy can be taught regionally or systemically; that is, respectively, studying anatomy by bodily regions such as the head and chest, or studying by specific systems, such as the nervous or respiratory systems. The major anatomy textbook, Gray's Anatomy, has been reorganized from a systems format to a regional format, in line with modern teaching methods. A thorough working knowledge of anatomy is required by physicians, especially surgeons and doctors working in some diagnostic specialties, such as histopathology and radiology.

Academic anatomists are usually employed by universities, medical schools or teaching hospitals. They are often involved in teaching anatomy, and research into certain systems, organs, tissues or cells.

Invertebrate Anatomy

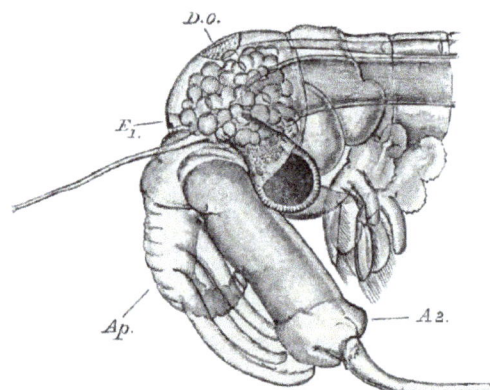

Head of a male *Daphnia*, a planktonic crustacean

Invertebrates constitute a vast array of living organisms ranging from the simplest unicellular eukaryotes such as *Paramecium* to such complex multicellular animals as the octopus, lobster and dragonfly. They constitute about 95% of the animal species. By definition, none of these creatures has a backbone. The cells of single-cell protozoans have the same basic structure as those of multicellular animals but some parts are specialized into the equivalent of tissues and organs. Locomotion is often provided by cilia or flagella or may proceed via the advance of pseudopodia, food may be gathered by phagocytosis, energy needs may be supplied by photosynthesis and the cell may be supported by an endoskeleton or an exoskeleton. Some protozoans can form multicellular colonies.

Metazoans are multicellular organism, different groups of cells of which have separate functions. The most basic types of metazoan tissues are epithelium and connective tissue, both of which are present in nearly all invertebrates. The outer surface of the epidermis is normally formed of epithelial cells and secretes an extracellular matrix which provides support to the organism. An endoskeleton derived from the mesoderm is present in echinoderms, sponges and some cephalopods. Exoskeletons are derived from the epidermis and is composed of chitin in arthropods (insects, spiders, ticks, shrimps, crabs, lobsters). Calcium carbonate constitutes the shells of molluscs, brachiopods and some tube-building polychaete worms and silica forms the exoskeleton of the microscopic diatoms and radiolaria. Other invertebrates may have no rigid structures but the epidermis may secrete a variety of surface coatings such as the pinacoderm of sponges, the gelatinous cuticle of cnidarians (polyps, sea anemones, jellyfish) and the collagenous cuticle of annelids. The outer epithelial layer may include cells of several types including sensory cells, gland cells and stinging cells. There may also be protrusions such as microvilli, cilia, bristles, spines and tubercles.

Marcello Malpighi, the father of microscopical anatomy, discovered that plants had tubules similar to those he saw in insects like the silk worm. He observed that when a ring-like portion of bark was removed on a trunk a swelling occurred in the tissues above the ring, and he unmistakably interpreted this as growth stimulated by food coming down from the leaves, and being captured above the ring.

Arthropod Anatomy

Arthropods comprise the largest phylum in the animal kingdom with over a million known invertebrate species.

Insects possess segmented bodies supported by a hard-jointed outer covering, the exoskeleton, made mostly of chitin. The segments of the body are organized into three distinct parts, a head, a thorax and an abdomen. The head typically bears a pair of sensory antennae, a pair of compound eyes, one to three simple eyes (ocelli) and three sets of modified appendages that form the mouthparts. The thorax has three pairs of segmented legs, one pair each for the three segments that compose the thorax and one or two pairs of wings. The abdomen is composed of eleven segments, some of which may be fused and houses the digestive, respiratory, excretory and reproductive systems. There is considerable variation between species and many adaptations to the body parts, especially wings, legs, antennae and mouthparts.

Spiders a class of arachnids have four pairs of legs; a body of two segments—a cephalothorax and an abdomen. Spiders have no wings and no antennae. They have mouthparts called chelicerae which are often connected to venom glands as most spiders are venomous. They have a second pair of appendages called pedipalps attached to the cephalothorax. These have similar segmentation to the legs and function as taste and smell organs. At the end of each male pedipalp is a spoon-shaped cymbium that acts to support the copulatory organ.

Other Branches of Anatomy

- Superficial or surface anatomy is important as the study of anatomical landmarks that can be readily seen from the exterior contours of the body. It enables physicians or veterinary surgeons to gauge the position and anatomy of the associated deeper structures. Superficial is a directional term that indicates that structures are located relatively close to the surface of the body.

- Comparative anatomy relates to the comparison of anatomical structures (both gross and microscopic) in different animals.

- Artistic anatomy relates to anatomic studies for artistic reasons.

History

Ancient

Ancient Greek anatomy and physiology underwent great changes and advances throughout the early medieval world. Over time, this medical practice expanded by a continually developing understanding of the functions of organs and structures in the body. Phenomenal anatomical observations of the human body were made, which have contributed towards the understanding of the brain, eye, liver, reproductive organs and the nervous system.

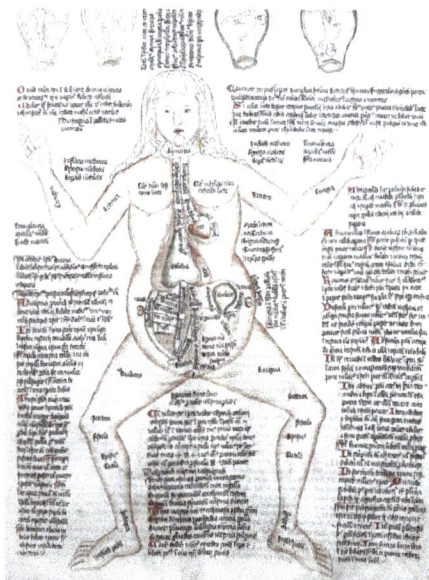

Image of early rendition of anatomy findings

The city of Alexandria was the stepping-stone for Greek anatomy and physiology. Alexandria not only housed the biggest library for medical records and books of the liberal arts in the world during the time of the Greeks, but was also home to many medical practitioners and philosophers. Great patronage of the arts and sciences from the Ptolemy rulers helped raise Alexandria up, further rivalling the cultural and scientific achievements of other Greek states.

Some of the most striking advances in early anatomy and physiology took place in Hellenistic Alexandria. Two of the most famous Greek anatomists and physiologists of the third century were Herophilus and Erasistratus. These two physicians helped pioneer human dissection for medical research. They also conducted vivisections on the cadavers of condemned criminals, which was considered taboo until the Renaissance – Herophilus was recognized as the first person to perform systematic dissections. Herophilus became known for his anatomical works making impressing contributions to many branches of anatomy and many other aspects of medicine. Some of the works included classifying the system of the pulse, the discovery that human arteries had thicker walls then veins, and that the atria were parts of the heart. Herophilus's knowledge of the human body has provided vital input towards understanding the brain, eye, liver, reproductive organs and nervous system, and characterizing the course of disease. Erasistratus accurately described the structure of the brain, including the cavities and membranes, and made a distinction between its cerebrum and cerebellum During his study in Alexandria, Erasistratus was particularly concerned with studies of the circulatory and nervous systems. He was able to distinguish the sensory and the motor nerves in the human body and believed that air entered the lungs and heart, which was then carried throughout the body. His distinction between the arteries and veins – the arteries carrying the air through the body, while the veins carried the blood from the heart was a great anatomical discovery. Erasistratus was also responsible for naming and describing the function of the epiglottis and the valves of the heart, including the tricuspid. During the third century, Greek physicians were able to differentiate nerves from blood vessels and tendons and to realize that the nerves convey neural impulses. It was Herophilus who made the point that damage to motor nerves induced paralysis. Herophilus named the meninges and ventricles in the brain, appreciated the division between cerebellum and cerebrum and recognized that the brain was the "seat of intellect" and not a "cooling chamber" as propounded by Aristotle Herophilus is also credited with describing the optic, oculomotor, motor division of the trigeminal, facial, vestibulocochlear and hypoglossal nerves.

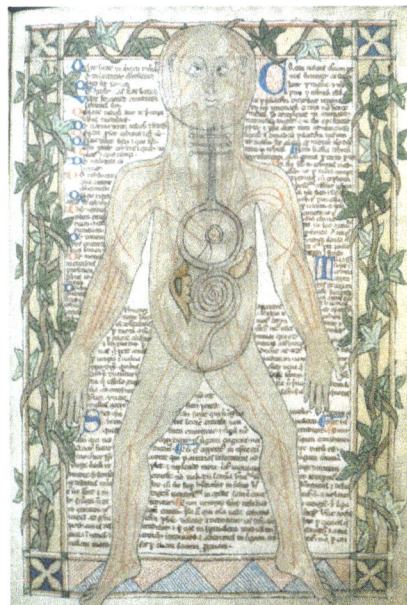

13th century anatomical illustration

Great feats were made during the third century in both the digestive and reproductive systems. Herophilus was able to discover and describe not only the salivary glands, but the small intestine and liver. He showed that the uterus is a hollow organ and described the ovaries and uterine tubes. He recognized that spermatozoa were produced by the testes and was the first to identify the prostate gland.

In 1600 BCE, the Edwin Smith Papyrus, an Ancient Egyptian medical text, described the heart, its vessels, liver, spleen, kidneys, hypothalamus, uterus and bladder, and showed the blood vessels diverging from the heart. The Ebers Papyrus (c. 1550 BCE) features a "treatise on the heart", with vessels carrying all the body's fluids to or from every member of the body.

The anatomy of the muscles and skeleton is described in the *Hippocratic Corpus*, an Ancient Greek medical work written by unknown authors. Aristotle described vertebrate anatomy based on animal dissection. Praxagoras identified the difference between arteries and veins. Also in the 4th century BCE, Herophilos and Erasistratus produced more accurate anatomical descriptions based on vivisection of criminals in Alexandria during the Ptolemaic dynasty.

In the 2nd century, Galen of Pergamum, an anatomist, clinician, writer and philosopher, wrote the final and highly influential anatomy treatise of ancient times. He compiled existing knowledge and studied anatomy through dissection of animals. He was one of the first experimental physiologists through his vivisection experiments on animals. Galen's drawings, based mostly on dog anatomy, became effectively the only anatomical textbook for the next thousand years. His work was known to Renaissance doctors only through Islamic Golden Age medicine until it was translated from the Greek some time in the 15th century.

Medieval to Early Modern

Anatomy developed little from classical times until the sixteenth century; as the historian Marie Boas writes, "Progress in anatomy before the sixteenth century is as mysteriously slow as its development after 1500 is startlingly rapid". Between 1275 and 1326, the anatomists Mondino de Luzzi, Alessandro Achillini and Antonio Benivieni at Bologna carried out the first systematic human dissections since ancient times. Mondino's *Anatomy* of 1316 was the first textbook in the medieval rediscovery of human anatomy. It describes the body in the order followed in Mondino's dissections, starting with the abdomen, then the thorax, then the head and limbs. It was the standard anatomy textbook for the next century.

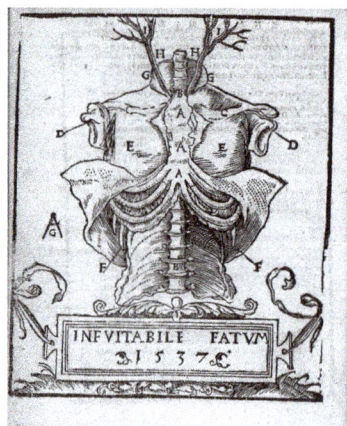

Mondino de Luzzi, *Anathomia*, 1541

Anatomical chart in Vesalius's *Epitome*, 1543

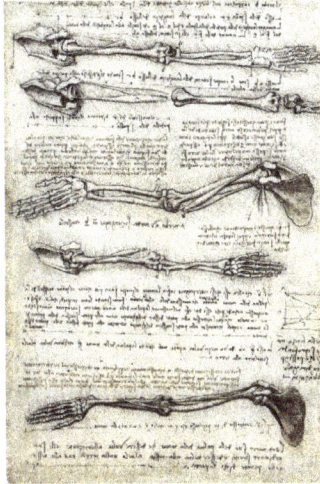

Anatomical study of the arm, by Leonardo da Vinci, (about 1510)

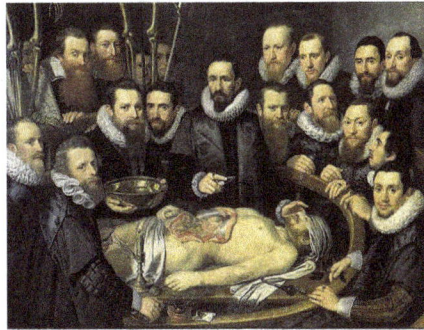

Michiel Jansz van Mierevelt – *Anatomy lesson of Dr. Willem van der Meer*, 1617

Leonardo da Vinci (1452–1519) was trained in anatomy by Andrea del Verrocchio. He made use of his anatomical knowledge in his artwork, making many sketches of skeletal structures, muscles and organs of humans and other vertebrates that he dissected.

Andreas Vesalius (1514–1564) (Latinized from Andries van Wezel), professor of anatomy at the University of Padua, is considered the founder of modern human anatomy. Originally from Brabant, Vesalius published the influential book *De humani corporis fabrica* ("the structure of the human body"), a large format book in seven volumes, in 1543. The accurate and intricately detailed illustrations, often in allegorical poses against Italianate landscapes, are thought to have been made by the artist Jan van Calcar, a pupil of Titian.

In England, anatomy was the subject of the first public lectures given in any science; these were given by the Company of Barbers and Surgeons in the 16th century, joined in 1583 by the Lumleian lectures in surgery at the Royal College of Physicians.

Late Modern

In the United States, medical schools began to be set up towards the end of the 18th century. Classes in anatomy needed a continual stream of cadavers for dissection and these were difficult to obtain. Philadelphia, Baltimore and New York were all renowned for body snatching activity as criminals raided graveyards at night, removing newly buried corpses from their coffins. A similar problem existed in Britain where demand for bodies became so great that grave-raiding and even anatomy murder were practised to obtain cadavers. Some graveyards were in consequence protected with watchtowers. The practice was halted in Britain by the Anatomy Act of 1832, while in the United States, similar legislation was enacted after the physician William S. Forbes of Jefferson Medical College was found guilty in 1882 of "complicity with resurrectionists in the despoliation of graves in Lebanon Cemetery".

The teaching of anatomy in Britain was transformed by Sir John Struthers, Regius Professor of Anatomy at the University of Aberdeen from 1863 to 1889. He was responsible for setting up the system of three years of "pre-clinical" academic teaching in the sciences underlying medicine, including especially anatomy. This system lasted until the reform of medical training in 1993 and 2003. As well as teaching, he collected many vertebrate skeletons for his museum of comparative anatomy, published over 70 research papers, and became famous for his public dissection of the Tay Whale. From 1822 the Royal College of Surgeons regulated the teaching of anatomy in medical schools. Medical museums provided examples in comparative anatomy, and were often used in teaching. Ignaz Semmelweis investigated puerperal fever and he discovered how it was caused. He noticed that the frequently fatal fever occurred more often in mothers examined by medical students than by midwives. The students went from the dissecting room to the hospital ward and examined women in childbirth. Semmelweis showed that when the trainees washed their hands in chlorinated lime before each clinical examination, the incidence of puerperal fever among the mothers could be reduced dramatically.

An electron microscope from 1973

Before the era of modern medical procedures, the main means for studying the internal structure of the body were palpation and dissection. It was the advent of microscopy that opened up an understanding of the building blocks that constituted living tissues. Technical advances in the development of achromatic lenses increased the resolving power of the microscope and around 1839, Matthias Jakob Schleiden and Theodor Schwann identified that cells were the fundamental unit of organization of all living things. Study of small structures involved passing light through them and the microtome was invented to provide sufficiently thin slices of tissue to examine. Staining techniques using artificial dyes were established to help distinguish between different types of tissue. The fields of cytology and histology developed from here in the late 19th century. The invention of the electron microscope brought a great advance in resolution power and allowed research into the ultrastructure of cells and the organelles and other structures within them. About the same time, in the 1950s, the use of X-ray diffraction for studying the crystal structures of proteins, nucleic acids and other biological molecules gave rise to a new field of molecular anatomy.

Short wavelength electromagnetic radiation such as X-rays can be passed through the body and used in medical radiography to view interior structures that have different degrees of opaqueness. Nowadays, modern techniques such as magnetic resonance imaging, computed tomography, fluoroscopy and ultrasound imaging have enabled researchers and practitioners to examine organs, living or dead, in unprecedented detail. They are used for diagnostic and therapeutic purposes and provide information on the internal structures and organs of the body to a degree far beyond the imagination of earlier generations.

Biomechanics

Biomechanics is the study of the structure and function of biological systems such as humans, animals, plants, organs, fungi, and cells by means of the methods of mechanics.

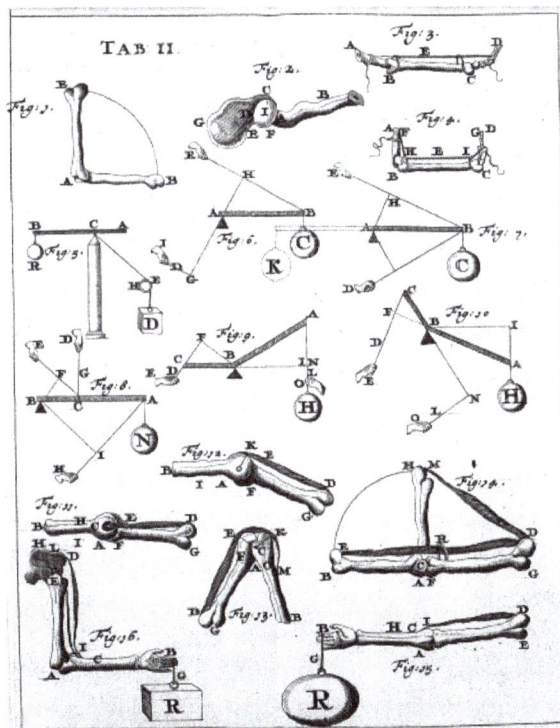

Page of one of the first works of Biomechanics (*De Motu Animalium* of Giovanni Alfonso Borelli) in the 17th century

Method

Biomechanics is closely related to engineering, because it often uses traditional engineering sciences to analyze biological systems. Some simple applications of Newtonian mechanics and/or materials sciences can supply correct approximations to the mechanics of many biological systems. Applied mechanics, most notably mechanical engineering disciplines such as continuum mechanics, mechanism analysis, structural analysis, kinematics and dynamics play prominent roles in the study of biomechanics.

Usually biological systems are much more complex than man-built systems. Numerical methods are hence applied in almost every biomechanical study. Research is done in an iterative process of hypothesis and verification, including several steps of modeling, computer simulation and experimental measurements.

Subfields

Applied subfields of biomechanics include:

- Soft body dynamics
- Kinesiology (kinetics + physiology)
- Animal locomotion & Gait analysis
- Musculoskeletal & orthopedic biomechanics
- Cardiovascular biomechanics
- Ergonomy
- Human factors engineering & occupational biomechanics
- Implant (medicine), Orthotics & Prosthesis
- Rehabilitation
- Sports biomechanics
- Allometry
- Injury biomechanics

Sports Biomechanics

In sports biomechanics, the laws of mechanics are applied to human movement in order to gain a greater understanding of athletic performance and to reduce sport injuries as well. It focuses on the application of the scientific principles of mechanical physics to understand movements of action of human bodies and sports implements such as cricket bat, hockey stick and javelin etc. Elements of mechanical engineering (e.g., strain gauges), electrical engineering (e.g., digital filtering), computer science (e.g., numerical methods), gait analysis (e.g., force platforms), and clinical neurophysiology (e.g., surface EMG) are common methods used in sports biomechanics.

Biomechanics in sports can be stated as the muscular, joint and skeletal actions of the body during the execution of a given task, skill and/or technique. Proper understanding of biomechanics relating to sports skill has the greatest implications on: sport's performance, rehabilitation and injury prevention, along with sport mastery. As noted by Doctor Michael Yessis, one could say that best athlete is the one that executes his or her skill the best.

Continuum Biomechanics

The mechanical analysis of biomaterials and biofluids is usually carried forth with the concepts of continuum mechanics. This assumption breaks down when the length scales of interest approach the order of the micro structural details of the material. One of the most remarkable characteristic of biomaterials is their hierarchical structure. In other words, the mechanical characteristics of these materials rely on physical phenomena occurring in multiple levels, from the molecular all the way up to the tissue and organ levels.

Biomaterials are classified in two groups, hard and soft tissues. Mechanical deformation of hard tissues (like wood, shell and bone) may be analysed with the theory of linear elasticity. On the other hand, soft tissues (like skin, tendon, muscle and cartilage) usually undergo large deformations and thus their analysis rely on the finite strain theory and computer simulations. The interest in continuum biomechanics is spurred by the need for realism in the development of medical simulation.

Biofluid Mechanics

Biological fluid mechanics, or biofluid mechanics, is the study of both gas and liquid fluid flows in or around biological organisms. An often studied liquid biofluids problem is that of blood flow in the human cardiovascular system. Under certain mathematical circumstances, blood flow can be modelled by the Navier–Stokes equations. *In vivo* whole blood is assumed to be an incompressible Newtonian fluid. However, this assumption fails when considering forward flow within arterioles. At the microscopic scale, the effects of individual red blood cells become significant, and whole blood can no longer be modelled as a continuum. When the diameter of the blood vessel is just slightly larger than the diameter of the red blood cell the Fahraeus–Lindquist effect occurs and there is a decrease in wall shear stress. However, as the diameter of the blood vessel decreases further, the red blood cells have to squeeze through the vessel and often can only pass in single file. In this case, the inverse Fahraeus–Lindquist effect occurs and the wall shear stress increases.

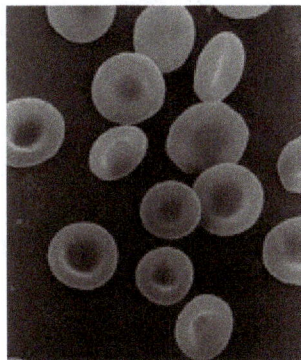

Red blood cells

An example of a gaseous biofluids problem is that of human respiration. Recently, respiratory systems in insects have been studied for bioinspiration for designing improved microfluidic devices.

Biotribology

The main aspects of Contact mechanics and tribology are related to friction, wear and lubrication. When the two surfaces come in contact during motion i.e. rub against each other, friction, wear and lubrication effects are very important to analyze in order to determine the performance of the material. Biotribology is a study of friction, wear and lubrication of biological systems especially human joints such as hips and knees. For example, femoral and tibial components of knee implant routinely rub against each other during daily activity such as walking or stair climbing. If the performance of tibial component needs to be analyzed, the principles of biotribology are used to determine the wear performance of the implant and lubrication effects of synovial fluid. In addition, the theory of contact mechanics also becomes very important for wear analysis. Additional aspects of biotribology can also include analysis of subsurface damage resulting from two surfaces coming in contact during motion, i.e. rubbing against each other, such as in the evaluation of tissue engineered cartilage.

Comparative Biomechanics

Comparative biomechanics is the application of biomechanics to non-human organisms, whether used to gain greater insights into humans (as in physical anthropology) or into the functions, ecology and adaptations of the organisms themselves. Common areas of investigation are Animal locomotion and feeding, as these have strong connections to the organism's fitness and impose high mechanical demands. Animal locomotion, has many manifestations, including running, jumping and flying. Locomotion requires energy to overcome friction, drag, inertia, and gravity, though which factor predominates varies with environment.

Chinstrap penguin leaping over water

Comparative biomechanics overlaps strongly with many other fields, including ecology, neurobiology, developmental biology, ethology, and paleontology, to the extent of commonly publishing papers in the journals of these other fields. Comparative biomechanics is often applied in medicine (with regards to common model organisms such as mice and rats) as well as in biomimetics, which looks to nature for solutions to engineering problems.

Plant Biomechanics

The application of biomechanical principles to plants and plant organs has developed into the subfield of plant biomechanics.

Computational Biomechanics

Over the past decade the Finite element method has become an established alternative to in vivo surgical assessment. The main advantage of Computational Biomechanics lies in its ability to determine the endo-anatomical response of an anatomy, without being subject to ethical restrictions. This has led FE modelling to the point of becoming ubiquitous in several fields of Biomechanics while several projects have even adopted an open source philosophy (e.g. BioSpine).

History

Antiquity

Aristotle wrote the first book on the motion of animals, *De Motu Animalium*, or On the Movement of Animals. He not only saw animals' bodies as mechanical systems, but pursued questions such as the physiological difference between imagining performing an action and actually doing it. In another work, *On the Parts of Animals*, he provided an accurate description of how the ureter uses peristalsis to carry urine from the kidneys to the bladder.

Renaissance

Leonardo da Vinci studied anatomy in the context of mechanics. He analyzed muscle forces as acting along lines connecting origins and insertions, and studied joint function. Da Vinci tended to mimic some animal features in his machines. For example, he studied the flight of birds to find means by which humans could fly; and because horses were the principal source of mechanical power in that time, he studied their muscular systems to design machines that would better benefit from the forces applied by this animal.

Galileo Galilei was interested in the strength of bones and suggested that bones are hollow because this affords maximum strength with minimum weight. He noted that animals' bone masses increased disproportionately to their size. Consequently, bones must also increase disproportionately in girth rather than mere size. This is because the bending strength of a tubular structure (such as a bone) is much more efficient relative to its weight. Mason suggests that this insight was one of the first grasps of the principles of biological optimization.

In the 16th century, Descartes suggested a philosophic system whereby all living systems, including the human body (but not the soul), are simply machines ruled by the same mechanical laws, an idea that did much to promote and sustain biomechanical study. Giovanni Alfonso Borelli embraced this idea and studied walking, running, jumping, the flight of birds, the swimming of fish, and even the piston action of the heart within a mechanical framework. He could determine the position of the human center of gravity, calculate and measured inspired and expired air volumes, and showed that inspiration is muscle-driven and expiration is due to tissue elasticity. Borelli was the first to understand that the levers of the musculoskeletal system magnify motion rather than force, so that muscles must produce much larger forces than those resisting the motion. Influenced by the work of Galileo, whom he personally knew, he had an intuitive understanding of static equilibrium in various joints of the human body well before Newton published the laws of motion.

Industrial Era

In the 19th century Étienne-Jules Marey used cinematography to scientifically investigate locomotion. He opened the field of modern 'motion analysis' by being the first to correlate ground reaction forces with movement. In Germany, the brothers Ernst Heinrich Weber and Wilhelm

Eduard Weber hypothesized a great deal about human gait, but it was Christian Wilhelm Braune who significantly advanced the science using recent advances in engineering mechanics. During the same period, the engineering mechanics of materials began to flourish in France and Germany under the demands of the industrial revolution. This led to the rebirth of bone biomechanics when the railroad engineer Karl Culmann and the anatomist Hermann von Meyer compared the stress patterns in a human femur with those in a similarly shaped crane. Inspired by this finding Julius Wolff proposed the famous Wolff's law of bone remodeling.

Applications

The study of biomechanics ranges from the inner workings of a cell to the movement and development of limbs, to the mechanical properties of soft tissue, and bones. Some simple examples of biomechanics research include the investigation of the forces that act on limbs, the aerodynamics of bird and insect flight, the hydrodynamics of swimming in fish, and locomotion in general across all forms of life, from individual cells to whole organisms. The biomechanics of human beings is a core part of kinesiology. As we develop a greater understanding of the physiological behavior of living tissues, researchers are able to advance the field of tissue engineering, as well as develop improved treatments for a wide array of pathologies.

Biomechanics is also applied to studying human musculoskeletal systems. Such research utilizes force platforms to study human ground reaction forces and infrared videography to capture the trajectories of markers attached to the human body to study human 3D motion. Research also applies electromyography (EMG) system to study the muscle activation. By this, it is feasible to investigate the muscle responses to the external forces as well as perturbations.

Biomechanics is widely used in orthopedic industry to design orthopedic implants for human joints, dental parts, external fixations and other medical purposes. Biotribology is a very important part of it. It is a study of the performance and function of biomaterials used for orthopedic implants. It plays a vital role to improve the design and produce successful biomaterials for medical and clinical purposes. One such example is in tissue engineered cartilage.

Synthetic Biology

Synthetic biology is an interdisciplinary branch of biology and engineering. The subject combines various disciplines from within these domains, such as biotechnology, evolutionary biology, genetic engineering, molecular biology, molecular engineering, systems biology, biophysics, and computer engineering.

Descriptions of synthetic biology depend on how the user approaches it, as a biologist or as an engineer. Originally seen as a subset of biology, in recent years the role of electrical and chemical engineering has become more important. For example, one description designates synthetic biology as "an emerging discipline that uses engineering principles to design and assemble biological components". Another description, by Jan Staman Director of the Rathenau Institute in The Hague in 2006, portrayed it as "a new emerging scientific field where ICT, biotechnology and nanotechnology meet and strengthen each other".

The definition of synthetic biology is debated, not only among natural scientists and engineers but also in the human sciences, arts and politics. One popular definition is "designing and constructing biological modules, biological systems, and biological machines for useful purposes." However, the functional aspects of this definition are rooted in molecular biology and biotechnology.

As usage of the term has expanded to many interdisciplinary fields, synthetic biology has been recently defined as the artificial design and engineering of biological systems and living organisms for purposes of improving applications for industry or biological research.

promoter	primer binding site
cds	restriction site
ribosome entry site	blunt restriction site
terminator	5' sticky restriction site
operator	3' sticky restriction site
insulator	5' overhang
ribonuclease site	3' overhang
rna stability element	assembly scar
protease site	signature
protein stability element	user defined
origin of replication	

Synthetic Biology Open Language (SBOL) standard visual symbols for use with BioBricks Standard

History

The first identifiable use of the term "synthetic biology" was in Stéphane Leduc's publication of *Théorie physico-chimique de la vie et générations spontanées*(1910) and his *La Biologie Synthétique* (1912).

Sixty-four years later, in 1974, the term gained its more modern usage when Polish geneticist Wacław Szybalski used the term "synthetic biology", writing:

Let me now comment on the question "what next". Up to now we are working on the descriptive phase of molecular biology. ... But the real challenge will start when we enter the synthetic phase of research in our field. We will then devise new control elements and add these new modules to the existing genomes or build up wholly new genomes. This would be a field with an unlimited expansion potential and hardly any limitations to building "new better control circuits" or ..... finally other "synthetic" organisms, like a "new better mouse". ... I am not concerned that we will run out of exciting and novel ideas, ... in the synthetic biology, in general.

When in 1978 Arber, Nathans and Smith won the Nobel Prize in Physiology or Medicine for the discovery of restriction enzymes, Wacław Szybalski wrote in an editorial comment in the journal *Gene*:

The work on restriction nucleases not only permits us easily to construct recombinant DNA molecules and to analyze individual genes, but also has led us into the new era of synthetic biology where not only existing genes are described and analyzed but also new gene arrangements can be constructed and evaluated.

A notable advance in synthetic biology occurred in 2000, when two articles in Nature by Michael B. Elowitz and Stanislas Leibler discussed the creation of synthetic biological circuit devices of a genetic toggle switch and a biological clock by combining genes within E. coli cells.

Perspectives

Engineering

Engineers view biology as a *technology* – the *systems biotechnology* or *systems biological engineering*. Synthetic biology includes the broad redefinition and expansion of biotechnology, with the ultimate goals of being able to design and build engineered biological systems that process information, manipulate chemicals, fabricate materials and structures, produce energy, provide food, and maintain and enhance human health and our environment.

Studies in synthetic biology can be subdivided into broad classifications according to the approach they take to the problem at hand: standardization of biological parts, biomolecular engineering, genome engineering. Biomolecular engineering includes approaches which aim to create a toolkit of functional units that can be introduced to present new technological functions in living cells. Genetic engineering includes approaches to construct synthetic chromosomes for whole or minimal organisms. Biomolecular design refers to the general idea of de novo design and additive combination of biomolecular components. Each of these approaches share a similar task: to develop a more synthetic entity at a higher level of complexity by inventively manipulating a simpler part at the preceding level.

Re-writing

Re-writers are synthetic biologists interested in testing the irreducibility of biological systems. Due to the complexity of natural biological systems, it would be simpler to re-build the natural systems of interest from the ground up; In order to provide engineered surrogates that are easier to comprehend, control and manipulate. Re-writers draw inspiration from refactoring, a process sometimes used to improve computer software.

Key Enabling Technologies

Several key enabling technologies are critical to the growth of synthetic biology. The key concepts include standardization of biological parts and hierarchical abstraction to permit using those parts in increasingly complex synthetic systems. Achieving this is greatly aided by basic technologies of reading and writing of DNA (sequencing and fabrication), which are improving in price/performance exponentially (Kurzweil 2001). Measurements under a variety of conditions are needed for accurate modeling and computer-aided-design (CAD).

Standardized DNA Parts

The most used standardized DNA parts are BioBrick plasmids invented by Tom Knight in 2003. Biobricks are stored at the Registry of Standard Biological Parts in Cambridge, Massachusetts and the BioBrick standard has been used by thousands of students worldwide in the international Genetically Engineered Machine (iGEM) competition.

DNA Synthesis

In 2007 it was reported that several companies were offering the synthesis of genetic sequences up to 2000 bp long, for a price of about $1 per base pair and a turnaround time of less than two weeks. Oligonucleotides harvested from a photolithographic or inkjet manufactured DNA chip combined with DNA mismatch error-correction allows inexpensive large-scale changes of codons in genetic systems to improve gene expression or incorporate novel amino-acids. This favors a synthesis-from-scratch approach.

Additionally, the CRISPR/Cas system has emerged as a promising technique for gene editing. It was hailed by The Washington Post as "the most important innovation in the synthetic biology space in nearly 30 years." While other methods take months or years to edit gene sequences, CRISPR speeds that time up to weeks. However, due to its ease of use and accessibility, it has raised a number of ethical concerns, especially surrounding its use in the biohacking space.

DNA Sequencing

DNA sequencing is determining the order of the nucleotide bases in a molecule of DNA. Synthetic biologists make use of DNA sequencing in their work in several ways. First, large-scale genome sequencing efforts continue to provide a wealth of information on naturally occurring organisms. This information provides a rich substrate from which synthetic biologists can construct parts and devices. Second, synthetic biologists use sequencing to verify that they fabricated their engineered system as intended. Third, fast, cheap, and reliable sequencing can also facilitate rapid detection and identification of synthetic systems and organisms.

Modular Protein Assembly

While DNA is most important for information storage, a large fraction of the cell's activities are carried out by proteins. Therefore, it is important to have tools to send proteins to specific regions of the cell and to link different proteins together, as desired. Ideally the interaction strength between protein partners should be tunable between a lifetime of seconds (desirable for dynamic signaling events) up to an irreversible interaction (desirable when building devices stable over days or resilient to harsh conditions). Interactions such as coiled coils, SH3 domain-peptide binding or SpyTag/SpyCatcher have helped to give such control. In addition it is important to be able to regulate protein-protein interactions in cells, such as with light (using Light-oxygen-voltage-sensing domains) or cell-permeable small molecules by Chemically induced dimerization.

Modeling

Models inform the design of engineered biological systems by allowing synthetic biologists to better predict system behavior prior to fabrication. Synthetic biology will benefit from better models of how biological molecules bind substrates and catalyze reactions, how DNA encodes the information needed to specify the cell and how multi-component integrated systems behave. Recently, multiscale models of gene regulatory networks have been developed that focus on synthetic biology applications. Simulations have been used that model all biomolecular interactions in transcription, translation, regulation, and induction of gene regulatory networks, guiding the design of synthetic systems.

Research Examples

Synthetic DNA

Driven by dramatic decreases in costs of making oligonucleotides ("oligos"), the sizes of DNA constructions from oligos have increased to the genomic level. For example, in 2000, researchers at Washington University reported synthesis of the 9.6 kbp (kilo base pair) Hepatitis C virus genome from chemically synthesized 60 to 80-mers. In 2002 researchers at SUNY Stony Brook succeeded in synthesizing the 7741 base poliovirus genome from its published sequence, producing the second synthetic genome. This took about two years of work. In 2003 the 5386 bp genome of the bacteriophage Phi X 174 was assembled in about two weeks. In 2006, the same team, at the J. Craig Venter Institute, had constructed and patented a synthetic genome of a novel minimal bacterium, *Mycoplasma laboratorium* and were working on getting it functioning in a living cell.

Synthetic Transcription Factors

Studies have also been performed on the components of the DNA translation mechanism. One desire of scientists creating synthetic biological circuits is to be able to control the translation of synthetic DNA in prokaryotes and eukaryotes. One study tested the adjustability of synthetic transcription factors (sTFs) in areas of transcription output and cooperative ability among multiple transcription factor complexes. Researchers were able to mutate zinc fingers, the DNA specific component of sTFs, to decrease their affinity for DNA, and thus decreasing the amount of translation. They were also able to use the zinc fingers as components of complex forming sTFs, which are the eukaryotic translation mechanisms.

Applications

Synthetic Life

One important topic in synthetic biology is *synthetic life*, that is, artificial life created *in vitro* from biomolecules and their component materials. Synthetic life experiments attempt to either probe the origins of life, study some of the properties of life, or more ambitiously to recreate life from non-living (abiotic) components. Synthetic biology attempts to create new biological molecules and even novel living species capable of carrying out a range of important medical and industrial functions, from manufacturing pharmaceuticals to detoxifying polluted land and water. In medicine, it offers prospects of using designer biological parts as a starting point for an entirely new class of therapies and diagnostic tools.

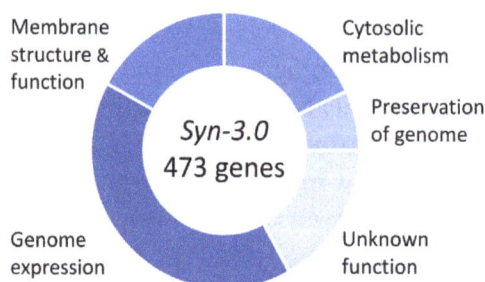

Gene functions in the minimal genome of the synthetic organism, *Syn 3*.

In the area of synthetic biology, a living "artificial cell" has been defined as a completely synthetically-made cell that can capture energy, maintain ion gradients, contain macromolecules as well as store information and have the ability to mutate. Nobody has been able to create such an artificial cell.

The first living organism with 'artificial' DNA was produced by scientists at the Scripps Research Institute as *E. coli* was engineered to replicate an expanded genetic alphabet.

A completely synthetic genome was produced by Craig Venter, and his team introduced it to genomically emptied bacterial host cells, and allowed the host cells to grow and replicate.

Cell Transformation

Currently, entire organisms are not being created from scratch, but instead living cells are being transformed with inserts of new DNA. There are several ways of constructing synthetic DNA components and even entire synthetic genomes, but once the desired genetic code is obtained, it is integrated into a living cell that is expected to manifest the desired new capabilities or phenotypes while growing and thriving. Cell transformation is used to create biological circuits, which can be manipulated to yield desired outputs.

Information Storage

Scientists can encode vast amounts of digital information onto a single strand of synthetic DNA. In 2012, George M. Church encoded one of his books about synthetic biology in DNA. The 5.3 Mb of data from the book is more than 1000 times greater than the previous largest amount of information to be stored in synthesized DNA. A similar project had encoded the complete sonnets of William Shakespeare in DNA.

Synthetic Genetic Pathways

Traditional metabolic engineering has been bolstered by the introduction of combinations of foreign genes and optimization by directed evolution. Perhaps the best known application of synthetic biology to date is engineering *E. coli* and yeast for commercial production of a precursor of the antimalarial drug, Artemisinin, by the laboratory of Jay Keasling

Unnatural Nucleotides

Many technologies have been developed for incorporating unnatural nucleotides and amino acids into nucleic acids and proteins, both in vitro and in vivo. For example, in May 2014, researchers announced that they had successfully introduced two new artificial nucleotides into bacterial DNA. By including individual artificial nucleotides in the culture media, were able to exchange the bacteria 24 times; they did not generate mRNA or proteins able to use the artificial nucleotides.

Unnatural Amino Acids

Another common topic of investigation is expansion of the normal repertoire of 20 amino acids. Excluding stop codons, there are 61 codons, but only 20 amino acids are coded generally in all organisms. Certain codons are engineered to code for alternative amino acids including: nonstandard amino acids such as O-methyl tyrosine; or exogenous amino acids such as 4-fluorophenylalanine. Typically, these projects make use of re-coded nonsense suppressor tRNA-Aminoacyl tRNA synthetase pairs from other organisms, though in most cases substantial engineering is still required.

Reduced Amino-acid Libraries

Instead of expanding the genetic code, other researchers have investigated the structure and function of proteins by reducing the normal set of 20 amino acids. Limited protein sequence libraries are made by generating proteins where certain groups of amino acids may be substituted with a single amino acid. For instance, several non-polar amino acids within a protein can all be replaced with a single non-polar amino acid. One project demonstrated that an engineered version of Chorismate mutase still had catalytic activity when only 9 amino acids were used.

Designed Proteins

While there are methods to engineer natural proteins such as by directed evolution, there are also projects to design novel protein structures that match or improve on the functionality of existing proteins. One group generated a helix bundle that was capable of binding oxygen with similar properties as hemoglobin, yet did not bind carbon monoxide. A similar protein structure was generated to support a variety of oxidoreductase activities. Another group generated a family of G-protein coupled receptors which could be activated by the inert small molecule clozapine-N-oxide but insensitive to the native ligand, acetylcholine.

The Top7 protein was one of the first proteins designed for a fold that had never been seen before in nature

Biosensors

A biosensor refers to an engineered organism, usually a bacterium, which is capable of reporting some ambient phenomenon such as the presence of heavy metals or toxins. In this capability, a very widely used system is the Lux operon of *Aliivibrio fischeri*. The Lux operon codes for an enzyme which is the source bacterial bioluminescence, and can be placed after a respondent promoter to express the luminescence genes in response to a specific environmental stimulus. One such sensor created in Oak Ridge National Laboratory, and named "critter on a chip", consisted of a bioluminescent bacterial coating on a photosensitive computer chip to detect certain petroleum pollutants. When the bacteria sense the pollutant, they begin to luminesce.

Materials Production

By integrating synthetic biology approaches with materials sciences, it would be possible to envision cells as microscopic molecular foundries to produce materials with properties that can be genetically encoded. Recent advances towards this include the re-engineering of curli fibers, the amyloid component of extracellular material of biofilms, as a platform for programmable nanomaterial. These nanofibers have been genetically constructed for specific functions, including: adhesion to substrates; nanoparticle templating; and protein immobilization.

Industrial Enzymes

Researchers and companies utilizing synthetic biology aim to synthesize enzymes with high activity, to produce products with optimal yields and effectiveness. These synthesized enzymes aim to improve products such as detergents and lactose-free dairy products, as well as make them more cost effective.

The improvements of metabolic engineering by synthetic biology is an example of a biotechnological technique utilized in industry to discover pharmaceuticals and fermentative chemicals. Synthetic biology may investigate modular pathway systems in biochemical production and increase yields of metabolic production. Artificial enzymatic activity and subsequent effects on metabolic reaction rates and yields may develop "efficient new strategies for improving cellular properties . . . for industrially important biochemical production."

Space Exploration

Synthetic biology raised NASA's interest as it could help to produce resources for astronauts from a restricted portfolio of compounds sent from Earth. On Mars, in particular, synthetic biology could also lead to production processes based on local resources, making it a powerful tool in the development of manned outposts with minimal dependence on Earth.

Bioethics and Security

In addition to numerous scientific and technical challenges, synthetic biology raises ethical issues and biosecurity issues. However, with the exception of regulating DNA synthesis companies, the issues are not seen as new because they were raised during the earlier recombinant DNA and genetically modified organism (GMO) debates and there were already extensive regulations of genetic engineering and pathogen research in place in the U.S.A., Europe and the rest of the world.

European Initiatives

The European Union funded project SYNBIOSAFE has issued several reports on how to manage the risks of synthetic biology. A 2007 paper identified key issues in safety, security, ethics and the science-society interface, which the project defined as public education and ongoing dialogue among scientists, businesses, government, and ethicists). The key security issues that SYNBIOSAFE identified involved engaging companies that sell synthetic DNA and the Biohacking community of amateur biologists. Key ethical issues concerned the creation of new life forms.

A subsequent report focused on biosecurity, especially the so-called dual-use challenge. For example, while synthetic biology may lead to more efficient production of medical treatments, for malaria for example, it may also lead to synthesis or redesign of harmful pathogens (e.g., smallpox). The bio-hacking community remains a source of special concern, as the distributed and diffuse nature of open-source biotechnology makes it difficult to track, regulate, or mitigate potential concerns over biosafety and biosecurity.

COSY, another European initiative, focuses on public perception and communication of synthetic biology. To better communicate synthetic biology and its societal ramifications to a broader public, COSY and SYNBIOSAFE published a 38-minute documentary film in October 2009.

The International Association Synthetic Biology has proposed an initiative for self-regulation. This suggests specific measures that the synthetic biology industry, especially DNA synthesis companies, should implement. In 2007, a group led by scientists from leading DNA-synthesis companies published a "practical plan for developing an effective oversight framework for the DNA-synthesis industry."

USA

In January 2009, the Alfred P. Sloan Foundation funded the Woodrow Wilson Center, the Hastings Center, and the J. Craig Venter Institute to examine the public perception, ethics, and policy implications of synthetic biology.

On July 9–10, 2009, the National Academies' Committee of Science, Technology & Law convened a symposium on "Opportunities and Challenges in the Emerging Field of Synthetic Biology".

After the publication of the first synthetic genome by Craig Venter's group and the accompanying media coverage about "life" being created, President Obama requested the Presidential Commission for the Study of Bioethical Issues to study synthetic biology. The commission convened a series of meetings, then issued a report in December 2010 titled "New Directions: The Ethics of Synthetic Biology and Emerging Technologies." The commission clarified that the "while Venter's achievement marked a significant technical advance in demonstrating that a relatively large genome could be accurately synthesized and substituted for another, it did not amount to the "creation of life". It also noted that synthetic biology is an emerging field, which creates potential risks and rewards. The commission did not recommend any changes to policy or oversight and called for continued funding of the research and new funding for monitoring, study of emerging ethical issues, and public education.

Synthetic biology, being a major tool for biological advances, results in the "potential for developing biological weapons, possible unforeseen negative impacts on human health . . . and any potential environmental impact." These security issues may be avoided by regulating industry uses of biotechnology through policy legislation. Federal guidelines on genetic manipulation are being proposed by "the President's Bioethics Commission . . . in response to the announced creation of a self-replicating cell from a chemically synthesized genome, put forward 18 recommendations not only for regulating the science . . . for educating the public."

Opposition

On March 13, 2012, over 100 environmental and civil society groups, including Friends of the Earth, the International Center for Technology Assessment and the ETC Group issued the manifesto *The Principles for the Oversight of Synthetic Biology*. This manifesto calls for a worldwide moratorium on the release and commercial use of synthetic organisms until more robust regulations and rigorous biosafety measures are established. The groups specifically call for an outright ban on the use of synthetic biology on the human genome or human microbiome. Richard Lewontin wrote that some of the safety tenets for oversight discussed in *The Principles for the Oversight of Synthetic Biology* are reasonable, but that the main problem with the recommendations in the manifesto is that "the public at large lacks the ability to enforce any meaningful realization of those recommendations."

Ethical Concerns

Synthetic biology brings to light a number of questions, including: who will have control and access to the products of synthetic biology, and who will gain from these innovations? Placing patents on living organisms and regulations on bioengineering of human embryos are large concerns in the bioethics field.

Evolutionary Biology

Evolutionary biology is the subfield of biology that studies the evolutionary processes that produced the diversity of life on Earth starting from a single origin of life. These processes include natural selection, the descent of species, and the origin of new species.

The discipline emerged through what Julian Huxley called the modern evolutionary synthesis (of the 1930s) of understanding from several previously unrelated fields of biological research, including genetics, ecology, systematics and palaeontology.

Current research has widened to cover the genetic architecture of adaptation, molecular evolution, and the different forces that contribute to evolution including sexual selection, genetic drift and biogeography. The newer field of evolutionary developmental biology ("evo-devo") investigates how embryonic development is controlled, thus creating a wider synthesis that integrates developmental biology with the fields covered by the earlier evolutionary synthesis.

Subfields

The study of evolution is the central unifying concept in biology. Biology can be divided in various ways. One way is by the level of biological organisation, from molecular to cell, organism to population. Another way is by taxonomic group, with fields such as zoology (all animals), ornithology (birds), and herpetology (reptiles and amphibians). A third way is by approach, such as field biology, theoretical biology, experimental evolution, and palaeontology. These alternative ways of dividing up the subject can be combined with evolutionary biology to create subfields like evolutionary ecology and evolutionary developmental biology.

History

Evolutionary biology, as an academic discipline in its own right, emerged during the period of the modern evolutionary synthesis in the 1930s and 1940s (Smocovitis, 1996). It was not until the 1980s that many universities had departments of evolutionary biology. In the United States, many universities have created departments of *molecular and cell biology* or *ecology and evolutionary biology*, in place of the older departments of botany and zoology. Palaeontology is often grouped with earth science.

The statistician Ronald Fisher (1890 – 1962) helped to form the modern evolutionary synthesis of Mendelian genetics and natural selection.

Microbiology too is becoming an evolutionary discipline, now that microbial physiology and genomics are better understood. The quick generation time of bacteria and viruses such as bacteriophages makes it possible to explore evolutionary questions.

J. B. S. Haldane (1892 – 1964) helped to create the field of population genetics.

Many biologists have contributed to our current understanding of evolution. Theodosius Dobzhansky and E. B. Ford established an empirical research programme. Ronald Fisher, Sewall Wright and J. S. Haldane created a sound theoretical framework. Ernst Mayr in systematics, George Gaylord Simpson in palaeontology and G. Ledyard Stebbins in botany helped to form the modern synthesis. James Crow, Richard Lewontin, Dan Hartl, Marcus Feldman, and Brian Charlesworth trained a generation of evolutionary biologists.

Current Research Topics

Current research in evolutionary biology covers diverse topics and incorporates ideas from diverse areas, such as molecular genetics and computer science.

First, some fields of evolutionary research try to explain phenomena that were poorly accounted for in the modern evolutionary synthesis. These include speciation, the evolution of sexual reproduction, the evolution of cooperation, the evolution of ageing, and evolvability.

Second, biologists ask the most straightforward evolutionary question: "what happened and when?". This includes fields such as palaeobiology, as well as systematics and phylogenetics.

Third, the modern evolutionary synthesis was devised at a time when nobody understood the molecular basis of genes. Today, evolutionary biologists try to determine the genetic architecture of interesting evolutionary phenomena such as adaptation and speciation. They seek answers to questions such as how many genes are involved, how large are the effects of each gene, how interdependent are the effects of different genes, what do the genes do, and what changes happen to them (e.g., point mutations vs. gene duplication or even genome duplication). They try to reconcile the high heritability seen in twin studies with the difficulty in finding which genes are responsible for this heritability using genome-wide association studies.

One challenge in studying genetic architecture is that the classical population genetics that catalysed the modern evolutionary synthesis must be updated to take into account modern molecular knowledge. This requires a great deal of mathematical development to relate DNA sequence data to evolutionary theory as part of a theory of molecular evolution. For example, biologists try to infer which genes have been under strong selection by detecting selective sweeps.

Fourth, the modern evolutionary synthesis involved agreement about which forces contribute to evolution, but not about their relative importance. Current research seeks to determine this. Evolutionary forces include natural selection, sexual selection, genetic drift, genetic draft, developmental constraints, mutation bias and biogeography.

An evolutionary approach is key to much current research in organismal biology and ecology, such as in life history theory. Annotation of genes and their function relies heavily on comparative, i.e., evolutionary, approaches. The field of evolutionary developmental biology ("evo-devo") investigates how developmental processes work, and compares them in different organisms determine how they evolved.

Astrobiology

Astrobiology is the study of the origin, evolution, distribution, and future of life in the universe:

extraterrestrial life and life on Earth. Astrobiology addresses the question of whether life exists beyond Earth, and how humans can detect it if it does (the term exobiology is similar but more specific—it covers the search for life beyond Earth, and the effects of extraterrestrial environments on living things).

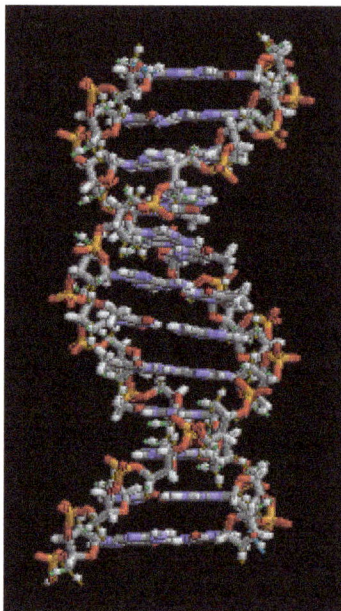

Nucleic acids may not be the only biomolecules in the Universe capable of coding for life processes.

Astrobiology makes use of physics, chemistry, astronomy, biology, molecular biology, ecology, planetary science, geography, and geology to investigate the possibility of life on other worlds and help recognize biospheres that might be different from that on Earth. The origin and early evolution of life is an inseparable part of the discipline of astrobiology. Astrobiology concerns itself with interpretation of existing scientific data; given more detailed and reliable data from other parts of the universe, the roots of astrobiology itself—physics, chemistry and biology—may have their theoretical bases challenged. Although speculation is entertained to give context, astrobiology concerns itself primarily with hypotheses that fit firmly into existing scientific theories.

This interdisciplinary field encompasses research on the origin and evolution of planetary systems, origins of organic compounds in space, rock-water-carbon interactions, abiogenesis on Earth, planetary habitability, research on biosignatures for life detection, and studies on the potential for life to adapt to challenges on Earth and in outer space.

The chemistry of life may have begun shortly after the Big Bang, 13.8 billion years ago, during a habitable epoch when the Universe was only 10–17 million years old. According to the panspermia hypothesis, microscopic life—distributed by meteoroids, asteroids and other small Solar System bodies—may exist throughout the universe. According to research published in August 2015, very large galaxies may be more favorable to the creation and development of habitable planets than smaller galaxies, like the Milky Way galaxy. Nonetheless, Earth is the only place in the universe known to harbor life. Estimates of habitable zones around other stars, along with the discovery of hundreds of extrasolar planets and new insights into the extreme habitats here on Earth, suggest that there may be many more habitable places in the universe than considered possible until very recently.

Current studies on the planet Mars by the *Curiosity* and *Opportunity* rovers are now searching for evidence of ancient life as well as plains related to ancient rivers or lakes that may have been habitable. The search for evidence of habitability, taphonomy (related to fossils), and organic molecules on the planet Mars is now a primary NASA and ESA objective on Mars.

Overview

The synonyms of astrobiology are diverse; however, the synonyms were structured in relation to the most important sciences implied in its development: astronomy and biology. The term exobiology was coined by molecular biologist Joshua Lederberg. Exobiology is considered to have a narrow scope limited to search of life external to Earth, whereas subject area of astrobiology is wider and investigates the link between life and the universe, which includes the search for extraterrestrial life, but also includes the study of life on Earth, its origin, evolution and limits.

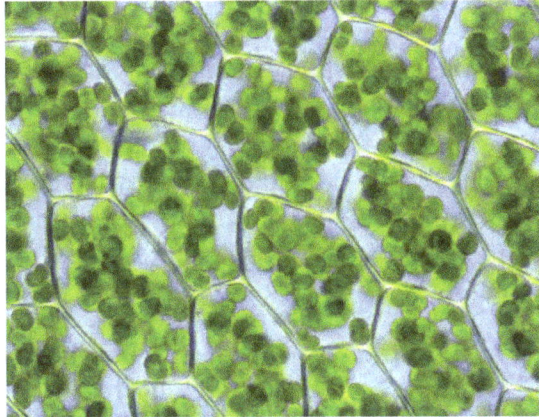

It is not known whether life elsewhere in the universe would utilize cell structures like those found on Earth. (Chloroplasts within plant cells shown here.)

Another term used in the past is xenobiology, ("biology of the foreigners") a word used in 1954 by science fiction writer Robert Heinlein in his work The Star Beast. The term *xenobiology* is now used in a more specialized sense, to mean "biology based on foreign chemistry", whether of extraterrestrial or terrestrial (possibly synthetic) origin. Since alternate chemistry analogs to some life-processes have been created in the laboratory, xenobiology is now considered as an extant subject.

While it is an emerging and developing field, the question of whether life exists elsewhere in the universe is a verifiable hypothesis and thus a valid line of scientific inquiry. Though once considered outside the mainstream of scientific inquiry, astrobiology has become a formalized field of study. Planetary scientist David Grinspoon calls astrobiology a field of natural philosophy, grounding speculation on the unknown, in known scientific theory. NASA's interest in exobiology first began with the development of the U.S. Space Program. In 1959, NASA funded its first exobiology project, and in 1960, NASA founded an Exobiology Program, which is now one of four main elements of NASA's current Astrobiology Program. In 1971, NASA funded the search for extraterrestrial intelligence (SETI) to search radio frequencies of the electromagnetic spectrum for interstellar communications transmitted by extraterrestrial life outside the Solar System. NASA's

Viking missions to Mars, launched in 1976, included three biology experiments designed to look for metabolism of present life on Mars.

In June 2014, the John W. Kluge Center of the Library of Congress held a seminar focusing on astrobiology. Panel members (L to R) Robin Lovin, Derek Malone-France, and Steven J. Dick

Advancements in the fields of astrobiology, observational astronomy and discovery of large varieties of extremophiles with extraordinary capability to thrive in the harshest environments on Earth, have led to speculation that life may possibly be thriving on many of the extraterrestrial bodies in the universe. A particular focus of current astrobiology research is the search for life on Mars due to its proximity to Earth and geological history. There is a growing body of evidence to suggest that Mars has previously had a considerable amount of water on its surface, water being considered an essential precursor to the development of carbon-based life.

Missions specifically designed to search for current life on Mars were the Viking program and Beagle 2 probes. The Viking results were inconclusive, and Beagle 2 failed minutes after landing. A future mission with a strong astrobiology role would have been the Jupiter Icy Moons Orbiter, designed to study the frozen moons of Jupiter—some of which may have liquid water—had it not been cancelled. In late 2008, the Phoenix lander probed the environment for past and present planetary habitability of microbial life on Mars, and to research the history of water there.

In November 2011, NASA launched the Mars Science Laboratory mission carrying the *Curiosity* rover, which landed on Mars at Gale Crater in August 2012. The *Curiosity* rover is currently probing the environment for past and present planetary habitability of microbial life on Mars. On 9 December 2013, NASA reported that, based on evidence from *Curiosity* studying Aeolis Palus, Gale Crater contained an ancient freshwater lake which could have been a hospitable environment for microbial life.

The European Space Agency is currently collaborating with the Russian Federal Space Agency (Roscosmos) and developing the ExoMars astrobiology rover, which is to be launched in 2018. While NASA is developing the Mars 2020 astrobiology rover and sample cacher for a later return to Earth.

Methodology

Planetary Habitability

When looking for life on other planets like Earth, some simplifying assumptions are useful to reduce the size of the task of the astrobiologist. One is the informed assumption that the vast majority of life forms in our galaxy are based on carbon chemistries, as are all life forms on Earth. Carbon is well known for the unusually wide variety of molecules that can be formed around it. Carbon is the fourth most abundant element in the universe and the energy required to make or break a bond is just at an appropriate level for building molecules which are not only stable, but also reactive. The fact that carbon atoms bond readily to other carbon atoms allows for the building of arbitrarily long and complex molecules.

The presence of liquid water is an assumed requirement, as it is a common molecule and provides an excellent environment for the formation of complicated carbon-based molecules that could eventually lead to the emergence of life. Some researchers posit environments of ammonia, or more likely, water-ammonia mixtures as possible solvents for hypothetical types of biochemistry.

A third assumption is to focus on planets orbiting Sun-like stars for increased probabilities of planetary habitability. Very large stars have relatively short lifetimes, meaning that life might not have time to emerge on planets orbiting them. Very small stars provide so little heat and warmth that only planets in very close orbits around them would not be frozen solid, and in such close orbits these planets would be tidally "locked" to the star. The long lifetimes of red dwarfs could allow the development of habitable environments on planets with thick atmospheres. This is significant, as red dwarfs are extremely common.

Since Earth is the only planet known to harbor life, there is no evident way to know if any of the simplifying assumptions are correct.

Communication Attempts

Research on communication with extraterrestrial intelligence (CETI) focuses on composing and deciphering messages that could theoretically be understood by another technological civilization. Communication attempts by humans have included broadcasting mathematical languages, pictorial systems such as the Arecibo message and computational approaches to detecting and deciphering 'natural' language communication. The SETI program, for example, uses both radio telescopes and optical telescopes to search for deliberate signals from an extraterrestrial intelligence.

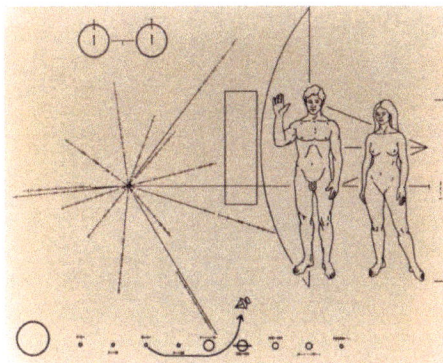

The illustration on the Pioneer plaque

While some high-profile scientists, such as Carl Sagan, have advocated the transmission of messages, scientist Stephen Hawking has warned against it, suggesting that aliens might simply raid Earth for its resources and then move on.

Elements of Astrobiology

Astronomy

Most astronomy-related astrobiology research falls into the category of extrasolar planet (exoplanet) detection, the hypothesis being that if life arose on Earth, then it could also arise on other planets with similar characteristics. To that end, a number of instruments designed to detect Earth-sized exoplanets have been considered, most notably NASA's Terrestrial Planet Finder (TPF) and ESA's Darwin programs, both of which have been cancelled. NASA launched the Kepler mission in March 2009, and the French Space Agency launched the COROT space mission in 2006. There are also several less ambitious ground-based efforts underway.

Artist's impression of the extrasolar planet OGLE-2005-BLG-390Lb orbiting its star 20,000 light-years from Earth; this planet was discovered with gravitational microlensing.

The goal of these missions is not only to detect Earth-sized planets, but also to directly detect light from the planet so that it may be studied spectroscopically. By examining planetary spectra, it would be possible to determine the basic composition of an extrasolar planet's atmosphere and/ or surface. Given this knowledge, it may be possible to assess the likelihood of life being found on that planet. A NASA research group, the Virtual Planet Laboratory, is using computer modeling to generate a wide variety of virtual planets to see what they would look like if viewed by TPF or Darwin. It is hoped that once these missions come online, their spectra can be cross-checked with these virtual planetary spectra for features that might indicate the presence of life.

The NASA Kepler mission, launched in March 2009, searches for extrasolar planets.

An estimate for the number of planets with intelligent *communicative* extraterrestrial life can be gleaned from the Drake equation, essentially an equation expressing the probability of intelligent life as the product of factors such as the fraction of planets that might be habitable and the fraction of planets on which life might arise:

where:

- N = The number of communicative civilizations

- R^* = The rate of formation of suitable stars (stars such as our Sun)

- f_p = The fraction of those stars with planets (current evidence indicates that planetary systems may be common for stars like the Sun)

- n_e = The number of Earth-sized worlds per planetary system

- f_l = The fraction of those Earth-sized planets where life actually develops

- f_i = The fraction of life sites where intelligence develops

- f_c = The fraction of communicative planets (those on which electromagnetic communications technology develops)

- L = The "lifetime" of communicating civilizations

However, whilst the rationale behind the equation is sound, it is unlikely that the equation will be constrained to reasonable error limits any time soon. The first term, R^*, number of stars, is generally constrained within a few orders of magnitude. The second and third terms, f_p, stars with planets and f_e, planets with habitable conditions, are being evaluated for the star's neighborhood. The problem with the formula is that it is not usable to generate or support hypotheses because it contains factors that can never be verified. Drake originally formulated the equation merely as an agenda for discussion at the Green Bank conference, but some applications of the formula had been taken literally and related to simplistic or pseudoscientific arguments. Another associated topic is the Fermi paradox, which suggests that if intelligent life is common in the universe, then there should be obvious signs of it.

Another active research area in astrobiology is planetary system formation. It has been suggested that the peculiarities of the Solar System (for example, the presence of Jupiter as a protective shield) may have greatly increased the probability of intelligent life arising on our planet.

Biology

Biology cannot state that a process or phenomenon, by being mathematically possible, has to exist forcibly in an extraterrestrial body. Biologists specify what is speculative and what is not.

Until the 1970s, life was thought to be entirely dependent on energy from the Sun. Plants on Earth's surface capture energy from sunlight to photosynthesize sugars from carbon dioxide and water, releasing oxygen in the process that is then consumed by oxygen-respiring organisms, passing their energy up the food chain. Even life in the ocean depths, where sunlight cannot reach, was thought to obtain its nourishment either from consuming organic detritus rained down from the surface waters or from eating animals that did. The world's ability to support life was thought to depend on its access to sunlight. However, in 1977, during an exploratory dive to the Galapagos Rift in the deep-sea exploration submersible *Alvin*, scientists discovered colonies of giant tube worms, clams, crustaceans, mussels, and other assorted creatures clustered around undersea volcanic features known as black smokers. These creatures thrive despite having no access to sunlight, and it was soon discovered that they comprise an entirely independent ecosystem. Although most of these multicellular lifeforms need dissolved oxygen (produced by oxygenic photosynthesis) for their aerobic cellular respiration and thus are not completely independent from sunlight by themselves, the basis for their food chain is a form of bacterium that derives its energy from oxidization of reactive chemicals, such as hydrogen or hydrogen sulfide, that bubble up from the Earth's interior. Other lifeforms entirely decoupled from the energy from sunlight are green sulphur bacteria which are capturing geothermal light for anoxygenic photosynthesis or bacteria running chemolithoautotrophy based on the radioactive decay of uranium. This chemosynthesis revolutionized the study of biology and astrobiology by revealing that life need not be sun-dependent; it only requires water and an energy gradient in order to exist.

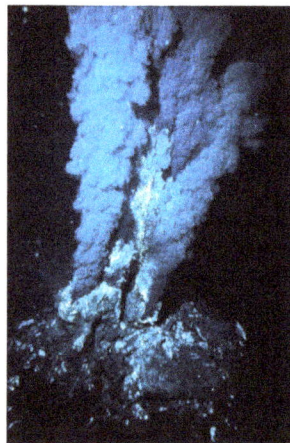

Hydrothermal vents are able to support extremophile bacteria on Earth and may also support life in other parts of the cosmos.

Extremophiles, organisms able to survive in extreme environments, are a core research element for astrobiologists. Such organisms include biota which are able to survive several kilometers below the ocean's surface near hydrothermal vents and microbes that thrive in highly acidic environments. It is now known that extremophiles thrive in ice, boiling water, acid, alkali, the water core of nuclear reactors, salt crystals, toxic waste and in a range of other extreme habitats that were previously thought to be inhospitable for life. It opened up a new avenue in astrobiology by massively expanding the number of possible extraterrestrial habitats. Characterization of these organisms, their environments and their evolutionary pathways, is considered a crucial component to understanding how life might evolve elsewhere in the universe. For example, some organisms able to withstand exposure to the vacuum and radiation of outer space include the lichen fungi *Rhizocarpon geographicum* and *Xanthoria elegans*, the bacterium *Bacillus safensis*, *Deinococcus radiodurans*, *Bacillus subtilis*, yeast *Saccharomyces cerevisiae*, seeds from *Arabidopsis thaliana* ('mouse-ear cress'), as well as the invertebrate animal Tardigrade.

Jupiter's moon, Europa, and Saturn's moon, Enceladus, are now considered the most likely locations for extant extraterrestrial life in the Solar System due to their subsurface water oceans where radiogenic and tidal heating enables liquid water to exist.

The origin of life, known as abiogenesis, distinct from the evolution of life, is another ongoing field of research. Oparin and Haldane postulated that the conditions on the early Earth were conducive to the formation of organic compounds from inorganic elements and thus to the formation of many of the chemicals common to all forms of life we see today. The study of this process, known as prebiotic chemistry, has made some progress, but it is still unclear whether or not life could have formed in such a manner on Earth. The alternative hypothesis of panspermia is that the first elements of life may have formed on another planet with even more favorable conditions (or even in interstellar space, asteroids, etc.) and then have been carried over to Earth — the panspermia hypothesis.

The cosmic dust permeating the universe contains complex organic matter ("amorphous organic solids with a mixed aromatic-aliphatic structure") that could be created naturally, and rapidly, by stars. Further, a scientist suggested that these compounds may have been related to the development of life on Earth and said that, "If this is the case, life on Earth may have had an easier time getting started as these organics can serve as basic ingredients for life." In September 2012, NASA scientists reported that polycyclic aromatic hydrocarbons (PAHs), subjected to interstellar medium conditions, are transformed through hydrogenation, oxygenation and hydroxylation, to more complex organics - "a step along the path toward amino acids and nucleotides, the raw materials of proteins and DNA, respectively".

More than 20% of the carbon in the universe may be associated with PAHs, possible starting materials for the formation of life. PAHs seem to have been formed shortly after the Big Bang, are widespread throughout the universe, and are associated with new stars and exoplanets.

Astroecology

Astroecology concerns the interactions of life with space environments and resources, in planets, asteroids and comets. On a larger scale, astroecology concerns resources for life about stars in the galaxy through the cosmological future. Astroecology attempts to quantify future life in space, addressing this area of astrobiology.

Experimental astroecology investigates resources in planetary soils, using actual space materials in meteorites. The results suggest that Martian and carbonaceous chondrite materials can support bacteria, algae and plant (asparagus, potato) cultures, with high soil fertilities. The results support that life could have survived in early aqueous asteroids and on similar materials imported to Earth by dust, comets and meteorites, and that such asteroid materials can be used as soil for future space colonies.

On the largest scale, cosmoecology concerns life in the universe over cosmological times. The main sources of energy may be red giant stars and white and red dwarf stars, sustaining life for 10^{20} years. Astroecologists suggest that their mathematical models may quantify the potential amounts of future life in space, allowing a comparable expansion in biodiversity, potentially leading to diverse intelligent life forms.

Astrogeology

Astrogeology is a planetary science discipline concerned with the geology of the celestial bodies such as the planets and their moons, asteroids, comets, and meteorites. The information gathered by this discipline allows the measure of a planet's or a natural satellite's potential to develop and sustain life, or planetary habitability.

An additional discipline of astrogeology is geochemistry, which involves study of the chemical composition of the Earth and other planets, chemical processes and reactions that govern the composition of rocks and soils, the cycles of matter and energy and their interaction with the hydrosphere and the atmosphere of the planet. Specializations include cosmochemistry, biochemistry and organic geochemistry.

The fossil record provides the oldest known evidence for life on Earth. By examining the fossil evidence, paleontologists are able to better understand the types of organisms that arose on the early Earth. Some regions on Earth, such as the Pilbara in Western Australia and the McMurdo Dry Valleys of Antarctica, are also considered to be geological analogs to regions of Mars, and as such, might be able to provide clues on how to search for past life on Mars.

The various organic functional groups, composed of hydrogen, oxygen, nitrogen, phosphorus, sulfur, and a host of metals, such as iron, magnesium, and zinc, provide the enormous diversity of chemical reactions necessarily catalyzed by a living organism. Silicon, in contrast, interacts with only a few other atoms, and the large silicon molecules are monotonous compared with the combinatorial universe of organic macromolecules. Indeed, it seems likely that the basic building blocks of life anywhere will be similar those on Earth, in the generality if not in the detail. Although terrestrial life and life that might arise independently of Earth are expected to use many similar, if not identical, building blocks, they also are expected to have some biochemical qualities that are unique. If life has had a comparable impact elsewhere in the Solar System, the relative abundances of chemicals key for its survival - whatever they may be - could betray its presence. Whatever extraterrestrial life may be, its tendency to chemically alter its environment might just give it away.

Life in The Solar System

People have long speculated about the possibility of life in settings other than Earth, however, speculation on the nature of life elsewhere often has paid little heed to constraints imposed by the nature of biochemistry. The likelihood that life throughout the universe is probably carbon-based is suggested by the fact that carbon is one of the most abundant of the higher elements. Only two of the natural atoms, carbon and silicon, are known to serve as the backbones of molecules sufficiently large to carry biological information. As the structural basis for life, one of carbon's important features is that unlike silicon, it can readily engage in the formation of chemical bonds with many other atoms, thereby allowing for the chemical versatility required to conduct the reactions of biological metabolism and propagation.

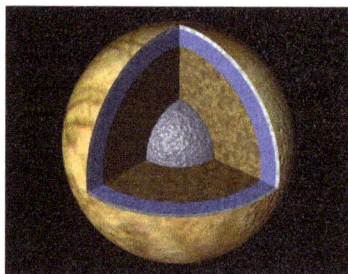

Europa, due to the ocean that exists under its icy surface, might host some form of microbial life.

Thought on where in the Solar System life might occur, was limited historically by the understanding that life relies ultimately on light and warmth from the Sun and, therefore, is restricted to the surfaces of planets. The three most likely candidates for life in the Solar System are the planet Mars, the Jovian moon Europa, and Saturn's moon Titan. More recently, Saturn's moon Enceladus may be considered a likely candidate as well.

Mars, Enceladus and Europa are considered likely candidates in the search for life primarily because they may have liquid water, a molecule essential for life as we know it for its use as a solvent in cells. Water on Mars is found in its polar ice caps, and newly carved gullies recently observed on Mars suggest that liquid water may exist, at least transiently, on the planet's surface. At the Martian low temperatures and low pressure, liquid water is likely to be highly saline. As for Europa, liquid water likely exists beneath the moon's icy outer crust. This water may be warmed to a liquid state by volcanic vents on the ocean floor, but the primary source of heat is probably tidal heating. On 11 December 2013, NASA reported the detection of "clay-like minerals" (specifically, phyllosilicates), often associated with organic materials, on the icy crust of Europa. The presence of the minerals may have been the result of a collision with an asteroid or comet according to the scientists.

Another planetary body that could potentially sustain extraterrestrial life is Saturn's largest moon, Titan. Titan has been described as having conditions similar to those of early Earth. On its surface, scientists have discovered the first liquid lakes outside Earth, but they seem to be composed of ethane and/or methane, not water. Some scientists think it possible that these liquid hydrocarbons might take the place of water in living cells different from those on Earth. After Cassini data was studied, it was reported on March 2008 that Titan may also have an underground ocean composed of liquid water and ammonia. Additionally, Saturn's moon Enceladus may have an ocean below its icy surface and, according to NASA scientists in May 2011, "is emerging as the most habitable spot beyond Earth in the Solar System for life as we know it".

Measuring the ratio of hydrogen and methane levels on Mars may help determine the likelihood of life on Mars. According to the scientists, "...low H_2/CH_4 ratios (less than approximately 40) indicate that life is likely present and active." Other scientists have recently reported methods of detecting hydrogen and methane in extraterrestrial atmospheres.

Complex organic compounds of life, including uracil, cytosine and thymine, have been formed in a laboratory under outer space conditions, using starting chemicals such as pyrimidine, found in meteorites. Pyrimidine, like polycyclic aromatic hydrocarbons (PAHs), the most carbon-rich chemical found in the universe.

Rare Earth Hypothesis

The Rare Earth hypothesis postulates that multicellular life forms found on Earth may actually be more of a rarity than scientists assume. It provides a possible answer to the Fermi paradox which suggests, "If extraterrestrial aliens are common, why aren't they obvious?" It is apparently in opposition to the principle of mediocrity, assumed by famed astronomers Frank Drake, Carl Sagan, and others. The Principle of Mediocrity suggests that life on Earth is not exceptional, but rather that life is more than likely to be found on innumerable other worlds.

The anthropic principle states that fundamental laws of the universe work specifically in a way that life would be possible. The anthropic principle supports the Rare Earth Hypothesis by arguing the overall elements that are needed to support life on Earth are so fine-tuned that it is nearly impossible for another just like it to exist by random chance.

Research

The systematic search for possible life outside Earth is a valid multidisciplinary scientific endeavor. However, hypotheses and predictions as to its existence and origin vary widely, and at the present, the development of hypotheses firmly grounded on science may be considered astrobiology's most concrete practical application. It has been proposed that viruses are likely to be encountered on other life-bearing planets.

Research Outcomes

As of 2015, no evidence of extraterrestrial life has been identified. Examination of the Allan Hills 84001 meteorite, which was recovered in Antarctica in 1984 and originated from Mars, is thought by David McKay, as well as few other scientists, to contain microfossils of extraterrestrial origin; this interpretation is controversial.

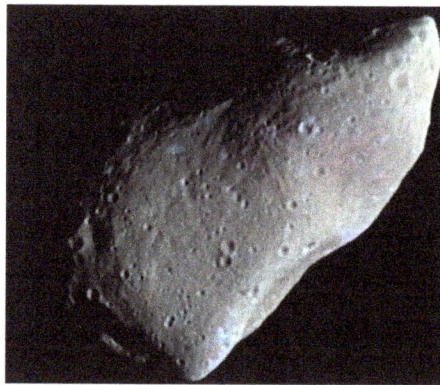

Asteroid(s) may have transported life to Earth.

Yamato 000593 is the second largest meteorite from Mars, and was found on Earth in 2000. At a microscopic level, spheres are found in the meteorite that are rich in carbon compared to surrounding areas that lack such spheres. The carbon-rich spheres may have been formed by biotic activity according to some NASA scientists.

On 5 March 2011, Richard B. Hoover, a scientist with the Marshall Space Flight Center, speculated on the finding of alleged microfossils similar to cyanobacteria in CI1 carbonaceous meteorites in the fringe *Journal of Cosmology*, a story widely reported on by mainstream media. However, NASA formally distanced itself from Hoover's claim. According to American astrophysicist Neil deGrasse Tyson: "At the moment, life on Earth is the only known life in the universe, but there are compelling arguments to suggest we are not alone."

Extreme Environments on Earth

On 17 March 2013, researchers reported that microbial life forms thrive in the Mariana Trench, the deepest spot on the Earth. Other researchers reported related studies that microbes thrive inside rocks up to 1900 feet below the sea floor under 8500 feet of ocean off the coast of the northwestern United States. According to one of the researchers, "You can find microbes everywhere — they're extremely adaptable to conditions, and survive wherever they are." These finds expand the potential habitability of certain niches of other planets.

Methane

In 2004, the spectral signature of methane (CH4) was detected in the Martian atmosphere by both Earth-based telescopes as well as by the Mars Express orbiter. Because of solar radiation and cosmic radiation, methane is predicted to disappear from the Martian atmosphere within several years, so the gas must be actively replenished in order to maintain the present concentration. The *Curiosity* rover will perform precision measurements of oxygen and carbon isotope ratios in carbon dioxide (CO_2) and methane (CH_4) in the atmosphere of Mars in order to distinguish between a geochemical and a biological origin.

Planetary Systems

It is possible that some exoplanets may have moons with solid surfaces or liquid oceans that are hospitable. Most of the planets so far discovered outside the Solar System are hot gas giants thought to be inhospitable to life, so it is not yet known whether the Solar System, with a warm, rocky, metal-rich inner planet such as Earth, is of an aberrant composition. Improved detection methods and increased observing time will undoubtedly discover more planetary systems, and possibly some more like ours. For example, NASA's Kepler Mission seeks to discover Earth-sized planets around other stars by measuring minute changes in the star's light curve as the planet passes between the star and the spacecraft. Progress in infrared astronomy and submillimeter astronomy has revealed the constituents of other star systems.

Planetary Habitability

Efforts to answer questions such as the abundance of potentially habitable planets in habitable zones and chemical precursors have had much success. Numerous extrasolar planets have been detected using the wobble method and transit method, showing that planets around other stars are more numerous than previously postulated. The first Earth-sized extrasolar planet to be discovered within its star's habitable zone is Gliese 581 c.

Missions

Research into the environmental limits of life and the workings of extreme ecosystems is ongoing, enabling researchers to better predict what planetary environments might be most likely to harbor life. Missions such as the Phoenix lander, Mars Science Laboratory, ExoMars, Mars 2020 rover to Mars, and the *Cassini* probe to Saturn's moons aim to further explore the possibilities of life on other planets in the Solar System.

Viking Program

Carl Sagan posing with a model of the Viking Lander.

The two Viking landers each carried four types of biological experiments to the surface of Mars in the late 1970s. These were the only Mars landers to carry out experiments to look specifically for metabolism by current microbial life on Mars. The landers used a robotic arm to collect soil samples into sealed test containers on the craft. The two landers were identical, so the same tests were carried out at two places on Mars' surface; Viking 1 near the equator and Viking 2 further north. The result was inconclusive, and is still disputed by some scientists.

Beagle 2

Beagle 2 was an unsuccessful British Mars lander that formed part of the European Space Agency's 2003 Mars Express mission. Its primary purpose was to search for signs of life on Mars, past or present. Although it landed safely, it was unable to correctly deploy its solar panels and telecom antenna.

Replica of the 33.2 kg *Beagle-2* lander

Mars Science Laboratory rover concept artwork

EXPOSE

EXPOSE is a multi-user facility mounted in 2008 outside the International Space Station dedicated to astrobiology. EXPOSE was developed by the European Space Agency (ESA) for long-term spaceflights that allows to expose organic chemicals and biological samples to outer space in low Earth orbit.

Mars Science Laboratory

The Mars Science Laboratory (MSL) mission landed a rover that is currently in operation on Mars. It was launched 26 November 2011, and landed at Gale Crater on 6 August 2012. Mission objectives are to help assess Mars' habitability and in doing so, determine whether Mars is or has ever been able to support life, collect data for a future human mission, study Martian geology, its climate, and further assess the role that water, an essential ingredient for life as we know it, played in forming minerals on Mars.

ExoMars rover

ExoMars is a robotic mission to Mars to search for possible biosignatures of Martian life, past or present. This astrobiological mission is currently under development by the European Space Agency (ESA) in partnership with the Russian Federal Space Agency (Roscosmos); it is planned for a 2018 launch.

ExoMars rover model

Red Dragon

Red Dragon is a planned series low-cost Mars lander missions that will utilize the SpaceX Falcon Heavy launch vehicle, and a modified Dragon V2 capsule to enter the Martian atmosphere and land using retrorockets. The lander's primary mission would be a technology demonstration, and to search for evidence of life on Mars (biosignatures), past or present. The concept had been meant to compete for funding on 2012/2013 as a NASA Discovery mission. On April 2016, SpaceX announced that they will proceed with the mission, with technical support from NASA, to be launched with a Falcon Heavy rocket in 2018. These Mars missions will also be pathfinders for the much larger SpaceX Mars colonization architecture that will be announced in September 2016.

Mars 2020

The 'Mars 2020' rover mission is a concept under development by NASA with a possible launch in 2020. It is intended to investigate environments on Mars relevant to astrobiology, investigate its surface geological processes and history, including the assessment of its past habitability and potential for preservation of biosignatures and biomolecules within accessible geological materials. The Science Definition Team is proposing the rover collect and package at least 31 samples of rock cores and soil for a later mission to bring back for more definitive analysis in laboratories on Earth. The rover could make measurements and technology demonstrations to help designers of a human expedition understand any hazards posed by Martian dust and demonstrate how to collect carbon dioxide (CO_2), which could be a resource for making molecular oxygen (O_2) and rocket fuel.

Proposed Concepts

Icebreaker Life

Icebreaker Life is a lander mission that is being proposed for NASA's Discovery Program for the 2018 launch opportunity. If selected and funded, the stationary lander would be a near copy of the successful 2008 *Phoenix* and it would carry an upgraded astrobiology scientific payload, including a 1-meter-long core drill to sample ice-cemented ground in the northern plains to conduct a search for organic molecules and evidence of current or past life on Mars. One of the key goals of the *Icebreaker Life* mission is to test the hypothesis that the ice-rich ground in the polar regions has significant concentrations of organics due to protection by the ice from oxidants and radiation.

Journey to Enceladus and Titan

Journey to Enceladus and Titan (JET) is an orbiter astrobiology mission concept to assess the habitability potential of Saturn's moons Enceladus and Titan.

Enceladus Life Finder

Enceladus Life Finder (ELF) is a proposed astrobiology mission concept for a space probe intended to assess the habitability of the internal aquatic ocean of Enceladus, Saturn's sixth-largest moon.

Life Investigation for Enceladus

Life Investigation For Enceladus (LIFE) is a proposed astrobiology sample-return mission concept for Enceladus. The spacecraft would enter into Saturn orbit and enable multiple flybys through Enceladus' icy plumes to collect icy plume particles and volatiles and return them to Earth on a capsule. The spacecraft may sample Enceladus' plumes, the E ring of Saturn, and the Titan upper atmosphere.

Europa Multiple-flyby Mission

Europa Multiple-Flyby Mission is a mission planned by NASA for a 2025 launch that will conduct detailed reconnaissance of Jupiter's moon Europa and will investigate whether the icy moon could harbor conditions suitable for life. It will also aid in the selection of future landing sites.

Cryobiology

Cryobiology is the branch of biology that studies the effects of low temperatures on living things within Earth's cryosphere or in science.In practice, cryobiology is the study of biological material or systems at temperatures below normal. Materials or systems studied may include proteins, cells, tissues, organs, or whole organisms. Temperatures may range from moderately hypothermic conditions to cryogenic temperatures.

Areas of Study

At least six major areas of cryobiology can be identified: 1) study of cold-adaptation of microorganisms, plants (cold hardiness), and animals, both invertebrates and vertebrates (including hibernation), 2) cryopreservation of cells, tissues, gametes, and embryos of animal and human origin for (medical) purposes of long-term storage by cooling to temperatures below the freezing point of water. This usually requires the addition of substances which protect the cells during freezing and thawing (cryoprotectants), 3) preservation of organs under hypothermic conditions for transplantation, 4) lyophilization (freeze-drying) of pharmaceuticals, 5) cryosurgery, a (minimally) invasive approach for the destruction of unhealthy tissue using cryogenic gases/fluids, and 6) physics of supercooling, ice nucleation/growth and mechanical engineering aspects of heat transfer during cooling and warming, as applied to biological systems. Cryobiology would include cryonics, the low temperature preservation of humans and mammals with the intention of future revival, although this is not part of mainstream cryobiology, depending heavily on speculative technology yet to be invented. Several of these areas of study rely on cryogenics, the branch of physics and engineering that studies the production and use of very low temperatures

Cryopreservation in Nature

Many living organisms are able to tolerate prolonged periods of time at temperatures below the freezing point of water. Most living organisms accumulate cryoprotectants such as antinucleating proteins, polyols, and glucose to protect themselves against frost damage by sharp ice crystals. Most plants, in particular, can safely reach temperatures of −4 °C to −12 °C.

Bacteria

Three species of bacteria, *Carnobacterium pleistocenium*, *Chryseobacterium greenlandensis*. and *Herminiimonas glaciei*, have reportedly been revived after surviving for thousands of years frozen in ice. Certain bacteria, notably *Pseudomonas syringae*, produce specialized proteins that serve as potent ice nucleators, which they use to force ice formation on the surface of various fruits and plants at about −2 °C. The freezing causes injuries in the epithelia and makes the nutrients in the underlying plant tissues available to the bacteria. *Listeria* grows slowly in temperatures as low as -1.5 °C and persists for some time in frozen foods.

Plants

Many plants undergo a process called hardening which allows them to survive temperatures below 0 °C for weeks to months.

Animals

Invertebrates

Nematodes that survive below 0 °C include *Trichostrongylus colubriformis* and *Panagrolaimus davidi*. Cockroach nymphs (*Periplaneta japonica*) survive short periods of freezing at -6 to -8 °C. The red flat bark beetle (*Cucujus clavipes*) can survive after being frozen to -150 °C. The fungus gnat *Exechia nugatoria* can survive after being frozen to -50 °C, by a unique mechanism whereby ice crystals form in the body but not the head. Another freeze-tolerant beetle is *Upis ceramboides*. See insect winter ecology and antifreeze protein. Another invertebrate that is briefly tolerant to temperatures down to -273 °C is the tardigrade.

The larvae of *Haemonchus contortus*, a nematode, can survive 44 weeks frozen at -196 °C.

Vertebrates

For the wood frog (*Rana sylvatica*), in the winter, as much as 45% of its body may freeze and turn to ice. "Ice crystals form beneath the skin and become interspersed among the body's skeletal muscles. During the freeze, the frog's breathing, blood flow, and heartbeat cease. Freezing is made possible by specialized proteins and glucose, which prevent intracellular freezing and dehydration." The wood frog can survive up to 11 days frozen at -4 °C.

Other vertebrates that survive at body temperatures below 0 °C include painted turtles (*Chrysemys picta*), gray tree frogs (*Hyla versicolor*), box turtles (*Terrapene carolina* - 48 hours at -2 °C), spring peeper (*Pseudacris crucifer*), garter snakes (*Thamnophis sirtalis* - 24 hours at -1.5 °C), the chorus frog (*Pseudacris triseriata*), Siberian salamander (*Salamandrella keyserlingii* - 24 hours at -15.3 °C), European common lizard (*Lacerta vivipara*) and Antarctic fish such as *Pagothenia borchgrevinki*. Antifreeze proteins cloned from such fish have been used to confer frost-resistance on transgenic plants.

Hibernating Arctic ground squirrels may have abdominal temperatures as low as −2.9 °C (26.8 °F), maintaining subzero abdominal temperatures for more than three weeks at a time, although the temperatures at the head and neck remain at 0 °C or above.

Applied Cryobiology

Historical Background

Cryobiology history can be traced back to antiquity. As early as in 2500 BC, low temperatures were used in Egypt in medicine. The use of cold was recommended by Hippocrates to stop bleeding and swelling. With the emergence of modern science, Robert Boyle studied the effects of low temperatures on animals.

Boyle

In 1949, bull semen was cryopreserved for the first time by a team of scientists led by Christopher Polge. This led to a much wider use of cryopreservation today, with many organs, tissues and cells routinely stored at low temperatures. Large organs such as hearts are usually stored and transported, for short times only, at cool but not freezing temperatures for transplantation. Cell suspensions (like blood and semen) and thin tissue sections can sometimes be stored almost indefinitely in liquid nitrogen temperature (cryopreservation). Human sperm, eggs, and embryos are routinely stored in fertility research and treatments. Controlled-rate and slow freezing are well established techniques pioneered in the early 1970s which enabled the first human embryo frozen birth (Zoe Leyland) in 1984. Since then, machines that freeze biological samples using programmable steps, or controlled rates, have been used all over the world for human, animal, and cell biology – 'freezing down' a sample to better preserve it for eventual thawing, before it is deep frozen, or cryopreserved, in liquid nitrogen. Such machines are used for freezing oocytes, skin, blood products, embryo, sperm, stem cells, and general tissue preservation in hospitals, veterinary practices, and research labs. The number of live births from 'slow frozen' embryos is some 300,000 to 400,000 or 20% of the estimated 3 million *in vitro* fertilized births. Dr Christopher Chen, Australia, reported the world's first pregnancy using slow-frozen oocytes from a British controlled-rate freezer in 1986.

Cryosurgery (intended and controlled tissue destruction by ice formation) was carried out by James Arnott in 1845 in an operation on a patient with cancer. Cryosurgery is not common.

Low temperature bank, Institute for Problems of Cryobiology and Cryomedicine of the National Academy of Sciences of Ukraine

Preservation Techniques

Cryobiology as an applied science is primarily concerned with low-temperature preservation. Hypothermic storage is typically above 0 °C but below normothermic (32 °C to 37 °C) mammalian temperatures. Storage by cryopreservation, on the other hand, will be in the −80 to −196 °C temperature range. Organs, and tissues are more frequently the objects of hypothermic storage, whereas single cells have been the most common objects cryopreserved.

A rule of thumb in hypothermic storage is that every 10 °C reduction in temperature is accompanied by a 50% decrease in oxygen consumption. Although hibernating animals have adapted mechanisms to avoid metabolic imbalances associated with hypothermia, hypothermic organs, and tissues being maintained for transplantation require special preservation solutions to counter acidosis, depressed sodium pump activity. and increased intracellular calcium. Special organ preservation solutions such as Viaspan (University of Wisconsin solution), HTK, and Celsior have been designed for this purpose. These solutions also contain ingredients to minimize damage by free radicals, prevent edema, compensate for ATP loss, etc.

Cryopreservation of cells is guided by the "two-factor hypothesis" of American cryobiologist Peter Mazur, which states that excessively rapid cooling kills cells by intracellular ice formation and excessively slow cooling kills cells by either electrolyte toxicity or mechanical crushing. During slow cooling, ice forms extracellularly, causing water to osmotically leave cells, thereby dehydrating them. Intracellular ice can be much more damaging than extracellular ice.

For red blood cells, the optimum cooling rate is very rapid (nearly 100 °C per second), whereas for stem cells the optimum cooling rate is very slow (1 °C per minute). Cryoprotectants, such as dimethyl sulfoxide and glycerol, are used to protect cells from freezing. A variety of cell types are protected by 10% dimethyl sulfoxide. Cryobiologists attempt to optimize cryoprotectant concentration (minimizing both ice formation and toxicity) and cooling rate. Cells may be cooled at an optimum rate to a temperature between −30 and −40 °C before being plunged into liquid nitrogen.

Slow cooling methods rely on the fact that cells contain few nucleating agents, but contain naturally occurring vitrifying substances that can prevent ice formation in cells that have been moderately dehydrated. Some cryobiologists are seeking mixtures of cryoprotectants for full vitrification (zero ice formation) in preservation of cells, tissues, and organs. Vitrification methods pose a challenge in the requirement to search for cryoprotectant mixtures that can minimize toxicity.

Cryobiology in Humans

Human gametes and two-, four- and eight-cell embryos can survive cryopreservation at -196 °C for 10 years under well-controlled laboratory conditions.

Cryopreservation in humans with regards to infertility involves preservation of embryos, sperm, or oocytes via freezing. Conception, in vitro, is attempted when the sperm is thawed and introduced to the 'fresh' eggs, the frozen eggs are thawed and sperm is placed with the eggs and together they are placed back into the uterus or a frozen embryo is introduced to the uterus. Vitrification has flaws and is not as reliable or proven as freezing fertilized sperm, eggs, or embryos as traditional slow freezing methods because eggs alone are extremely sensitive to temperature. Many researchers are also freezing ovarian tissue in conjunction with the eggs in hopes that the ovarian tissue can be transplanted back into the uterus, stimulating normal ovulation cycles. In 2004, Donnez of Louvain in Belgium reported the first successful ovarian birth from frozen ovarian tissue. In 1997, samples of ovarian cortex were taken from a woman with Hodgkin's lymphoma and cryopreserved in a (Planer, UK) controlled-rate freezer and then stored in liquid nitrogen. Chemotherapy was initiated after the patient had premature ovarian failure. In 2003, after freeze-thawing, orthotopic autotransplantation of ovarian cortical tissue was done by laparoscopy and after five months, re-implantation signs indicated recovery of regular ovulatory cycles. Eleven months after reimplantation, a viable intrauterine pregnancy was confirmed, which resulted in the first such live birth – a girl named Tamara.

Therapeutic hypothermia, e.g. during heart surgery on a "cold" heart (generated by cold perfusion without any ice formation) allows for much longer operations and improves recovery rates for patients.

Scientific Societies

The Society for Cryobiology was founded in 1964 to bring together those from the biological, medical, and physical sciences who have a common interest in the effects of low temperatures on biological systems. As of 2007, the Society for Cryobiology had about 280 members from around the world, and one-half of them are US-based. The purpose of the Society is to promote scientific research in low temperature biology, to improve scientific understanding in this field, and to disseminate and apply this knowledge to the benefit of mankind. The Society requires of all its members the highest ethical and scientific standards in the performance of their professional activities. According to the Society's bylaws, membership may be refused to applicants whose conduct is deemed detrimental to the Society; in 1982, the bylaws were amended explicitly to exclude "any practice or application of freezing deceased persons in the anticipation of their reanimation", over the objections of some members who were cryonicists, such as Jerry Leaf. The Society organizes an annual scientific meeting dedicated to all aspects of low-temperature biology. This international meeting offers opportunities for presentation and discussion of the most up-to-date research in cryobiology, as well as reviewing specific aspects through symposia and workshops. Members are also kept informed of news and forthcoming meetings through the Society newsletter, *News Notes*. The 2011-2012 president of the Society for Cryobiology was John H. Crowe.

The Society for Low Temperature Biology was founded in 1964 and became a registered charity in 2003 with the purpose of promoting research into the effects of low temperatures on all types of organisms and their constituent cells, tissues, and organs. As of 2006, the society had around 130 (mostly British and European) members and holds at least one annual general meeting. The program usually includes both a symposium on a topical subject and a session of free communications on any aspect of low-temperature biology. Recent symposia have included long-term stability, preservation of aquatic organisms, cryopreservation of embryos and gametes, preservation of plants, low-temperature microscopy, vitrification (glass formation of aqueous systems during cooling), freeze drying and tissue banking. Members are informed through the Society Newsletter, which is presently published three times a year.

Developmental Biology

Developmental biology is the study of the process by which animals and plants grow and develop, their ontogeny. In animals most development occurs in embryonic life, but it is also found in regeneration, asexual reproduction and metamorphosis, and in the growth and differentiation of stem cells in the adult organism. In plants, development occurs in embryos, during vegetative reproduction, and in the normal outgrowth of roots, shoots and flowers.

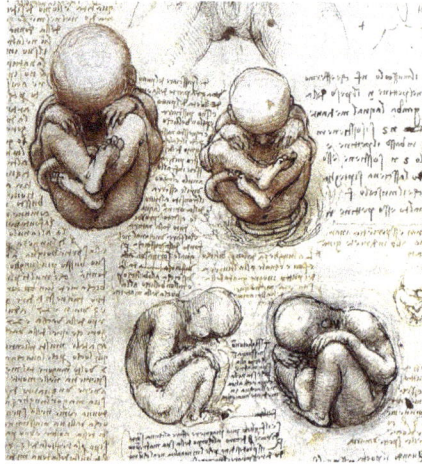

Views of a Fetus in the Womb, Leonardo da Vinci, c. 1510 - 1512. The subject of prenatal development is a major subset of developmental biology.

Practical outcomes from the study of animal developmental biology have included in vitro fertilization, now widely used in fertility treatment, the understanding of risks from substances that can damage the fetus (teratogens), and the creation of various animal models for human disease which are useful in research. Developmental biology has helped to generate modern stem cell biology which promises practical benefits for human health.

Developmental biology, along with molecular genetics, is a central part of evolutionary developmental biology ("evo-devo"), which seeks to explain evolution and the development of body plans at a molecular level.

Perspectives

The main processes involved in the embryonic development of animals are: regional specification, morphogenesis, cell differentiation, growth, and the overall control of timing explored in evolutionary developmental biology. Regional specification refers to the processes that create spatial pattern in a ball or sheet of initially similar cells. This generally involves the action of cytoplasmic determinants, located within parts of the fertilized egg, and of inductive signals emitted from signaling centers in the embryo. The early stages of regional specification do not generate functional differentiated cells, but cell populations committed to develop to a specific region or part of the organism. These are defined by the expression of specific combinations of transcription factors. Morphogenesis relates to the formation of three-dimensional shape. It mainly involves the orchestrated movements of cell sheets and of individual cells. Morphogenesis is important for creating the three germ layers of the early embryo (ectoderm, mesoderm and endoderm) and for building up complex structures during organ development. Cell differentiation relates specifically to the formation of functional cell types such as nerve, muscle, secretory epithelia etc. Differentiated cells contain large amounts of specific proteins associated with the cell function. Growth involves both an overall increase in size, and also the differential growth of parts (allometry) which contributes to morphogenesis. Growth mostly occurs through cell division but also through changes of cell size and the deposition of extracellular materials. The control of timing of events and the integration of the various processes with one another is the least well understood area of the subject. It remains unclear whether animal embryos contain a master clock mechanism or not.

The development of plants involves similar processes to that of animals. However plant cells are mostly immotile so morphogenesis is achieved by differential growth, without cell movements. Also, the inductive signals and the genes involved in plant development are different from those that control animal development.

Developmental Processes

Cell Differentiation

Cell differentiation is the process whereby different functional cell types arise in development. For example, neurons, muscle fibers and hepatocytes (liver cells) are well known types of differentiated cell. Differentiated cells usually produce large amounts of a few proteins that are required for their specific function and this gives them the characteristic appearance that enables them to be recognized under the light microscope. The genes encoding these proteins are highly active. Typically their chromatin structure is very open, allowing access for the transcription enzymes, and specific transcription factors bind to regulatory sequences in the DNA in order to activate gene expression. For example, NeuroD is a key transcription factor for neuronal differentiation, myogenin for muscle differentiation, and HNF4 for hepatocyte differentiation.

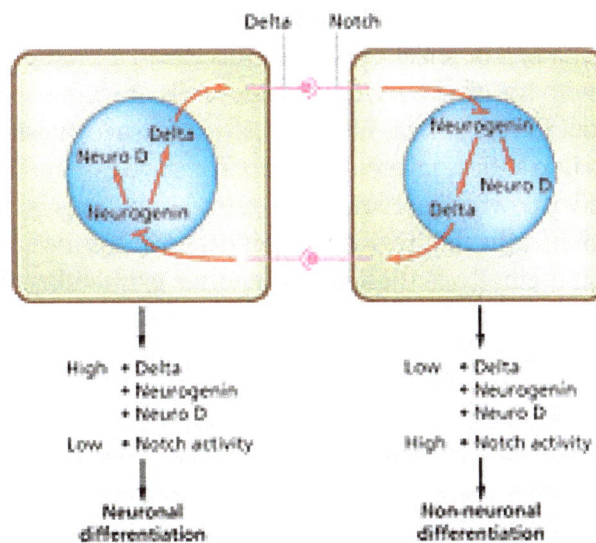

The Notch-delta system in neurogenesis.(Slack Essential Dev Biol Fig 14.12a)

Cell differentiation is usually the final stage of development, preceded by several states of commitment which are not visibly differentiated. A single tissue, formed from a single type of progenitor cell or stem cell, often consists of several differentiated cell types. Control of their formation involves a process of lateral inhibition, based on the properties of the Notch-Delta signaling system. For example, in the neural plate of the embryo this system operates to generate a population of neuronal precursor cells in which NeuroD is highly expressed.

Regeneration

Regeneration indicates the ability to regrow a missing part. This is very prevalent amongst plants, which show continuous growth, and also among colonial animals such as hydroids and ascidians. But most interest by developmental biologists has been shown in the regeneration of parts in free living animals. In particular four models have been the subject of much investigation. Two of these have the ability to regenerate whole bodies: *Hydra*, which can regenerate any part of the polyp from a small fragment, and planarian worms, which can usually regenerate both heads and tails. Both of these examples have continuous cell turnover fed by stem cells and, at least in planaria, at least some of the stem cells have been shown to be pluripotent. The other two models show only distal regeneration of appendages. These are the insect appendages, usually the legs of hemimetabolous insects such as the cricket, and the limbs of urodele amphibians. Considerable information is now available about amphibian limb regeneration and it is known that each cell type regenerates itself, except for connective tissues where there is considerable interconversion between cartilage, dermis and tendons. In terms of the pattern of structures, this is controlled by a re-activation of signals active in the embryo. There is still debate about the old question of whether regeneration is a "pristine" or an "adaptive" property. If the former is the case, with improved knowledge, we might expect to be able to improve regenerative ability in humans. If the latter, then each instance of regeneration is presumed to have arisen by natural selection in circumstances particular to the species, so no general rules would be expected.

Embryonic Development of Animals

The sperm and egg fuse in the process of fertilization to form a fertilized egg, or zygote. This undergoes a period of divisions to form a ball or sheet of similar cells called a blastula or blastoderm. These cell divisions are usually rapid with no growth so the daughter cells are half the size of the mother cell and the whole embryo stays about the same size. They are called cleavage divisions. Morphogenetic movements convert the cell mass into a three layered structure consisting of multicellular sheets called ectoderm, mesoderm and endoderm, which are known as germ layers. This is the process of gastrulation. During cleavage and gastrulation the first regional specification events occur. In addition to the formation of the three germ layers themselves, these often generate extraembryonic structures, such as the mammalian placenta, needed for support and nutrition of the embryo, and also establish differences of commitment along the anteroposterior axis (head, trunk and tail).

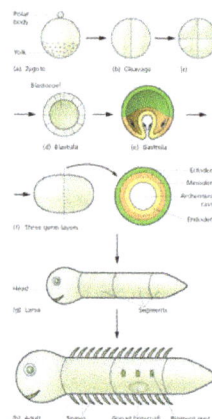

Generalized scheme of embryonic development. Slack "Essential Developmental Biology" Fig.2.8

Regional specification is initiated by the presence of cytoplasmic determinants in one part of the zygote. The cells that contain the determinant become a signaling center and emit an inducing factor. Because the inducing factor is produced in one place, diffuses away, and decays, it forms a concentration gradient, high near the source cells and low further away. The remaining cells of the embryo, which do not contain the determinant, are competent to respond to different concentrations by upregulating specific developmental control genes. This results in a series of zones becoming set up, arranged at progressively greater distance from the signaling center. In each zone a different combination of developmental control genes is upregulated. These genes encode transcription factors which upregulate new combinations of gene activity in each region. Among other functions, these transcription factors control expression of genes conferring specific adhesive and motility properties on the cells in which they are active. Because of these different morphogenetic properties, the cells of each germ layer move to form sheets such that the ectoderm ends up on the outside, mesoderm in the middle, and endoderm on the inside. Morphogenetic movements not only change the shape and structure of the embryo, but by bringing cell sheets into new spatial relationships they also make possible new phases of signaling and response between them.

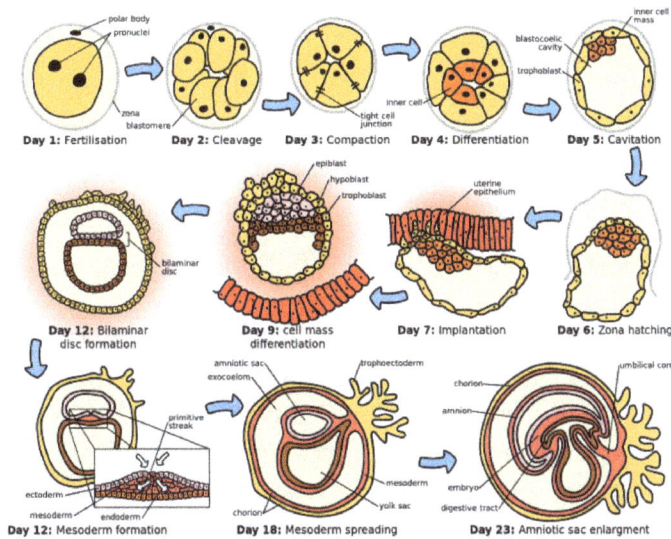

The initial stages of human embryogenesis.

Growth in embryos is mostly autonomous. For each territory of cells the growth rate is controlled by the combination of genes that are active. Free living embryos do not grow in mass as they have no external food supply. But embryos fed by a placenta or extraembryonic yolk supply can grow very fast, and changes to relative growth rate between parts in these organisms help to produce the final overall anatomy.

The whole process needs to be coordinated in time and how this is controlled is not understood. There may be a master clock able to communicate with all parts of the embryo that controls the course of events, or timing may depend simply on local causal sequences of events.

Metamorphosis

Developmental processes are very evident during the process of metamorphosis. This occurs in various types of animal. Well known are the examples of the frog, which usually hatches as a tadpole and metamorphoses to an adult frog, and certain insects which hatch as a larva and then become remodeled to the adult form during a pupal stage.

All the developmental processes listed above occur during metamorphosis. Examples that have been especially well studied include tail loss and other changes in the tadpole of the frog *Xenopus*, and the biology of the imaginal discs, which generate the adult body parts of the fly *Drosophila melanogaster*.

Plant Development

Plant development is the process by which structures originate and mature as a plant grows. It is studied in plant anatomy and plant physiology as well as plant morphology.

Plants constantly produce new tissues and structures throughout their life from meristems located at the tips of organs, or between mature tissues. Thus, a living plant always has embryonic tissues. By contrast, an animal embryo will very early produce all of the body parts that it will ever have in its life. When the animal is born (or hatches from its egg), it has all its body parts and from that point will only grow larger and more mature.

The properties of organization seen in a plant are emergent properties which are more than the sum of the individual parts. "The assembly of these tissues and functions into an integrated multicellular organism yields not only the characteristics of the separate parts and processes but also quite a new set of characteristics which would not have been predictable on the basis of examination of the separate parts."

Growth

A vascular plant begins from a single celled zygote, formed by fertilisation of an egg cell by a sperm cell. From that point, it begins to divide to form a plant embryo through the process of embryogenesis. As this happens, the resulting cells will organize so that one end becomes the first root, while the other end forms the tip of the shoot. In seed plants, the embryo will develop one or more "seed leaves" (cotyledons). By the end of embryogenesis, the young plant will have all the parts necessary to begin in its life.

Once the embryo germinates from its seed or parent plant, it begins to produce additional organs (leaves, stems, and roots) through the process of organogenesis. New roots grow from root meristems located at the tip of the root, and new stems and leaves grow from shoot meristems located at the tip of the shoot. Branching occurs when small clumps of cells left behind by the meristem, and which have not yet undergone cellular differentiation to form a specialized tissue, begin to grow as the tip of a new root or shoot. Growth from any such meristem at the tip of a root or shoot is termed primary growth and results in the lengthening of that root or shoot. Secondary growth results in widening of a root or shoot from divisions of cells in a cambium.

In addition to growth by cell division, a plant may grow through cell elongation. This occurs when individual cells or groups of cells grow longer. Not all plant cells will grow to the same length. When cells on one side of a stem grow longer and faster than cells on the other side, the stem will bend to the side of the slower growing cells as a result. This directional growth can occur via a plant's response to a particular stimulus, such as light (phototropism), gravity (gravitropism), water, (hydrotropism), and physical contact (thigmotropism).

Plant growth and development are mediated by specific plant hormones and plant growth regulators (PGRs) (Ross et al. 1983). Endogenous hormone levels are influenced by plant age, cold hardiness, dormancy, and other metabolic conditions; photoperiod, drought, temperature, and other external environmental conditions; and exogenous sources of PGRs, e.g., externally applied and of rhizospheric origin.

Morphological Variation

Plants exhibit natural variation in their form and structure. While all organisms vary from individual to individual, plants exhibit an additional type of variation. Within a single individual, parts are repeated which may differ in form and structure from other similar parts. This variation is most easily seen in the leaves of a plant, though other organs such as stems and flowers may show similar variation. There are three primary causes of this variation: positional effects, environmental effects, and juvenility.

Evolution of Plant Morphology

Transcription factors and transcriptional regulatory networks play key roles in plant morphogenesis and their evolution. During plant landing, many novel transcription factor families emerged and are preferentially wired into the networks of multicellular development, reproduction, and organ development, contributing to more complex morphogenesis of land plants.

Developmental Model Organisms

Much of developmental biology research in recent decades has focused on the use of a small number of model organisms. It has turned out that there is much conservation of developmental mechanisms across the animal kingdom. In early development different vertebrate species all use essentially the same inductive signals and the same genes encoding regional identity. Even invertebrates use a similar repertoire of signals and genes although the body parts formed are significantly different. Model organisms each have some particular experimental advantages which have enabled them to become popular among researchers. In one sense they are "models" for the whole animal kingdom, and in another sense they are "models" for human development, which is difficult to study directly for both ethical and practical reasons. Model organisms have been most useful for elucidating the broad nature of developmental mechanisms. The more detail is sought, the more they differ from each other and from humans.

Plants:

- Thale cress (*Arabidopsis thaliana*)

Vertebrates:

- Frog: *Xenopus* (*X.laevis and tropicalis*). Good embryo supply. Especially suitable for microsurgery.
- Zebrafish: *Danio rerio*. Good embryo supply. Well developed genetics.
- Chicken: *Gallus gallus*. Early stages similar to mammal, but microsurgery easier. Low cost.

- Mouse: *Mus musculus*. A mammal with well developed genetics.

Invertebrates:

- Fruit fly: *Drosophila melanogaster*. Good embryo supply. Well developed genetics.

- Nematode: *Caenorhabditis elegans*. Good embryo supply. Well developed genetics. Low cost.

Also popular for some purposes have been sea urchins and ascidians. For studies of regeneration urodele amphibians such as the axolotl *Ambystoma mexicanum* are used, and also planarian worms such as *Schmidtea mediterranea*. Plant development has focused on the thale cress *Arabidopsis thaliana* as a model organism.

Genetics

Genetics is the study of genes, genetic variation, and heredity in living organisms. It is generally considered a field of biology, but it intersects frequently with many of the life sciences and is strongly linked with the study of information systems.

The father of genetics is Gregor Mendel, a late 19th-century scientist and Augustinian friar. Mendel studied "trait inheritance," patterns in the way traits are handed down from parents to offspring. He observed that organisms (pea plants) inherit traits by way of discrete "units of inheritance." This term, still used today, is a somewhat ambiguous definition of what is referred to as a gene.

Trait inheritance and molecular inheritance mechanisms of genes are still primary principles of genetics in the 21st century, but modern genetics has expanded beyond inheritance to studying the function and behavior of genes. Gene structure and function, variation, and distribution are studied within the context of the cell, the organism (e.g. dominance), and within the context of a population. Genetics has given rise to a number of sub-fields, including epigenetics and population genetics. Organisms studied within the broad field span the domain of life, including bacteria, plants, animals, and humans.

Genetic processes work in combination with an organism's environment and experiences to influence development and behavior, often referred to as nature versus nurture. The intra- or extra-cellular environment of a cell or organism may switch gene transcription on or off. A classic example is two seeds of genetically identical corn, one placed in a temperate climate and one in an arid climate. While the average height of the two corn stalks may be genetically determined to be equal, the one in the arid climate only grows to half the height of the one in the temperate climate due to lack of water and nutrients in its environment.

History

The observation that living things inherit traits from their parents has been used since prehistoric times to improve crop plants and animals through selective breeding. The modern science of genetics, seeking to understand this process, began with the work of Gregor Mendel in the mid-19th century.

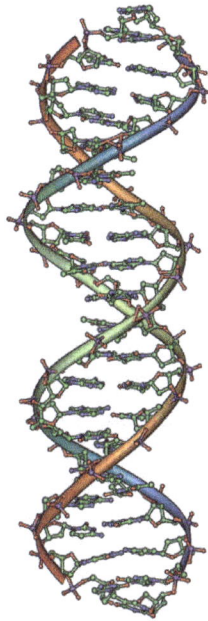

DNA, the molecular basis for biological inheritance. Each strand of DNA is a chain of nucleotides, matching each other in the center to form what look like rungs on a twisted ladder.

Prior to Mendel, Imre Festetics, a Hungarian noble, who lived in Kőszeg before Mendel, was the first who used the word "genetics." He described several rules of genetic inheritance in his work *The genetic law of the Nature* (Die genetische Gesätze der Natur, 1819). His second law is the same as what Mendel published. In his third law, he developed the basic principles of mutation (he can be considered a forerunner of Hugo de Vries.)

Other theories of inheritance preceded his work. A popular theory during Mendel's time was the concept of blending inheritance: the idea that individuals inherit a smooth blend of traits from their parents. Mendel's work provided examples where traits were definitely not blended after hybridization, showing that traits are produced by combinations of distinct genes rather than a continuous blend. Blending of traits in the progeny is now explained by the action of multiple genes with quantitative effects. Another theory that had some support at that time was the inheritance of acquired characteristics: the belief that individuals inherit traits strengthened by their parents. This theory (commonly associated with Jean-Baptiste Lamarck) is now known to be wrong—the experiences of individuals do not affect the genes they pass to their children, although evidence in the field of epigenetics has revived some aspects of Lamarck's theory. Other theories included the pangenesis of Charles Darwin (which had both acquired and inherited aspects) and Francis Galton's reformulation of pangenesis as both particulate and inherited.

Mendelian and Classical Genetics

Modern genetics started with Gregor Johann Mendel, a scientist and Augustinian friar who studied the nature of inheritance in plants. In his paper *"Versuche über Pflanzenhybriden"* ("Experiments on Plant Hybridization"), presented in 1865 to the *Naturforschender Verein* (Society for Research in Nature) in Brünn, Mendel traced the inheritance patterns of certain traits in pea plants and described them mathematically. Although this pattern of inheritance could only be observed for a few traits, Mendel's work suggested that heredity was particulate, not acquired, and that the inheritance patterns of many traits could be explained through simple rules and ratios.

The importance of Mendel's work did not gain wide understanding until the 1890s, after his death, when other scientists working on similar problems re-discovered his research. William Bateson, a proponent of Mendel's work, coined the word *genetics* in 1905. Bateson both acted as a mentor and was aided significantly by the work of female scientists from Newnham College at Cambridge, specifically the work of Becky Saunders, Nora Darwin Barlow, and Muriel Wheldale Onslow. Bateson popularized the usage of the word *genetics* to describe the study of inheritance in his inaugural address to the Third International Conference on Plant Hybridization in London, England, in 1906.

After the rediscovery of Mendel's work, scientists tried to determine which molecules in the cell were responsible for inheritance. In 1911, Thomas Hunt Morgan argued that genes are on chromosomes, based on observations of a sex-linked white eye mutation in fruit flies. In 1913, his student Alfred Sturtevant used the phenomenon of genetic linkage to show that genes are arranged linearly on the chromosome.

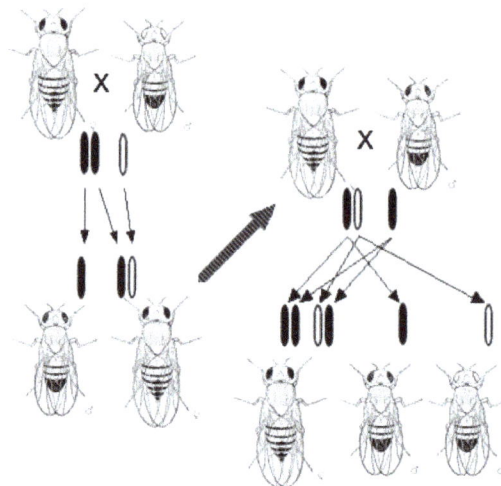

Morgan's observation of sex-linked inheritance of a mutation causing white eyes in *Drosophila* led him to the hypothesis that genes are located upon chromosomes.

Molecular Genetics

Although genes were known to exist on chromosomes, chromosomes are composed of both protein and DNA, and scientists did not know which of the two is responsible for inheritance. In 1928, Frederick Griffith discovered the phenomenon of transformation: dead bacteria could transfer genetic material to "transform" other still-living bacteria. Sixteen years later, in 1944, the Avery–MacLeod–McCarty experiment identified DNA as the molecule responsible for transformation. The role of the nucleus as the repository of genetic information in eukaryotes had been established by Hämmerling in 1943 in his work on the single celled alga *Acetabularia*. The Hershey–Chase experiment in 1952 confirmed that DNA (rather than protein) is the genetic material of the viruses that infect bacteria, providing further evidence that DNA is the molecule responsible for inheritance.

James Watson and Francis Crick determined the structure of DNA in 1953, using the X-ray crystallography work of Rosalind Franklin and Maurice Wilkins that indicated DNA has a helical structure (i.e., shaped like a corkscrew). Their double-helix model had two strands of DNA with the nucleotides pointing inward, each matching a complementary nucleotide on the other strand to form what look like rungs on a twisted ladder. This structure showed that genetic information exists in the sequence of nucleotides on each strand of DNA. The structure also suggested a simple method for replication: if the strands are separated, new partner strands can be reconstructed for each based on the sequence of the old strand. This property is what gives DNA its semi-conservative nature where one strand of new DNA is from an original parent strand.

Although the structure of DNA showed how inheritance works, it was still not known how DNA influences the behavior of cells. In the following years, scientists tried to understand how DNA controls the process of protein production. It was discovered that the cell uses DNA as a template to create matching messenger RNA, molecules with nucleotides very similar to DNA. The nucleotide sequence of a messenger RNA is used to create an amino acid sequence in protein; this translation between nucleotide sequences and amino acid sequences is known as the genetic code.

With the newfound molecular understanding of inheritance came an explosion of research. A notable theory arose from Tomoko Ohta in 1973 with her amendment to the neutral theory of molecular evolution through publishing the nearly neutral theory of molecular evolution. In this theory, Ohta stressed the importance of natural selection and the environment to the rate at which genetic evolution occurs. One important development was chain-termination DNA sequencing in 1977 by Frederick Sanger. This technology allows scientists to read the nucleotide sequence of a DNA molecule. In 1983, Kary Banks Mullis developed the polymerase chain reaction, providing a quick way to isolate and amplify a specific section of DNA from a mixture. The efforts of the Human Genome Project, Department of Energy, NIH, and parallel private efforts by Celera Genomics led to the sequencing of the human genome in 2003.

Features of Inheritance

Discrete Inheritance and Mendel's Laws

At its most fundamental level, inheritance in organisms occurs by passing discrete heritable units, called genes, from parents to offspring. This property was first observed by Gregor Mendel, who studied the segregation of heritable traits in pea plants. In his experiments studying the trait for flower color, Mendel observed that the flowers of each pea plant were either purple or white—but never an intermediate between the two colors. These different, discrete versions of the same gene are called alleles.

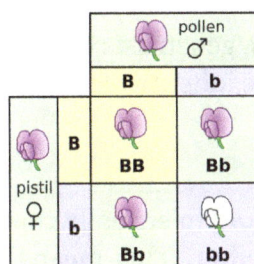

A Punnett square depicting a cross between two pea plants heterozygous for purple (B) and white (b) blossoms.

In the case of the pea, which is a diploid species, each individual plant has two copies of each gene, one copy inherited from each parent. Many species, including humans, have this pattern of inheritance. Diploid organisms with two copies of the same allele of a given gene are called homozygous at that gene locus, while organisms with two different alleles of a given gene are called heterozygous.

The set of alleles for a given organism is called its genotype, while the observable traits of the organism are called its phenotype. When organisms are heterozygous at a gene, often one allele is called dominant as its qualities dominate the phenotype of the organism, while the other allele is called recessive as its qualities recede and are not observed. Some alleles do not have complete dominance and instead have incomplete dominance by expressing an intermediate phenotype, or codominance by expressing both alleles at once.

When a pair of organisms reproduce sexually, their offspring randomly inherit one of the two alleles from each parent. These observations of discrete inheritance and the segregation of alleles are collectively known as Mendel's first law or the Law of Segregation.

Notation and Diagrams

Geneticists use diagrams and symbols to describe inheritance. A gene is represented by one or a few letters. Often a "+" symbol is used to mark the usual, non-mutant allele for a gene.

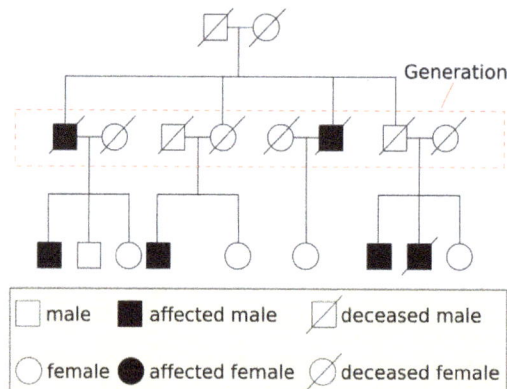

Genetic pedigree charts help track the inheritance patterns of traits.

In fertilization and breeding experiments (and especially when discussing Mendel's laws) the parents are referred to as the "P" generation and the offspring as the "F1" (first filial) generation. When the F1 offspring mate with each other, the offspring are called the "F2" (second filial) generation. One of the common diagrams used to predict the result of cross-breeding is the Punnett square.

When studying human genetic diseases, geneticists often use pedigree charts to represent the inheritance of traits. These charts map the inheritance of a trait in a family tree.

Multiple Gene Interactions

Organisms have thousands of genes, and in sexually reproducing organisms these genes generally assort independently of each other. This means that the inheritance of an allele for yellow or green pea color is unrelated to the inheritance of alleles for white or purple flowers.

This phenomenon, known as "Mendel's second law" or the "law of independent assortment," means that the alleles of different genes get shuffled between parents to form offspring with many different combinations.

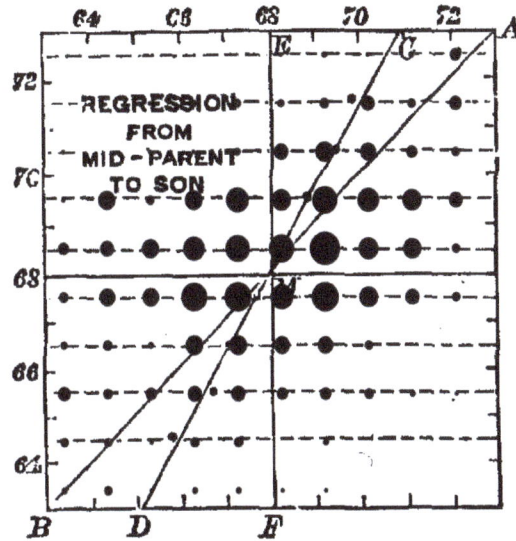

Human height is a trait with complex genetic causes. Francis Galton's data from 1889 shows the relationship between offspring height as a function of mean parent height.

Often different genes can interact in a way that influences the same trait. In the Blue-eyed Mary (*Omphalodes verna*), for example, there exists a gene with alleles that determine the color of flowers: blue or magenta. Another gene, however, controls whether the flowers have color at all or are white. When a plant has two copies of this white allele, its flowers are white—regardless of whether the first gene has blue or magenta alleles. This interaction between genes is called epistasis, with the second gene epistatic to the first.

Many traits are not discrete features (e.g. purple or white flowers) but are instead continuous features (e.g. human height and skin color). These complex traits are products of many genes. The influence of these genes is mediated, to varying degrees, by the environment an organism has experienced. The degree to which an organism's genes contribute to a complex trait is called heritability. Measurement of the heritability of a trait is relative—in a more variable environment, the environment has a bigger influence on the total variation of the trait. For example, human height is a trait with complex causes. It has a heritability of 89% in the United States. In Nigeria, however, where people experience a more variable access to good nutrition and health care, height has a heritability of only 62%.

Molecular Basis for Inheritance

DNA and Chromosomes

The molecular basis for genes is deoxyribonucleic acid (DNA). DNA is composed of a chain of nucleotides, of which there are four types: adenine (A), cytosine (C), guanine (G), and thymine (T). Genetic information exists in the sequence of these nucleotides, and genes exist as stretches of sequence along the DNA chain. Viruses are the only exception to this rule—sometimes viruses use the very similar molecule RNA instead of DNA as their genetic material. Viruses cannot reproduce without a host and are unaffected by many genetic processes, so tend not to be considered living organisms.

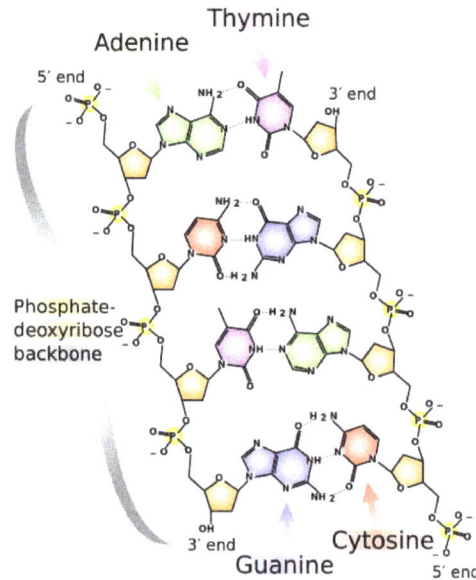

The molecular structure of DNA. Bases pair through the arrangement of hydrogen bonding between the strands.

DNA normally exists as a double-stranded molecule, coiled into the shape of a double helix. Each nucleotide in DNA preferentially pairs with its partner nucleotide on the opposite strand: A pairs with T, and C pairs with G. Thus, in its two-stranded form, each strand effectively contains all necessary information, redundant with its partner strand. This structure of DNA is the physical basis for inheritance: DNA replication duplicates the genetic information by splitting the strands and using each strand as a template for synthesis of a new partner strand.

Genes are arranged linearly along long chains of DNA base-pair sequences. In bacteria, each cell usually contains a single circular genophore, while eukaryotic organisms (such as plants and animals) have their DNA arranged in multiple linear chromosomes. These DNA strands are often extremely long; the largest human chromosome, for example, is about 247 million base pairs in length. The DNA of a chromosome is associated with structural proteins that organize, compact, and control access to the DNA, forming a material called chromatin; in eukaryotes, chromatin is usually composed of nucleosomes, segments of DNA wound around cores of histone proteins. The full set of hereditary material in an organism (usually the combined DNA sequences of all chromosomes) is called the genome.

While haploid organisms have only one copy of each chromosome, most animals and many plants are diploid, containing two of each chromosome and thus two copies of every gene. The two alleles for a gene are located on identical loci of the two homologous chromosomes, each allele inherited from a different parent.

Walther Flemming's 1882 diagram of eukaryotic cell division. Chromosomes are copied, condensed, and organized. Then, as the cell divides, chromosome copies separate into the daughter cells.

Many species have so-called sex chromosomes that determine the gender of each organism. In humans and many other animals, the Y chromosome contains the gene that triggers the development of the specifically male characteristics. In evolution, this chromosome has lost most of its content and also most of its genes, while the X chromosome is similar to the other chromosomes and contains many genes. The X and Y chromosomes form a strongly heterogeneous pair.

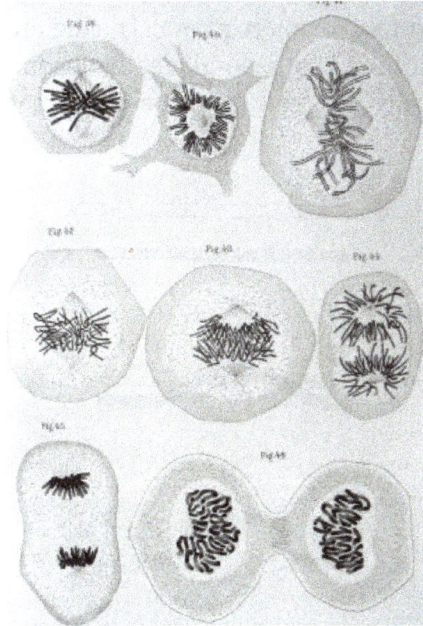

Reproduction

When cells divide, their full genome is copied and each daughter cell inherits one copy. This process, called mitosis, is the simplest form of reproduction and is the basis for asexual reproduction. Asexual reproduction can also occur in multicellular organisms, producing offspring that inherit their genome from a single parent. Offspring that are genetically identical to their parents are called clones.

Eukaryotic organisms often use sexual reproduction to generate offspring that contain a mixture of genetic material inherited from two different parents. The process of sexual reproduction alternates between forms that contain single copies of the genome (haploid) and double copies (diploid). Haploid cells fuse and combine genetic material to create a diploid cell with paired chromosomes. Diploid organisms form haploids by dividing, without replicating their DNA, to create daughter cells that randomly inherit one of each pair of chromosomes. Most animals and many plants are diploid for most of their lifespan, with the haploid form reduced to single cell gametes such as sperm or eggs.

Although they do not use the haploid/diploid method of sexual reproduction, bacteria have many methods of acquiring new genetic information. Some bacteria can undergo conjugation, transferring a small circular piece of DNA to another bacterium. Bacteria can also take up raw DNA fragments found in the environment and integrate them into their genomes, a phenomenon known as transformation. These processes result in horizontal gene transfer, transmitting fragments of genetic information between organisms that would be otherwise unrelated.

Recombination and Genetic Linkage

The diploid nature of chromosomes allows for genes on different chromosomes to assort independently or be separated from their homologous pair during sexual reproduction wherein haploid gametes are formed. In this way new combinations of genes can occur in the offspring of a mating pair. Genes on the same chromosome would theoretically never recombine. However, they do, via the cellular process of chromosomal crossover. During crossover, chromosomes exchange stretches of DNA, effectively shuffling the gene alleles between the chromosomes. This process of chromosomal crossover generally occurs during meiosis, a series of cell divisions that creates haploid cells.

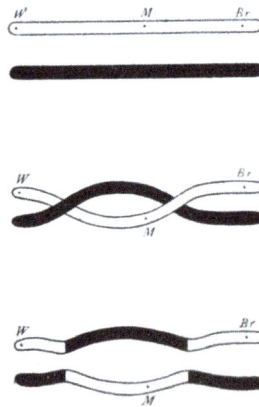

Thomas Hunt Morgan's 1916 illustration of a double crossover between chromosomes.

The first cytological demonstration of crossing over was performed by Harriet Creighton and Barbara McClintock in 1931. Their research and experiments on corn provided cytological evidence for the genetic theory that linked genes on paired chromosomes do in fact exchange places from one homolog to the other.

The probability of chromosomal crossover occurring between two given points on the chromosome is related to the distance between the points. For an arbitrarily long distance, the probability of crossover is high enough that the inheritance of the genes is effectively uncorrelated. For genes that are closer together, however, the lower probability of crossover means that the genes demonstrate genetic linkage; alleles for the two genes tend to be inherited together. The amounts of linkage between a series of genes can be combined to form a linear linkage map that roughly describes the arrangement of the genes along the chromosome.

Gene Expression

Genetic Code

Genes generally express their functional effect through the production of proteins, which are complex molecules responsible for most functions in the cell. Proteins are made up of one or more polypeptide chains, each of which is composed of a sequence of amino acids, and the DNA sequence of a gene (through an RNA intermediate) is used to produce a specific amino acid sequence. This process begins with the production of an RNA molecule with a sequence matching the gene's DNA sequence, a process called transcription.

This messenger RNA molecule is then used to produce a corresponding amino acid sequence through a process called translation. Each group of three nucleotides in the sequence, called a codon, corresponds either to one of the twenty possible amino acids in a protein or an instruction to end the amino acid sequence; this correspondence is called the genetic code. The flow of information is unidirectional: information is transferred from nucleotide sequences into the amino acid sequence of proteins, but it never transfers from protein back into the sequence of DNA—a phenomenon Francis Crick called the central dogma of molecular biology.

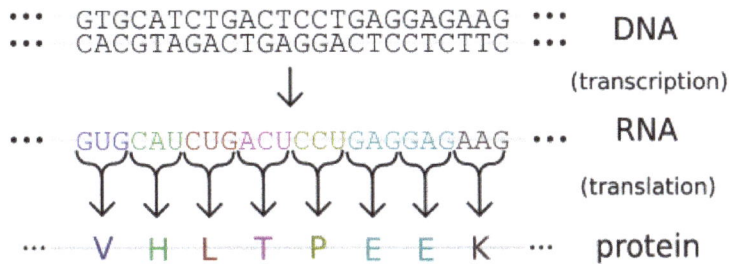

```
···   GTGCATCTGACTCCTGAGGAGAAG  ···   DNA
···   CACGTAGACTGAGGACTCCTCTTC  ···
                  ↓                          (transcription)

···   GUGCAUCUGACUCCUGAGGAGAAG  ···   RNA
      ᶜᶜᶜᶜᶜᶜᶜᶜ                            (translation)
      ↓  ↓  ↓  ↓  ↓  ↓  ↓  ↓
···   V  H  L  T  P  E  E  K   ···   protein
```

The genetic code: Using a triplet code, DNA, through a messenger RNA intermediary, specifies a protein.

The specific sequence of amino acids results in a unique three-dimensional structure for that protein, and the three-dimensional structures of proteins are related to their functions. Some are simple structural molecules, like the fibers formed by the protein collagen. Proteins can bind to other proteins and simple molecules, sometimes acting as enzymes by facilitating chemical reactions within the bound molecules (without changing the structure of the protein itself). Protein structure is dynamic; the protein hemoglobin bends into slightly different forms as it facilitates the capture, transport, and release of oxygen molecules within mammalian blood.

A single nucleotide difference within DNA can cause a change in the amino acid sequence of a protein. Because protein structures are the result of their amino acid sequences, some changes can dramatically change the properties of a protein by destabilizing the structure or changing the surface of the protein in a way that changes its interaction with other proteins and molecules. For example, sickle-cell anemia is a human genetic disease that results from a single base difference within the coding region for the β-globin section of hemoglobin, causing a single amino acid change that changes hemoglobin's physical properties. Sickle-cell versions of hemoglobin stick to themselves, stacking to form fibers that distort the shape of red blood cells carrying the protein. These sickle-shaped cells no longer flow smoothly through blood vessels, having a tendency to clog or degrade, causing the medical problems associated with this disease.

Some DNA sequences are transcribed into RNA but are not translated into protein products—such RNA molecules are called non-coding RNA. In some cases, these products fold into structures which are involved in critical cell functions (e.g. ribosomal RNA and transfer RNA). RNA can also have regulatory effects through hybridization interactions with other RNA molecules (e.g. microRNA).

Nature and Nurture

Although genes contain all the information an organism uses to function, the environment plays an important role in determining the ultimate phenotypes an organism displays. This is the complementary relationship often referred to as "nature and nurture." The phenotype of an organism depends on the interaction of genes and the environment. An interesting example is the coat coloration of the Siamese cat. In this case, the body temperature of the cat plays the role of the environment. The cat's genes code for dark hair, thus the hair-producing cells in the cat make cellular proteins resulting in dark hair. But these dark hair-producing proteins are sensitive to temperature (i.e. have a mutation causing temperature-sensitivity) and denature in higher-temperature environments, failing to produce dark-hair pigment in areas where the cat has a higher body temperature. In a low-temperature environment, however, the protein's structure is stable and produces dark-hair pigment normally. The protein remains functional in areas of skin that are colder – such as its legs, ears, tail and face – so the cat has dark-hair at its extremities.

Siamese cats have a temperature-sensitive pigment-production mutation.

Environment plays a major role in effects of the human genetic disease phenylketonuria. The mutation that causes phenylketonuria disrupts the ability of the body to break down the amino acid phenylalanine, causing a toxic build-up of an intermediate molecule that, in turn, causes severe symptoms of progressive intellectual disability and seizures. However, if someone with the phenylketonuria mutation follows a strict diet that avoids this amino acid, they remain normal and healthy.

A popular method for determining how genes and environment ("nature and nurture") contribute to a phenotype involves studying identical and fraternal twins, or other siblings of multiple births. Because identical siblings come from the same zygote, they are genetically the same. Fraternal twins are as genetically different from one another as normal siblings. By comparing how often a certain disorder occurs in a pair of identical twins to how often it occurs in a pair of fraternal twins, scientists can determine whether that disorder is caused by genetic or postnatal environmental factors – whether it has "nature" or "nurture" causes. One famous example is the multiple birth study of the Genain quadruplets, who were identical quadruplets all diagnosed with schizophrenia. However such tests cannot separate genetic factors from environmental factors affecting fetal development.

Gene Regulation

The genome of a given organism contains thousands of genes, but not all these genes need to be active at any given moment. A gene is expressed when it is being transcribed into mRNA and there exist many cellular methods of controlling the expression of genes such that proteins are produced only when needed by the cell. Transcription factors are regulatory proteins that bind to DNA, either promoting or inhibiting the transcription of a gene. Within the genome of *Escherichia coli* bacteria, for example, there exists a series of genes necessary for the synthesis of the amino acid tryptophan. However, when tryptophan is already available to the cell, these genes for tryptophan synthesis are no longer needed. The presence of tryptophan directly affects the activity of the genes—tryptophan molecules bind to the tryptophan repressor (a transcription factor), changing the repressor's structure such that the repressor binds to the genes. The tryptophan repressor blocks the transcription and expression of the genes, thereby creating negative feedback regulation of the tryptophan synthesis process.

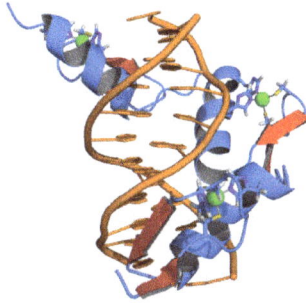

Transcription factors bind to DNA, influencing the transcription of associated genes.

Differences in gene expression are especially clear within multicellular organisms, where cells all contain the same genome but have very different structures and behaviors due to the expression of different sets of genes. All the cells in a multicellular organism derive from a single cell, differentiating into variant cell types in response to external and intercellular signals and gradually establishing different patterns of gene expression to create different behaviors. As no single gene is responsible for the development of structures within multicellular organisms, these patterns arise from the complex interactions between many cells.

Within eukaryotes, there exist structural features of chromatin that influence the transcription of genes, often in the form of modifications to DNA and chromatin that are stably inherited by daughter cells. These features are called "epigenetic" because they exist "on top" of the DNA sequence and retain inheritance from one cell generation to the next. Because of epigenetic features, different cell types grown within the same medium can retain very different properties. Although epigenetic features are generally dynamic over the course of development, some, like the phenomenon of paramutation, have multigenerational inheritance and exist as rare exceptions to the general rule of DNA as the basis for inheritance.

Genetic Change

Mutations

During the process of DNA replication, errors occasionally occur in the polymerization of the second strand. These errors, called mutations, can affect the phenotype of an organism, especially if they occur within the protein coding sequence of a gene. Error rates are usually very low—1 error in every 10–100 million bases—due to the "proofreading" ability of DNA polymerases. Processes that increase the rate of changes in DNA are called mutagenic: mutagenic chemicals promote errors in

DNA replication, often by interfering with the structure of base-pairing, while UV radiation induces mutations by causing damage to the DNA structure. Chemical damage to DNA occurs naturally as well and cells use DNA repair mechanisms to repair mismatches and breaks. The repair does not, however, always restore the original sequence.

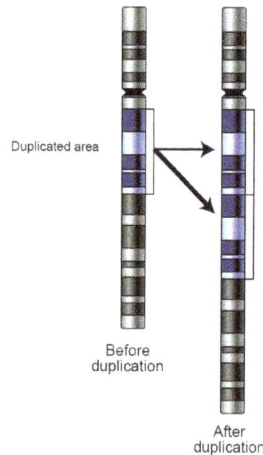

Gene duplication allows diversification by providing redundancy: one gene can mutate and lose its original function without harming the organism.

In organisms that use chromosomal crossover to exchange DNA and recombine genes, errors in alignment during meiosis can also cause mutations. Errors in crossover are especially likely when similar sequences cause partner chromosomes to adopt a mistaken alignment; this makes some regions in genomes more prone to mutating in this way. These errors create large structural changes in DNA sequence – duplications, inversions, deletions of entire regions – or the accidental exchange of whole parts of sequences between different chromosomes (chromosomal translocation).

Natural Selection and Evolution

Mutations alter an organism's genotype and occasionally this causes different phenotypes to appear. Most mutations have little effect on an organism's phenotype, health, or reproductive fitness. Mutations that do have an effect are usually detrimental, but occasionally some can be beneficial. Studies in the fly *Drosophila melanogaster* suggest that if a mutation changes a protein produced by a gene, about 70 percent of these mutations will be harmful with the remainder being either neutral or weakly beneficial.

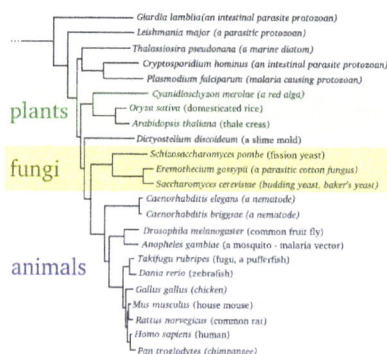

An evolutionary tree of eukaryotic organisms, constructed by the comparison of several orthologous gene sequences.

Population genetics studies the distribution of genetic differences within populations and how these distributions change over time. Changes in the frequency of an allele in a population are mainly influenced by natural selection, where a given allele provides a selective or reproductive advantage to the organism, as well as other factors such as mutation, genetic drift, genetic draft, artificial selection and migration.

Over many generations, the genomes of organisms can change significantly, resulting in evolution. In the process called adaptation, selection for beneficial mutations can cause a species to evolve into forms better able to survive in their environment. New species are formed through the process of speciation, often caused by geographical separations that prevent populations from exchanging genes with each other. The application of genetic principles to the study of population biology and evolution is known as the "modern evolutionary synthesis."

By comparing the homology between different species' genomes, it is possible to calculate the evolutionary distance between them and when they may have diverged. Genetic comparisons are generally considered a more accurate method of characterizing the relatedness between species than the comparison of phenotypic characteristics. The evolutionary distances between species can be used to form evolutionary trees; these trees represent the common descent and divergence of species over time, although they do not show the transfer of genetic material between unrelated species (known as horizontal gene transfer and most common in bacteria).

Model Organisms

Although geneticists originally studied inheritance in a wide range of organisms, researchers began to specialize in studying the genetics of a particular subset of organisms. The fact that significant research already existed for a given organism would encourage new researchers to choose it for further study, and so eventually a few model organisms became the basis for most genetics research. Common research topics in model organism genetics include the study of gene regulation and the involvement of genes in development and cancer.

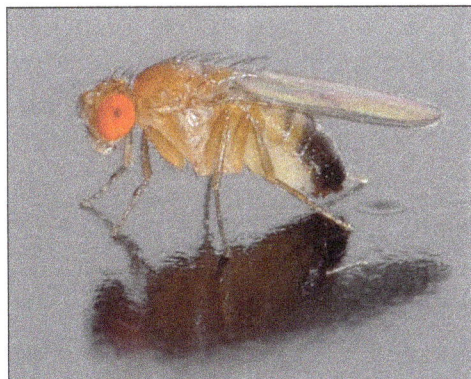

The common fruit fly (*Drosophila melanogaster*) is a popular model organism in genetics research.

Organisms were chosen, in part, for convenience—short generation times and easy genetic manipulation made some organisms popular genetics research tools. Widely used model organisms include the gut bacterium *Escherichia coli*, the plant *Arabidopsis thaliana*, baker's yeast (*Saccharomyces cerevisiae*), the nematode *Caenorhabditis elegans*, the common fruit fly (*Drosophila melanogaster*), and the common house mouse (*Mus musculus*).

Medicine

Medical genetics seeks to understand how genetic variation relates to human health and disease. When searching for an unknown gene that may be involved in a disease, researchers commonly use genetic linkage and genetic pedigree charts to find the location on the genome associated with the disease. At the population level, researchers take advantage of Mendelian randomization to look for locations in the genome that are associated with diseases, a method especially useful for multigenic traits not clearly defined by a single gene. Once a candidate gene is found, further research is often done on the corresponding (or homologous) genes of model organisms. In addition to studying genetic diseases, the increased availability of genotyping methods has led to the field of pharmacogenetics: the study of how genotype can affect drug responses.

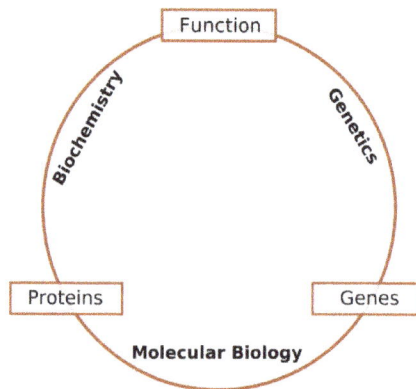

Schematic relationship between biochemistry, genetics and molecular biology.

Individuals differ in their inherited tendency to develop cancer, and cancer is a genetic disease. The process of cancer development in the body is a combination of events. Mutations occasionally occur within cells in the body as they divide. Although these mutations will not be inherited by any offspring, they can affect the behavior of cells, sometimes causing them to grow and divide more frequently. There are biological mechanisms that attempt to stop this process; signals are given to inappropriately dividing cells that should trigger cell death, but sometimes additional mutations occur that cause cells to ignore these messages. An internal process of natural selection occurs within the body and eventually mutations accumulate within cells to promote their own growth, creating a cancerous tumor that grows and invades various tissues of the body.

Normally, a cell divides only in response to signals called growth factors and stops growing once in contact with surrounding cells and in response to growth-inhibitory signals. It usually then divides a limited number of times and dies, staying within the epithelium where it is unable to migrate to other organs. To become a cancer cell, a cell has to accumulate mutations in a number of genes (three to seven) that allow it to bypass this regulation: it no longer needs growth factors to divide, continues growing when making contact to neighbor cells, ignores inhibitory signals, keeps growing indefinitely and is immortal, escapes from the epithelium and ultimately may be able to escape from the primary tumor, cross the endothelium of a blood vessel, be transported by the bloodstream and colonize a new organ, forming deadly metastasis. Although there are some genetic predispositions in a small fraction of cancers, the major fraction is due to a set of new genetic mutations that originally appear and accumulate in one or a small number of cells that will divide to form the tumor and are not transmitted to the progeny (somatic mutations). The most frequent mutations are a loss of function of p53 protein, a tumor suppressor, or in the p53 pathway, and gain of function mutations in the Ras proteins, or in other oncogenes.

Research Methods

DNA can be manipulated in the laboratory. Restriction enzymes are commonly used enzymes that cut DNA at specific sequences, producing predictable fragments of DNA. DNA fragments can be visualized through use of gel electrophoresis, which separates fragments according to their length.

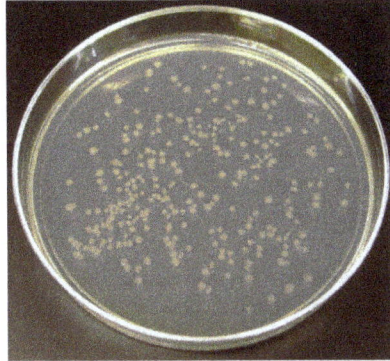

Colonies of *E. coli* produced by cellular cloning. A similar methodology is often used in molecular cloning.

The use of ligation enzymes allows DNA fragments to be connected. By binding ("ligating") fragments of DNA together from different sources, researchers can create recombinant DNA, the DNA often associated with genetically modified organisms. Recombinant DNA is commonly used in the context of plasmids: short circular DNA molecules with a few genes on them. In the process known as molecular cloning, researchers can amplify the DNA fragments by inserting plasmids into bacteria and then culturing them on plates of agar (to isolate clones of bacteria cells). ("Cloning" can also refer to the various means of creating cloned ("clonal") organisms.)

DNA can also be amplified using a procedure called the polymerase chain reaction (PCR). By using specific short sequences of DNA, PCR can isolate and exponentially amplify a targeted region of DNA. Because it can amplify from extremely small amounts of DNA, PCR is also often used to detect the presence of specific DNA sequences.

DNA Sequencing and Genomics

DNA sequencing, one of the most fundamental technologies developed to study genetics, allows researchers to determine the sequence of nucleotides in DNA fragments. The technique of chain-termination sequencing, developed in 1977 by a team led by Frederick Sanger, is still routinely used to sequence DNA fragments. Using this technology, researchers have been able to study the molecular sequences associated with many human diseases.

As sequencing has become less expensive, researchers have sequenced the genomes of many organisms using a process called genome assembly, which utilizes computational tools to stitch together sequences from many different fragments. These technologies were used to sequence the human genome in the Human Genome Project completed in 2003. New high-throughput sequencing technologies are dramatically lowering the cost of DNA sequencing, with many researchers hoping to bring the cost of resequencing a human genome down to a thousand dollars.

Next generation sequencing (or high-throughput sequencing) came about due to the ever-increasing demand for low-cost sequencing. These sequencing technologies allow the production of potentially millions of sequences concurrently. The large amount of sequence data available has created the field of genomics, research that uses computational tools to search for and analyze patterns in the full genomes of organisms. Genomics can also be considered a subfield of bioinformatics, which uses computational approaches to analyze large sets of biological data. A common problem to these fields of research is how to manage and share data that deals with human subject and personally identifiable information.

Society and Culture

On 19 March 2015, a leading group of biologists urged a worldwide ban on clinical use of methods, particularly the use of CRISPR and zinc finger, to edit the human genome in a way that can be inherited. In April 2015, Chinese researchers reported results of basic research to edit the DNA of non-viable human embryos using CRISPR.

Marine Biology

Marine biology is the scientific study of organisms in the ocean or other marine bodies of water. Given that in biology many phyla, families and genera have some species that live in the sea and others that live on land, marine biology classifies species based on the environment rather than on taxonomy. Marine biology differs from marine ecology as marine ecology is focused on how organisms interact with each other and the environment, while biology is the study of the organisms themselves.

Only 29 percent of the Earth's surface is land. The rest is ocean, home to marine life. The oceans average nearly four kilometres in depth and are fringed with coastlines that run for 360,000 kilometres.

A large proportion of all life on Earth lives in the ocean. Exactly how large the proportion is unknown, since many ocean species are still to be discovered. The ocean is a complex three-dimensional world covering approximately 71% of the Earth's surface. The habitats studied in marine biology include everything from the tiny layers of surface water in which organisms and abiotic items may be trapped in surface tension between the ocean and atmosphere, to the depths of the oceanic trenches, sometimes 10,000 meters or more beneath the surface of the ocean. Specific habitats include coral reefs, kelp forests, seagrass meadows, the surrounds of seamounts and thermal vents, tidepools, muddy, sandy and rocky bottoms, and the open ocean (pelagic) zone, where solid objects are rare and the surface of the water is the only visible boundary. The organisms studied range from microscopic phytoplankton and zooplankton to huge cetaceans (whales) 30 meters (98 feet) in length.

Marine life is a vast resource, providing food, medicine, and raw materials, in addition to helping to support recreation and tourism all over the world. At a fundamental level, marine life helps determine the very nature of our planet. Marine organisms contribute significantly to the oxygen cycle, and are involved in the regulation of the Earth's climate. Shorelines are in part shaped and protected by marine life, and some marine organisms even help create new land.

Many species are economically important to humans, including both finfish and shellfish. It is also becoming understood that the well-being of marine organisms and other organisms are linked in fundamental ways. The human body of knowledge regarding the relationship between life in the sea and important cycles is rapidly growing, with new discoveries being made nearly every day. These cycles include those of matter (such as the carbon cycle) and of air (such as Earth's respiration, and movement of energy through ecosystems including the ocean). Large areas beneath the ocean surface still remain effectively unexplored.

History

Early instances of the study of marine biology trace back to Aristotle (384–322 BC) who made several contributions which laid the foundation for many future discoveries and were the first big step in the early exploration period of the ocean and marine life. In 1768, Samuel Gottlieb Gmelin (1744–1774) published the *Historia Fucorum*, the first work dedicated to marine algae and the first book on marine biology to use the then new binomial nomenclature of Linnaeus. It included elaborate illustrations of seaweed and marine algae on folded leaves. The British naturalist Edward Forbes (1815–1854) is generally regarded as the founder of the science of marine biology. The pace of oceanographic and marine biology studies quickly accelerated during the course of the 19th century.

H.M.S. CHALLENGER UNDER SAIL, 1874.

HMS *Challenger* during its pioneer expedition of 1872–76

The observations made in the first studies of marine biology fueled the age of discovery and exploration that followed. During this time, a vast amount of knowledge was gained about the life that exists in the oceans of the world. Many voyages contributed significantly to this pool of knowledge. Among the most significant were the voyages of the HMS *Beagle* where Charles Darwin came up with his theories of evolution and on the formation of coral reefs. Another important expedition was undertaken by HMS *Challenger*, where findings were made of unexpectedly high species diversity among fauna stimulating much theorizing by population ecologists on how such varieties of life could be maintained in what was thought to be such a hostile environment. This era was important for the history of marine biology but naturalists were still limited in their studies because they lacked technology that would allow them to adequately examine species that lived in deep parts of the oceans.

The creation of marine laboratories was important because it allowed marine biologists to conduct research and process their specimens from expeditions. The oldest marine laboratory in the world, Station biologique de Roscoff, was established in France in 1872. In the United States, the Scripps Institution of Oceanography dates back to 1903, while the prominent Woods Hole Oceanographic Institute was founded in 1930. The development of technology such as sound navigation ranging, scuba diving gear, submersibles and remotely operated vehicles allowed marine biologists to discover and explore life in deep oceans that was once thought to not exist.

Marine Life

Microscopic Life

As inhabitants of the largest environment on Earth, microbial marine systems drive changes in every global system. Microbes are responsible for virtually all the photosynthesis that occurs in the ocean, as well as the cycling of carbon, nitrogen, phosphorus and other nutrients and trace elements.

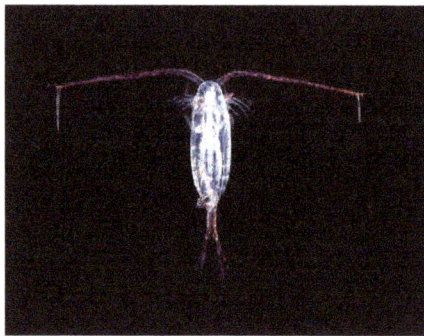

Copepod

Microscopic life undersea is incredibly diverse and still poorly understood. For example, the role of viruses in marine ecosystems is barely being explored even in the beginning of the 21st century.

The role of phytoplankton is better understood due to their critical position as the most numerous primary producers on Earth. Phytoplankton are categorized into cyanobacteria (also called blue-green algae/bacteria), various types of algae (red, green, brown, and yellow-green), diatoms, dinoflagellates, euglenoids, coccolithophorids, cryptomonads, chrysophytes, chlorophytes, prasinophytes, and silicoflagellates.

Zooplankton tend to be somewhat larger, and not all are microscopic. Many Protozoa are zooplankton, including dinoflagellates, zooflagellates, foraminiferans, and radiolarians. Some of these (such as dinoflagellates) are also phytoplankton; the distinction between plants and animals often breaks down in very small organisms. Other zooplankton include cnidarians, ctenophores, chaetognaths, molluscs, arthropods, urochordates, and annelids such as polychaetes. Many larger animals begin their life as zooplankton before they become large enough to take their familiar forms. Two examples are fish larvae and sea stars (also called starfish).

Plants and Algae

Microscopic algae and plants provide important habitats for life, sometimes acting as hiding places for larval forms of larger fish and foraging places for invertebrates.

Algal life is widespread and very diverse under the ocean. Microscopic photosynthetic algae contribute a larger proportion of the world's photosynthetic output than all the terrestrial forests combined. Most of the niche occupied by sub plants on land is actually occupied by macroscopic algae in the ocean, such as *Sargassum* and kelp, which are commonly known as seaweeds that create kelp forests.

Plants that survive in the sea are often found in shallow waters, such as the seagrasses (examples of which are eelgrass, *Zostera*, and turtle grass, *Thalassia*). These plants have adapted to the high salinity of the ocean environment. The intertidal zone is also a good place to find plant life in the sea, where mangroves or cordgrass or beach grass might grow. Microscopic algae and plants provide important habitats for life, sometimes acting as hiding and foraging places for larval forms of larger fish and invertebrates.

Invertebrates

As on land, invertebrates make up a huge portion of all life in the sea. Invertebrate sea life includes Cnidaria such as jellyfish and sea anemones; Ctenophora; sea worms including the phyla Platyhelminthes, Nemertea, Annelida, Sipuncula, Echiura, Chaetognatha, and Phoronida; Mollusca including shellfish, squid, octopus; Arthropoda including Chelicerata and Crustacea; Porifera; Bryozoa; Echinodermata including starfish; and Urochordata including sea squirts or tunicates. Invertebrates have no backbone.There are over a million species

Crown-of-thorns starfish

Fungi

Over 1500 species of fungi are known from marine environments. These are parasitic on marine algae or animals, or are saprobes on algae, corals, protozoan cysts, sea grasses, wood and other substrata, and can also be found in sea foam. Spores of many species have special appendages which facilitate attachment to the substratum. A very diverse range of unusual secondary metabolites is produced by marine fungi.

Vertebrates

Fish

Fish anatomy includes a two-chambered heart, operculum, swim bladder, scales, eyes adapted to seeing underwater, and secretory cells that produce mucous. Fish breathe by extracting oxygen from water through gills. Fins propel and stabilize the fish in the water. Fish fall into two main groups: fish with bony skeletons and fish with cartilaginous skeletons.

A reported 32,700 species of fish have been described (as of December 2013), more than the combined total of all other vertebrates. About 60% of fish species are saltwater fish.

Reptiles

Reptiles which inhabit or frequent the sea include sea turtles, sea snakes, terrapins, the marine iguana, and the saltwater crocodile. Most extant marine reptiles, except for some sea snakes, are oviparous and need to return to land to lay their eggs. Thus most species, excepting sea turtles, spend most of their lives on or near land rather than in the ocean. Despite their marine adaptations, most sea snakes prefer shallow waters nearby land, around islands, especially waters that are somewhat sheltered, as well as near estuaries. Some extinct marine reptiles, such as ichthyosaurs, evolved to be viviparous and had no requirement to return to land.

Green turtle

Birds

Birds adapted to living in the marine environment are often called seabirds. Examples include albatross, penguins, gannets, and auks. Although they spend most of their lives in the ocean, species such as gulls can often be found thousands of miles inland.

Mammals

Sea otters

There are five main types of marine mammals.

- Cetaceans include toothed whales (suborder Odontoceti), such as the sperm whale, dolphins, and porpoises such as the Dall's porpoise. Cetaceans also include baleen whales (suborder Mysticeti), such as the gray whale, humpback whale, and blue whale.

- Sirenians include manatees, the dugong, and the extinct Steller's sea cow.

- Seals (family Phocidae), sea lions (family Otariidae - which also include the fur seals), and the walrus (family Odobenidae) are all considered pinnipeds.

- The sea otter is a member of the family Mustelidae, which includes weasels and badgers.

- The polar bear is a member of the family Ursidae.

Marine Habitats

Marine habitats can be divided into coastal and open ocean habitats. Coastal habitats are found in the area that extends from the shoreline to the edge of the continental shelf. Most marine life is found in coastal habitats, even though the shelf area occupies only seven percent of the total ocean area. Open ocean habitats are found in the deep ocean beyond the edge of the continental shelf

Alternatively, marine habitats can be divided into pelagic and demersal habitats. Pelagic habitats are found near the surface or in the open water column, away from the bottom of the ocean. Demersal habitats are near or on the bottom of the ocean. An organism living in a pelagic habitat is said to be a pelagic organism, as in pelagic fish. Similarly, an organism living in a demersal habitat is said to be a demersal organism, as in demersal fish. Pelagic habitats are intrinsically shifting and ephemeral, depending on what ocean currents are doing.

Marine habitats can be modified by their inhabitants. Some marine organisms, like corals, kelp and seagrasses, are ecosystem engineers which reshape the marine environment to the point where they create further habitat for other organisms.

Intertidal and Near Shore

Intertidal zones, those areas close to shore, are constantly being exposed and covered by the ocean's tides. A huge array of life lives within this zone.

Tide pools with sea stars and sea anemone

Shore habitats span from the upper intertidal zones to the area where land vegetation takes prominence. It can be underwater anywhere from daily to very infrequently. Many species here are scavengers, living off of sea life that is washed up on the shore. Many land animals also make much use of the shore and intertidal habitats. A subgroup of organisms in this habitat bores and grinds exposed rock through the process of bioerosion.

Estuaries

Estuaries are also near shore and influenced by the tides. An estuary is a partially enclosed coastal body of water with one or more rivers or streams flowing into it and with a free connection to the open sea. Estuaries form a transition zone between freshwater river environments and saltwater maritime environments. They are subject both to marine influences—such as tides, waves, and the influx of saline water—and to riverine influences—such as flows of fresh water and sediment. The shifting flows of both sea water and fresh water provide high levels of nutrients both in the water column and in sediment, making estuaries among the most productive natural habitats in the world.

Estuaries have shifting flows of sea water and fresh water.

Reefs

Reefs comprise some of the densest and most diverse habitats in the world. The best-known types of reefs are tropical coral reefs which exist in most tropical waters; however, reefs can also exist in cold water. Reefs are built up by corals and other calcium-depositing animals, usually on top of a rocky outcrop on the ocean floor. Reefs can also grow on other surfaces, which has made it possible to create artificial reefs. Coral reefs also support a huge community of life, including the corals themselves, their symbiotic zooxanthellae, tropical fish and many other organisms.

Coral reefs form complex marine ecosystems with tremendous biodiversity.

Much attention in marine biology is focused on coral reefs and the El Niño weather phenomenon. In 1998, coral reefs experienced the most severe mass bleaching events on record, when vast expanses of reefs across the world died because sea surface temperatures rose well above normal. Some reefs are recovering, but scientists say that between 50% and 70% of the world's coral reefs are now endangered and predict that global warming could exacerbate this trend.

Open Ocean

The open ocean is relatively unproductive because of a lack of nutrients, yet because it is so vast, in total it produces the most primary productivity. The open ocean is separated into different zones, and the different zones each have different ecologies. Zones which vary according to their depth include the epipelagic, mesopelagic, bathypelagic, abyssopelagic, and hadopelagic zones. Zones which vary by the amount of light they receive include the photic and aphotic zones. Much of the aphotic zone's energy is supplied by the open ocean in the form of detritus.

The open ocean is the area of deep sea beyond the continental shelves

Deep Sea and Trenches

The deepest recorded oceanic trench measured to date is the Mariana Trench, near the Philippines, in the Pacific Ocean at 10,924 m (35,840 ft). At such depths, water pressure is extreme and there is no sunlight, but some life still exists. A white flatfish, a shrimp and a jellyfish were seen by the American crew of the bathyscaphe *Trieste* when it dove to the bottom in 1960.

Other notable oceanic trenches include Monterey Canyon, in the eastern Pacific, the Tonga Trench in the southwest at 10,882 m (35,702 ft), the Philippine Trench, the Puerto Rico Trench at 8,605 m (28,232 ft), the Romanche Trench at 7,760 m (25,460 ft), Fram Basin in the Arctic Ocean at 4,665 m (15,305 ft), the Java Trench at 7,450 m (24,440 ft), and the South Sandwich Trench at 7,235 m (23,737 ft).

In general, the deep sea is considered to start at the aphotic zone, the point where sunlight loses its power of transference through the water. Many life forms that live at these depths have the ability to create their own light known as bio-luminescence.

Marine life also flourishes around seamounts that rise from the depths, where fish and other sea life congregate to spawn and feed. Hydrothermal vents along the mid-ocean ridge spreading centers act as oases, as do their opposites, cold seeps. Such places support unique biomes and many new microbes and other lifeforms have been discovered at these locations .

Subfields

The marine ecosystem is large, and thus there are many sub-fields of marine biology. Most involve studying specializations of particular animal groups, such as phycology, invertebrate zoology and ichthyology.

Other subfields study the physical effects of continual immersion in sea water and the ocean in general, adaptation to a salty environment, and the effects of changing various oceanic properties on marine life. A subfield of marine biology studies the relationships between oceans and ocean life, and global warming and environmental issues (such as carbon dioxide displacement).

Recent marine biotechnology has focused largely on marine biomolecules, especially proteins, that may have uses in medicine or engineering. Marine environments are the home to many exotic biological materials that may inspire biomimetic materials.

Related Fields

Marine biology is a branch of biology. It is closely linked to oceanography and may be regarded as a sub-field of marine science. It also encompasses many ideas from ecology. Fisheries science and marine conservation can be considered partial offshoots of marine biology (as well as environmental studies). Marine Chemistry, Physical oceanography and Atmospheric sciences are closely related to this field.

Distribution Factors

An active research topic in marine biology is to discover and map the life cycles of various species and where they spend their time. Technologies that aid in this discovery include pop-up satellite archival tags, acoustic tags, and a variety of other data loggers. Marine biologists study how the ocean currents, tides and many other oceanic factors affect ocean life forms, including their growth, distribution and well-being. This has only recently become technically feasible with advances in GPS and newer underwater visual devices.

Most ocean life breeds in specific places, nests or not in others, spends time as juveniles in still others, and in maturity in yet others. Scientists know little about where many species spend different parts of their life cycles especially in the infant and juvenile years. For example, it is still largely unknown where juvenile sea turtles and some year-1 sharks travel. Recent advances in underwater tracking devices are illuminating what we know about marine organisms that live at great Ocean depths. The information that pop-up satellite archival tags give aids in certain time of the year fishing closures and development of a marine protected area. This data is important to both scientists and fishermen because they are discovering that by restricting commercial fishing in one small area they can have a large impact in maintaining a healthy fish population in a much larger area.

Microbiology

Microbiology is the study of microscopic organisms, those being unicellular (single cell), multicellular (cell colony), or acellular (lacking cells). Microbiology encompasses numerous sub-disciplines including virology, mycology, parasitology, and bacteriology.

Eukaryotic micro-organisms possess membrane-bound cell organelles and include fungi and protists, whereas prokaryotic organisms—all of which are microorganisms—are conventionally classified as lacking membrane-bound organelles and include eubacteria and archaebacteria. Microbiologists traditionally relied on culture, staining, and microscopy. However, less than 1% of the microorganisms present in common environments can be cultured in isolation using current means. Microbiologists often rely on extraction or detection of nucleic acid, either DNA or RNA sequences.

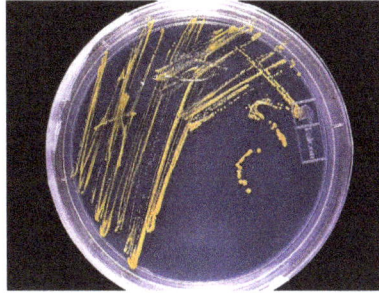

An agar plate streaked with microorganisms

Viruses have been variably classified as organisms, as they have been considered either as very simple microorganisms or very complex molecules. Prions, never considered microorganisms, have been investigated by virologists, however, as the clinical effects traced to them were originally presumed due to chronic viral infections, and virologists took search—discovering "infectious proteins".

As an application of microbiology, medical microbiology is often introduced with medical principles of immunology as *microbiology and immunology*. Otherwise, microbiology, virology, and immunology as basic sciences have greatly exceeded the medical variants, applied sciences.

Branches

The branches of microbiology can be classified into pure and applied sciences. Microbiology can be also classified based on taxonomy, in the cases of bacteriology, mycology, protozoology, and phycology. There is considerable overlap between the specific branches of microbiology with each other and with other disciplines, and certain aspects of these branches can extend beyond the traditional scope of microbiology.

Pure Microbiology

- Bacteriology: The study of bacteria.

- Mycology: The study of fungi.

- Protozoology: The study of protozoa.

- Phycology/algology: The study of algae.

- Parasitology: The study of parasites.

- Immunology: The study of the immune system.

- Virology: The study of viruses.

- Nematology: The study of nematodes.

- Microbial cytology: The study of microscopic and submicroscopic details of microorganisms.

- Microbial physiology: The study of how the microbial cell functions biochemically. Includes the study of microbial growth, microbial metabolism and microbial cell structure.

- Microbial ecology: The relationship between microorganisms and their environment.

- Microbial genetics: The study of how genes are organized and regulated in microbes in relation to their cellular functions. Closely related to the field of molecular biology.

- Cellular microbiology: A discipline bridging microbiology and cell biology.

- Evolutionary microbiology: The study of the evolution of microbes. This field can be subdivided into:

  o Microbial taxonomy: The naming and classification of microorganisms.

  o Microbial systematic: The study of the diversity and genetic relationship of microorganisms.

- Generation microbiology: The study of those microorganisms that have the same characters as their parents.

- Systems microbiology: A discipline bridging systems biology and microbiology.

- Molecular microbiology: The study of the molecular principles of the physiological processes in microorganisms.

Other

- Nano microbiology: The study of those organisms on nano level.

- Exo microbiology (or Astro microbiology): The study of microorganisms in outer space.

- Biological agent: The study of those microorganisms which are being used in weapon industries.

- Predictive microbiology: The quantification of relations between controlling factors in foods and responses of pathogenic and spoilage microorganisms using mathematical modelling

Applied Microbiology

- Medical microbiology: The study of the pathogenic microbes and the role of microbes in human illness. Includes the study of microbial pathogenesis and epidemiology and is related to the study of disease pathology and immunology. This area of microbiology also covers the study of human microbiota, cancer, and the tumor microenvironment.

- Pharmaceutical microbiology: The study of microorganisms that are related to the production of antibiotics, enzymes, vitamins, vaccines, and other pharmaceutical products and that cause pharmaceutical contamination and spoil.

- Industrial microbiology: The exploitation of microbes for use in industrial processes. Examples include industrial fermentation and wastewater treatment. Closely linked to the biotechnology industry. This field also includes brewing, an important application of microbiology.

- Microbial biotechnology: The manipulation of microorganisms at the genetic and molecular level to generate useful products.

Food microbiology laboratory at the Faculty of Food Technology, Latvia University of Agriculture.

- Food microbiology: The study of microorganisms causing food spoilage and foodborne illness. Using microorganisms to produce foods, for example by fermentation.

- Agricultural microbiology: The study of agriculturally relevant microorganisms. This field can be further classified into the following:

  o Plant microbiology and Plant pathology: The study of the interactions between microorganisms and plants and plant pathogens.

  o Soil microbiology: The study of those microorganisms that are found in soil.

- Veterinary microbiology: The study of the role of microbes in veterinary medicine or animal taxonomy.

- Environmental microbiology: The study of the function and diversity of microbes in their natural environments. This involves the characterization of key bacterial habitats such as the rhizosphere and phyllosphere, soil and groundwater ecosystems, open oceans or extreme environments (extremophiles). This field includes other branches of microbiology such as:

  o Microbial ecology

  o Microbially mediated nutrient cycling

  o Geomicrobiology

  o Microbial diversity

  o Bioremediation

- Water microbiology (or Aquatic microbiology): The study of those microorganisms that are found in water.

- Aeromicrobiology (or Air microbiology): The study of airborne microorganisms.

Benefits

While some fear microbes due to the association of some microbes with various human illnesses, many microbes are also responsible for numerous beneficial processes such as industrial fermentation (e.g. the production of alcohol, vinegar and dairy products), antibiotic production and as vehicles for cloning in more complex organisms such as plants. Scientists have also exploited their knowledge of microbes to produce biotechnologically important enzymes such as Taq polymerase, reporter genes for use in other genetic systems and novel molecular biology techniques such as the yeast two-hybrid system.

Fermenting tanks with yeast being used to brew beer

Bacteria can be used for the industrial production of amino acids. *Corynebacterium glutamicum* is one of the most important bacterial species with an annual production of more than two million tons of amino acids, mainly L-glutamate and L-lysine. Since some bacteria have the ability to synthesize antibiotics, they are used for medicinal purposes, such as Streptomyces to make aminoglycoside antibiotics.

A variety of biopolymers, such as polysaccharides, polyesters, and polyamides, are produced by microorganisms. Microorganisms are used for the biotechnological production of biopolymers with tailored properties suitable for high-value medical application such as tissue engineering and drug delivery. Microorganisms are used for the biosynthesis of xanthan, alginate, cellulose, cyanophycin, poly(gamma-glutamic acid), levan, hyaluronic acid, organic acids, oligosaccharides and polysaccharide, and polyhydroxyalkanoates.

Microorganisms are beneficial for microbial biodegradation or bioremediation of domestic, agricultural and industrial wastes and subsurface pollution in soils, sediments and marine environments. The ability of each microorganism to degrade toxic waste depends on the nature of each contaminant. Since sites typically have multiple pollutant types, the most effective approach to microbial biodegradation is to use a mixture of bacterial and fungal species and strains, each specific to the biodegradation of one or more types of contaminants.

Symbiotic microbial communities are known to confer various benefits to their human and animal hosts health including aiding digestion, production of beneficial vitamins and amino acids, and suppression of pathogenic microbes. Some benefit may be conferred by consuming fermented foods, probiotics (bacteria potentially beneficial to the digestive system) and/or prebiotics (substances consumed to promote the growth of probiotic microorganisms). The ways the microbiome influences human and animal health, as well as methods to influence the microbiome are active areas of research.

Research has suggested that microorganisms could be useful in the treatment of cancer. Various strains of non-pathogenic clostridia can infiltrate and replicate within solid tumors. Clostridial vectors can be safely administered and their potential to deliver therapeutic proteins has been demonstrated in a variety of preclinical models.

History

Ancient Times

The existence of microorganisms was hypothesized for many centuries before their actual discovery. The existence of unseen microbiological life was postulated by Jainism which is based on Mahavira's teachings as early as 6th century BCE. Paul Dundas notes that Mahavira asserted existence of unseen microbiological creatures living in earth, water, air and fire. Jain scriptures also describe nigodas which are sub-microscopic creatures living in large clusters and having a very short life and are said to pervade each and every part of the universe, even in tissues of plants and flesh of animals. The Roman Marcus Terentius Varro made references to microbes when he warned against locating a homestead in the vicinity of swamps "because there are bred certain minute creatures which cannot be seen by the eyes, which float in the air and enter the body through the mouth and nose and there by cause serious diseases."

Medieval Islamic World

At the golden age of Islamic civilization, some scientists had knowledge about microorganisms, such as Ibn Sina in his book *The Canon of Medicine*, Ibn Zuhr (also known as Avenzoar) who discovered scabies mites, and Al-Razi who gave the earliest known description of smallpox in his book *The Virtuous Life* (al-Hawi).

Avicenna "ibn Sina"

In 1546, Girolamo Fracastoro proposed that epidemic diseases were caused by transferable seedlike entities that could transmit infection by direct or indirect contact, or vehicle transmission.

However, early claims about the existence of microorganisms were speculative, and not based on microscopic observation. Actual observation and discovery of microbes had to await the invention of the microscope in the 17th century.

Modern Times

In 1676, Anton van Leeuwenhoek, who lived most of his life in Delft, Holland, observed bacteria and other microorganisms using a single-lens microscope of his own design. While Van Leeuwenhoek is often cited as the first to observe microbes, Robert Hooke made the first recorded microscopic observation, of the fruiting bodies of moulds, in 1665. It has, however, been suggested that a Jesuit priest called Athanasius Kircher was the first to observe micro-organisms.

Antonie van Leeuwenhoek, is considered to be the one of the first to observe microorganisms using a microscope.

He was among the first to design magic lanterns for projection purposes, so he must have been well acquainted with the properties of lenses. One of his books contains a chapter in Latin, which reads in translation – "Concerning the wonderful structure of things in nature, investigated by Microscope." Here, he wrote "who would believe that vinegar and milk abound with an innumerable multitude of worms." He also noted that putrid material is full of innumerable creeping animalcule. These observations antedate Robert Hooke's *Micrographia* by nearly 20 years and were published some 29 years before van Leeuwenhoek saw protozoa and 37 years before he described having seen bacteria. Joseph Lister was the first person who said infectious diseases are caused by micro-organism and was first person who used phenol as disinfectant on the open wounds of patients.

Innovative laboratory glassware and experimental methods developed by Louis Pasteur and other biologists contributed to the young field of bacteriology in the late 19th century.

The field of bacteriology (later a subdiscipline of microbiology) was founded in the 19th century by Ferdinand Cohn, a botanist whose studies on algae and photosynthetic bacteria led him to describe several bacteria including *Bacillus* and *Beggiatoa*. Cohn was also the first to formulate a scheme for the taxonomic classification of bacteria and discover spores. Louis Pasteur and Robert Koch were contemporaries of Cohn's and are often considered to be the father of microbiology and medical microbiology, respectively. Pasteur is most famous for his series of experiments designed to disprove the then widely held theory of spontaneous generation, thereby solidifying microbiology's identity as a biological science. Pasteur also designed methods for food preservation (pasteurization) and vaccines against several diseases such as anthrax, fowl cholera and rabies. Koch is best known for his contributions to the germ theory of disease, proving that specific diseases were caused by specific pathogenic micro-organisms. He developed a series of criteria that have become known as the Koch's postulates. Koch was one of the first scientists to focus on the isolation of bacteria in pure culture resulting in his description of several novel bacteria including *Mycobacterium tuberculosis*, the causative agent of tuberculosis.

While Pasteur and Koch are often considered the founders of microbiology, their work did not accurately reflect the true diversity of the microbial world because of their exclusive focus on micro-organisms having direct medical relevance. It was not until the late 19th century and the work of Martinus Beijerinck and Sergei Winogradsky, the founders of general microbiology (an older term encompassing aspects of microbial physiology, diversity and ecology), that the true breadth of microbiology was revealed. Beijerinck made two major contributions to microbiology: the discovery of viruses and the development of enrichment culture techniques. While his work on the Tobacco Mosaic Virus established the basic principles of virology, it was his development of enrichment culturing that had the most immediate impact on microbiology by allowing for the cultivation of a wide range of microbes with wildly different physiologies. Winogradsky was the first to develop the concept of chemolithotrophy and to thereby reveal the essential role played by micro-organisms in geochemical processes. He was responsible for the first isolation and description of both nitrifying and nitrogen-fixing bacteria. French-Canadian microbiologist Felix d'Herelle co-discovered bacteriophages and was one of the earliest applied microbiologists.

Mycology

Mycology is the branch of biology concerned with the study of fungi, including their genetic and biochemical properties, their taxonomy and their use to humans as a source for tinder, medicine, food, and entheogens, as well as their dangers, such as poisoning or infection. A biologist specializing in mycology is called a mycologist.

Mushrooms are a kind of fungal reproductive structure.

From mycology arose the field of phytopathology, the study of plant diseases, and the two disciplines remain closely related because the vast majority of "plant" pathogens are fungi.

Historically, mycology was a branch of botany because, although fungi are evolutionarily more closely related to animals than to plants, this was not recognized until a few decades ago. Pioneer *mycologists* included Elias Magnus Fries, Christian Hendrik Persoon, Anton de Bary, and Lewis David von Schweinitz.

Many fungi produce toxins, antibiotics, and other secondary metabolites. For example, the cosmopolitan (worldwide) genus *Fusarium* and their toxins associated with fatal outbreaks of alimentary toxic aleukia in humans were extensively studied by Abraham Joffe.

Fungi are fundamental for life on earth in their roles as symbionts, e.g. in the form of mycorrhizae, insect symbionts, and lichens. Many fungi are able to break down complex organic biomolecules such as lignin, the more durable component of wood, and pollutants such as xenobiotics, petroleum, and polycyclic aromatic hydrocarbons. By decomposing these molecules, fungi play a critical role in the global carbon cycle.

Fungi and other organisms traditionally recognized as fungi, such as oomycetes and myxomycetes (slime molds), often are economically and socially important, as some cause diseases of animals (such as histoplasmosis) as well as plants (such as Dutch elm disease and Rice blast).

Field meetings to find interesting species of fungi are known as 'forays', after the first such meeting organized by the Woolhope Naturalists' Field Club in 1868 and entitled "A foray among the funguses"[*sic*].

Some fungi can cause disease in humans or other organisms. The study of pathogenic fungi is referred to as medical mycology.

History

It is presumed that humans started collecting mushrooms as food in Prehistoric times. Mushrooms were first written about in the works of Euripides (480-406 B.C.). The Greek philosopher Theophrastos of Eressos (371-288 B.C.) was perhaps the first to try to systematically classify plants; mushrooms were considered to be plants missing certain organs. It was later Pliny the elder (23–79 A.D.), who wrote about truffles in his encyclopedia Naturalis historia.

The Middle Ages saw little advancement in the body of knowledge about fungi. Rather, the invention of the printing press allowed some authors to disseminate superstitions and misconceptions about the fungi that had been perpetuated by the classical authors.

> 66 *Fungi and truffles are neither herbs, nor roots, nor flowers, nor seeds, but merely the superfluous moisture or earth, of trees, or rotten wood, and of other rotting things. This is plain from the fact that all fungi and truffles, especially those that are used for eating, grow most commonly in thundery and wet weather.* 99

— *Jerome Bock (Hieronymus Tragus), 1552*

The start of the modern age of mycology begins with Pier Antonio Micheli's 1737 publication of *Nova plantarum genera*. Published in Florence, this seminal work laid the foundations for the systematic classification of grasses, mosses and fungi. The term *mycology* and the complementary *mycologist* were first used in 1836 by M.J. Berkeley.

Medical Mycology

For centuries, certain mushrooms have been documented as a folk medicine in China, Japan, and Russia. Although the use of mushrooms in folk medicine is centered largely on the Asian continent, people in other parts of the world like the Middle East, Poland, and Belarus have been documented using mushrooms for medicinal purposes. Certain mushrooms, especially polypores like Reishi were thought to be able to benefit a wide variety of health ailments. Medicinal mushroom research in the United States is currently active, with studies taking place at City of Hope National Medical Center, as well as the Memorial Sloan–Kettering Cancer Center.

Current research focuses on mushrooms that may have hypoglycemic activity, anti-cancer activity, anti-pathogenic activity, and immune system-enhancing activity. Recent research has found that the oyster mushroom naturally contains the cholesterol-lowering drug lovastatin, mushrooms produce large amounts of vitamin D when exposed to ultraviolet (UV) light, and that certain fungi may be a future source of taxol. To date, penicillin, lovastatin, ciclosporin, griseofulvin, cephalosporin, ergometrine, and statins are the most famous pharmaceuticals that have been isolated from the fungi kingdom.

References

- Dorit, R. L.; Walker, W. F.; Barnes, R. D. (1991). Zoology. Saunders College Publishing. pp. 547–549. ISBN 978-0-03-030504-7.

- Ruppert, Edward E.; Fox, Richard, S.; Barnes, Robert D. (2004). Invertebrate Zoology, 7th edition. Cengage Learning. pp. 59–60. ISBN 81-315-0104-3.

- McGrath, J.A.; Eady, R.A.; Pope, F.M. (2004). Rook's Textbook of Dermatology (7th ed.). Blackwell Publishing. pp. 3.1–3.6. ISBN 978-0-632-06429-8.

- Ruppert, Edward E.; Fox, Richard, S.; Barnes, Robert D. (2004). Invertebrate Zoology, 7th edition. Cengage Learning. pp. 105–107. ISBN 81-315-0104-3.

- Liem, Karel F.; Warren Franklin Walker (2001). Functional anatomy of the vertebrates: an evolutionary perspective. Harcourt College Publishers. p. 277. ISBN 978-0-03-022369-3.

- Kotpal, R. L. (2010). Modern Text Book of Zoology: Vertebrates. Rastogi Publications. p. 193. ISBN 978-81-7133-891-7.

- Stebbins, Robert C.; Cohen, Nathan W. (1995). A Natural History of Amphibians. Princeton University Press. pp. 24–25. ISBN 0-691-03281-5.

Reproduction: A Comprehensive Study

The process by which organisms/humans are born is known as reproduction. In asexual reproduction the offspring is reproduced without the participation of another organism. The types of reproduction explained are budding, vegetative reproduction, fission and apomixes. This section on reproduction offers an insightful focus, keeping in mind the theme of the chapter.

Reproduction

Production of new individuals along a leaf margin of the miracle leaf plant (*Kalanchoe pinnata*). The small plant in front is about 1 cm (0.4 in) tall. The concept of "individual" is obviously stretched by this asexual reproductive process.

Reproduction (or procreation, breeding) is the biological process by which new individual organisms – "offspring" – are produced from their "parents". Reproduction is a fundamental feature of all known life; each individual organism exists as the result of reproduction. There are two forms of reproduction: asexual and sexual.

In asexual reproduction, an organism can reproduce without the involvement of another organism. Asexual reproduction is not limited to single-celled organisms. The cloning of an organism is a form of asexual reproduction. By asexual reproduction, an organism creates a genetically similar or identical copy of itself. The evolution of sexual reproduction is a major puzzle for biologists. The two-fold cost of sex is that only 50% of organisms reproduce and organisms only pass on 50% of their genes.

Sexual reproduction typically requires the sexual interaction of two specialized organisms, called gametes, which contain half the number of chromosomes of normal cells and are created by meiosis, with typically a male fertilizing a female of the same species to create a fertilized zygote. This produces offspring organisms whose genetic characteristics are derived from those of the two parental organisms.

Asexual

Asexual reproduction is a process by which organisms create genetically similar or identical copies of themselves without the contribution of genetic material from another organism. Bacteria divide asexually via binary fission; viruses take control of host cells to produce more viruses; Hydras (invertebrates of the order *Hydroidea*) and yeasts are able to reproduce by budding. These organisms often do not possess different sexes, and they are capable of "splitting" themselves into two or more copies of themselves. Most plants have the ability to reproduce asexually and the ant species Mycocepurus smithii is thought to reproduce entirely by asexual means.

Some species that are capable of reproducing asexually, like hydra, yeast and jellyfish, may also reproduce sexually. For instance, most plants are capable of vegetative reproduction—reproduction without seeds or spores—but can also reproduce sexually. Likewise, bacteria may exchange genetic information by conjugation.

Other ways of asexual reproduction include parthenogenesis, fragmentation and spore formation that involves only mitosis. Parthenogenesis is the growth and development of embryo or seed without fertilization by a male. Parthenogenesis occurs naturally in some species, including lower plants (where it is called apomixis), invertebrates (e.g. water fleas, aphids, some bees and parasitic wasps), and vertebrates (e.g. some reptiles, fish, and, very rarely, birds and sharks). It is sometimes also used to describe reproduction modes in hermaphroditic species which can self-fertilize.

Sexual

Hoverflies mating in midair flight

Sexual reproduction is a biological process that creates a new organism by combining the genetic material of two organisms in a process that starts with meiosis, a specialized type of cell division. Each of two parent organisms contributes half of the offspring's genetic makeup by creating haploid gametes. Most organisms form two different types of gametes. In these *anisogamous* species, the two sexes are referred to as male (producing sperm or microspores) and female (producing ova or megaspores). In *isogamous species*, the gametes are similar or identical in form (isogametes), but may have separable properties and then may be given other different names . For example, in the green alga, *Chlamydomonas reinhardtii*, there are so-called "plus" and "minus" gametes. A few types of organisms, such as ciliates, *Paramecium aurelia*, have more than two types of "sex", called syngens.

Most animals (including humans) and plants reproduce sexually. Sexually reproducing organisms have different sets of genes for every trait (called alleles). Offspring inherit one allele for each trait from each parent, thereby ensuring that offspring have a combination of the parents' genes. Diploid having two copies of every gene within an organism, it is believed that "the masking of deleterious alleles favors the evolution of a dominant diploid phase in organisms that alternate between haploid and diploid phases" where recombination occurs freely.

Bryophyte reproduces sexually but its commonly seen life forms are all haploid, which produce gametes. The zygotes of the gametes develop into sporangium, which produces haploid spores. The diploid stage is relatively short compared with that of haploid stage, i.e. *haploid dominance*. The advantage of diploid, e.g. heterosis, only takes place in diploid life stage. Bryophyte still maintains the sexual reproduction during its evolution despite the fact that the haploid stage does not benefit from heterosis at all. This may be an example that the sexual reproduction has a bigger advantage by itself, since it allows gene shuffling (hybrid or recombination between multiple loci) among different members of the species, that permits natural selection of the fit over these new hybrids or recombinants that are haploid forms.

Allogamy

Allogamy is the fertilization of an ovum from one individual with the spermatozoa of another.

Autogamy

Self-fertilization, also known as autogamy, occurs in hermaphroditic organisms where the two gametes fused in fertilization come from the same individual, e.g., some foraminiferans, some ciliates. The term "autogamy" is sometimes substituted for autogamous pollination (not necessarily leading to successful fertilization) and describes self-pollination within the same flower, distinguished from geitonogamous pollination, transfer of pollen to a different flower on the same flowering plant, or within a single monoecious Gymnosperm plant. For example, species *Helonias bullata* suffer from low genetic diversity due to self-fertilization.

Mitosis and Meiosis

Mitosis and meiosis are types of cell division. Mitosis occurs in somatic cells, while meiosis occurs in gametes.

Mitosis The resultant number of cells in mitosis is twice the number of original cells. The number of chromosomes in the offspring cells is the same as that of the parent cell.

Meiosis The resultant number of cells is four times the number of original cells. This results in cells with half the number of chromosomes present in the parent cell. A diploid cell duplicates itself, then undergoes two divisions (tetraploid to diploid to haploid), in the process forming four haploid cells. This process occurs in two phases, meiosis I and meiosis II.

Same-sex

In recent decades, developmental biologists have been researching and developing techniques to facilitate same-sex reproduction. The obvious approaches, subject to a growing amount of activity, are female sperm and male eggs, with female sperm closer to being a reality for humans, given that Japanese scientists have already created female sperm for chickens. "However, the ratio of produced W chromosome-bearing (W-bearing) spermatozoa fell substantially below expectations. It is therefore concluded that most of the W-bearing PGC could not differentiate into spermatozoa because of restricted spermatogenesis." In 2004, by altering the function of a few genes involved with imprinting, other Japanese scientists combined two mouse eggs to produce daughter mice.

Strategies

There are a wide range of reproductive strategies employed by different species. Some animals, such as the human and northern gannet, do not reach sexual maturity for many years after birth and even then produce few offspring. Others reproduce quickly; but, under normal circumstances, most offspring do not survive to adulthood. For example, a rabbit (mature after 8 months) can produce 10–30 offspring per year, and a fruit fly (mature after 10–14 days) can produce up to 900 offspring per year. These two main strategies are known as K-selection (few offspring) and r-selection (many offspring). Which strategy is favoured by evolution depends on a variety of circumstances. Animals with few offspring can devote more resources to the nurturing and protection of each individual offspring, thus reducing the need for many offspring. On the other hand, animals with many offspring may devote fewer resources to each individual offspring; for these types of animals it is common for many offspring to die soon after birth, but enough individuals typically survive to maintain the population. Some organisms such as honey bees and fruit flies retain sperm in a process called sperm storage thereby increasing the duration of their fertility.

Other Types

- Polycyclic animals reproduce intermittently throughout their lives.

- Semelparous organisms reproduce only once in their lifetime, such as annual plants (including all grain crops), and certain species of salmon, spider, bamboo and century plant. Often, they die shortly after reproduction. This is often associated with r-strategists.

- Iteroparous organisms produce offspring in successive (e.g. annual or seasonal) cycles, such as perennial plants. Iteroparous animals survive over multiple seasons (or periodic condition changes). This is more associated with K-strategists.

Asexual Vs. Sexual Reproduction

Organisms that reproduce through asexual reproduction tend to grow in number exponentially. However, because they rely on mutation for variations in their DNA, all members of the species have similar vulnerabilities. Organisms that reproduce sexually yield a smaller number of offspring, but the large amount of variation in their genes makes them less susceptible to disease.

Many organisms can reproduce sexually as well as asexually. Aphids, slime molds, sea anemones, some species of starfish (by fragmentation), and many plants are examples. When environmental factors are favorable, asexual reproduction is employed to exploit suitable conditions for survival such as an abundant food supply, adequate shelter, favorable climate, disease, optimum pH or a proper mix of other lifestyle requirements. Populations of these organisms increase exponentially via asexual reproductive strategies to take full advantage of the rich supply resources.

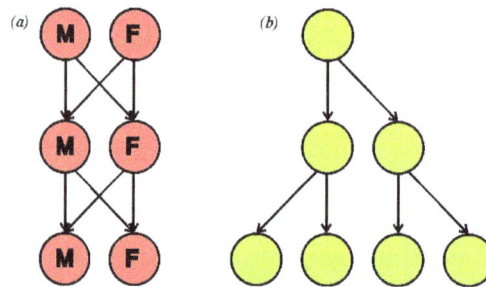

Illustration of the *twofold cost of sexual reproduction*. If each organism were to contribute to the same number of offspring (two), *(a)* the population remains the same size each generation, where the *(b)* asexual population doubles in size each generation.

When food sources have been depleted, the climate becomes hostile, or individual survival is jeopardized by some other adverse change in living conditions, these organisms switch to sexual forms of reproduction. Sexual reproduction ensures a mixing of the gene pool of the species. The variations found in offspring of sexual reproduction allow some individuals to be better suited for survival and provide a mechanism for selective adaptation to occur. The meiosis stage of the sexual cycle also allows especially effective repair of DNA damages. In addition, sexual reproduction usually results in the formation of a life stage that is able to endure the conditions that threaten the offspring of an asexual parent. Thus, seeds, spores, eggs, pupae, cysts or other "over-wintering" stages of sexual reproduction ensure the survival during unfavorable times and the organism can "wait out" adverse situations until a swing back to suitability occurs.

Life Without

The existence of life without reproduction is the subject of some speculation. The biological study of how the origin of life produced reproducing organisms from non-reproducing elements is called abiogenesis. Whether or not there were several independent abiogenetic events, biologists believe that the last universal ancestor to all present life on Earth lived about 3.5 billion years ago.

Scientists have speculated about the possibility of creating life non-reproductively in the laboratory. Several scientists have succeeded in producing simple viruses from entirely non-living materials. However, viruses are often regarded as not alive. Being nothing more than a bit of RNA or DNA in a protein capsule, they have no metabolism and can only replicate with the assistance of a hijacked cell's metabolic machinery.

The production of a truly living organism (e.g. a simple bacterium) with no ancestors would be a much more complex task, but may well be possible to some degree according to current biological knowledge. A synthetic genome has been transferred into an existing bacterium where it replaced the native DNA, resulting in the artificial production of a new *M. mycoides* organism.

There is some debate within the scientific community over whether this cell can be considered completely synthetic on the grounds that the chemically synthesized genome was an almost 1:1 copy of a naturally occurring genome and, the recipient cell was a naturally occurring bacterium. The Craig Venter Institute maintains the term "synthetic bacterial cell" but they also clarify "... we do not consider this to be "creating life from scratch" but rather we are creating new life out of already existing life using synthetic DNA". Venter plans to patent his experimental cells, stating that "they are pretty clearly human inventions". Its creators suggests that building 'synthetic life' would allow researchers to learn about life by building it, rather than by tearing it apart. They also propose to stretch the boundaries between life and machines until the two overlap to yield "truly programmable organisms". Researchers involved stated that the creation of "true synthetic biochemical life" is relatively close in reach with current technology and cheap compared to the effort needed to place man on the Moon.

Lottery Principle

Sexual reproduction has many drawbacks, since it requires far more energy than asexual reproduction and diverts the organisms from other pursuits, and there is some argument about why so many species use it. George C. Williams used lottery tickets as an analogy in one explanation for the widespread use of sexual reproduction. He argued that asexual reproduction, which produces little or no genetic variety in offspring, was like buying many tickets that all have the same number, limiting the chance of "winning" - that is, producing surviving offspring. Sexual reproduction, he argued, was like purchasing fewer tickets but with a greater variety of numbers and therefore a greater chance of success. The point of this analogy is that since asexual reproduction does not produce genetic variations, there is little ability to quickly adapt to a changing environment. The lottery principle is less accepted these days because of evidence that asexual reproduction is more prevalent in unstable environments, the opposite of what it predicts.

Plant Reproduction

Plant reproduction is the production of new individuals or offspring in plants, which can be accomplished by sexual or asexual reproduction. Sexual reproduction produces offspring by the fusion of gametes, resulting in offspring genetically different from the parent or parents. Asexual reproduction produces new individuals without the fusion of gametes, genetically identical to the parent plants and each other, except when mutations occur. In seed plants, the offspring can be packaged in a protective seed, which is used as an agent of dispersal.

Bryophyllum, a plant that reproduces asexually via new shoots from the leaves

Asexual Reproduction

Plants have two main types of asexual reproduction in which new plants are produced that are genetically identical clones of the parent individual. Vegetative reproduction involves a vegetative piece of the original plant (budding, tillering, etc.) and is distinguished from *apomixis*, which is a replacement for sexual reproduction, and in some cases involves seeds. Apomixis occurs in many plant species and also in some non-plant organisms. For apomixis and similar processes in non-plant organisms.

Natural vegetative reproduction is mostly a process found in herbaceous and woody perennial plants, and typically involves structural modifications of the stem or roots and in a few species leaves. Most plant species that employ vegetative reproduction do so as a means to perennialize the plants, allowing them to survive from one season to the next and often facilitating their expansion in size. A plant that persists in a location through vegetative reproduction of individuals constitutes a clonal colony; a single ramet, or apparent individual, of a clonal colony is genetically identical to all others in the same colony. The distance that a plant can move during vegetative reproduction is limited, though some plants can produce ramets from branching rhizomes or stolons that cover a wide area, often in only a few growing seasons. In a sense, this process is not one of reproduction but one of survival and expansion of biomass of the individual. When an individual organism increases in size via cell multiplication and remains intact, the process is called vegetative growth. However, in vegetative reproduction, the new plants that result are new individuals in almost every respect except genetic. A major disadvantage to vegetative reproduction, is the transmission of pathogens from parent to offspring; it is uncommon for pathogens to be transmitted from the plant to its seeds (in sexual reproduction or in apomixis), though there are occasions when it occurs.

Seeds generated by apomixis are a means of asexual reproduction, involving the formation and dispersal of seeds that do not originate from the fertilization of the embryos. Hawkweed (*Hieracium*), dandelion (*Taraxacum*), some Citrus (*Citrus*) and Kentucky blue grass (*Poa pratensis*) all use this form of asexual reproduction. Pseudogamy occurs in some plants that have apomictic seeds, where pollination is often needed to initiate embryo growth, though the pollen contributes no genetic material to the developing offspring. Other forms of apomixis occur in plants also, including the generation of a plantlet in replacement of a seed or the generation of bulbils instead of flowers, where new cloned individuals are produced.

Structures

A rhizome is a modified underground stem serving as an organ of vegetative reproduction; the growing tips of the rhizome can separate as new plants, e.g., Polypody, Iris, Couch Grass and Nettles.

Prostrate aerial stems, called runners or stolons are important vegetative reproduction organs in some species, such as the strawberry, numerous grasses, and some ferns.

Adventitious buds form on roots near the ground surface, on damaged stems (as on the stumps of cut trees), or on old roots. These develop into above-ground stems and leaves. A form of budding called suckering is the reproduction or regeneration of a plant by shoots that arise from an existing root system. Species that characteristically produce suckers include Elm (*Ulmus*), Dandelion (*Taraxacum*), and many members of the Rose family such as *Rosa* and *Rubus*.

Plants like onion (*Allium cepa*), hyacinth (*Hyacinth*), narcissus (*Narcissus*) and tulips (*Tulipa*) reproduce by dividing their underground bulbs into more bulbs. Other plants like potatoes (*Solanum tuberosum*) and dahlia (*Dahlia*) reproduce by a similar method involving underground tubers. Gladioli and crocuses (*Crocus*) reproduce in a similar way with corms.

Usage

The most common form of plant reproduction utilized by people is seeds, but a number of asexual methods are utilized which are usually enhancements of natural processes, including: cutting, grafting, budding, layering, division, sectioning of rhizomes, roots, tubers, bulbs, stolons, tillers (suckers), etc., and artificial propagation by laboratory tissue cloning. Asexual methods are most often used to propagate cultivars with individual desirable characteristics that do not come true from seed. Fruit tree propagation is frequently performed by budding or grafting desirable cultivars (clones), onto rootstocks that are also clones, propagated by stooling.

In horticulture, a "cutting" is a branch that has been cut off from a mother plant below an internode and then rooted, often with the help of a rooting liquid or powder containing hormones. When a full root has formed and leaves begin to sprout anew, the clone is a self-sufficient plant, genetically identical to the mother plant. Examples include cuttings from the stems of blackberries (*Rubus occidentalis*), African violets (*Saintpaulia*), verbenas (*Verbena*) to produce new plants. A related use of cuttings is grafting, where a stem or bud is joined onto a different stem. Nurseries offer for sale trees with grafted stems that can produce four or more varieties of related fruits, including apples. The most common usage of grafting is the propagation of cultivars onto already rooted plants, sometimes the rootstock is used to dwarf the plants or protect them from root damaging pathogens.

Since vegetatively propagated plants are clones, they are important tools in plant research. When a clone is grown in various conditions, differences in growth can be ascribed to environmental effects instead of genetic differences.

Sexual Reproduction

Sexual reproduction involves two fundamental processes: meiosis, which rearranges the genes and reduces the number of chromosomes, and fertilization, which restores the chromosome to a complete diploid number. In between these two processes, different types of plants and algae vary, but many of them, including all land plants, undergo alternation of generations, with two different multicellular structures (phases), a gametophyte and a sporophyte. The evolutionary origin and adaptive significance of sexual reproduction are discussed in the pages "Evolution of sexual reproduction" and "Origin and function of meiosis."

The gametophyte is the multicellular structure (plant) that is haploid, containing a single set of chromosomes in each cell. The gametophyte produces male or female gametes (or both), by a process of cell division called mitosis. In vascular plants with separate gametophytes, female gametophytes are known as mega gametophytes (mega=large, they produce the large egg cells) and the male gametophytes are called micro gametophytes (micro=small, they produce the small sperm cells).

The fusion of male and female gametes (fertilization) produces a diploid zygote, which develops by mitotic cell divisions into a multicellular sporophyte.

The mature sporophyte produces spores by meiosis, sometimes referred to as "reduction division" because the chromosome pairs are separated once again to form single sets.

In mosses and liverworts the gametophyte is relatively large, and the sporophyte is a much smaller structure that is never separated from the gametophyte. In ferns, gymnosperms, and flowering plants (angiosperms), the gametophytes are relatively small and the sporophyte is much larger. In gymnosperms and flowering plants the mega gametophyte is contained within the ovule (that may develop into a seed) and the micro gametophyte is contained within a pollen grain.

History of Sexual Reproduction

Unlike animals, plants are immobile, and cannot seek out sexual partners for reproduction. In the evolution of early plants, abiotic means, including water and wind, transported sperm for reproduction. The first plants were aquatic, as described in the page "Evolutionary history of plants", and released sperm freely into the water to be carried with the currents. Primitive land plants like liverworts and mosses had motile sperm that swam in a thin film of water or were splashed in water droplets from the male reproduction organs onto the female organs. As taller and more complex plants evolved, modifications in the alternation of generations evolved; in the Paleozoic era progymnosperms reproduced by using spores dispersed on the wind. The seed plants including seed ferns, conifers and cordaites, which were all gymnosperms, evolved 350 million years ago; they had pollen grains that contained the male gametes for protection of the sperm during the process of transfer from the male to female parts. It is believed that insects fed on the pollen, and plants thus evolved to use insects to actively carry pollen from one plant to the next. Seed producing plants, which include the angiosperms and the gymnosperms, have heteromorphic alternation of generations with large sporophytes containing much reduced gametophytes. Angiosperms have distinctive reproductive organs called flowers, with carpels, and the female gametophyte is greatly reduced to a female embryo sac, with as few as eight cells. The male gametophyte consists of the pollen grains. The sperm of seed plants are non-motile, except for two older groups of plants, the Cycadophyta and the Ginkgophyta, which have flagellated sperm.

Flowering Plants

Flowering plants are the dominant plant form on land and they reproduce by sexual and asexual means. Often their most distinguishing feature is their reproductive organs, commonly called flowers. Sexual reproduction in flowering plants involves the production of male and female gametes, the transfer of the male gametes to the female ovules in a process called pollination. After pollination occurs, fertilization happens and the ovules grow into seeds within a fruit. After the seeds are ready for dispersal, the fruit ripens and by various means the seeds are freed from the fruit and after varying amounts of time and under specific conditions the seeds germinate and grow into the next generation.

The anther produces male gametophytes which are pollen grains, which attach to the stigma on top of a carpel, in which the female gametophytes (inside ovules) are located. After the pollen tube grows through the carpel's style, the sperm from the pollen grain migrate into the ovule to fertilize the egg cell and central cell within the female gametophyte in a process termed double fertilization. The resulting zygote develops into an embryo, while the triploid endosperm (one sperm cell plus a binucleate female cell) and female tissues of the ovule give rise to the surrounding tissues in the developing seed. The ovary, which produced the female gametophyte(s), then grows into a fruit, which surrounds the seed(s). Plants may either self-pollinate or cross-pollinate.

Pollination

In plants that use insects or other animals to move pollen from one flower to the next, plants have developed greatly modified flower parts to attract pollinators and to facilitate the movement of pollen from one flower to the insect and from the insect back to the next flower. Flowers of wind pollinated plants tend to lack petals and or sepals; typically large amounts of pollen are produced and pollination often occurs early in the growing season before leaves can interfere with the dispersal of the pollen. Many trees and all grasses and sedges are wind pollinated, as such they have no need for large fancy flowers.

An orchid flower

Plants have a number of different means to attract pollinators including colour, scent, heat, nectar glands, edible pollen and flower shape. Along with modifications involving the above structures two other conditions play a very important role in the sexual reproduction of flowering plants, the first is timing of flowering and the other is the size or number of flowers produced. Often plant species have a few large, very showy flowers while others produce many small flowers, often flowers are collected together into large inflorescences to maximize their visual effect, becoming more noticeable to passing pollinators. Flowers are attraction strategies and sexual expressions are functional strategies used to produce the next generation of plants, with pollinators and plants having co-evolved, often to some extraordinary degrees, very often rendering mutual benefit.

Flower heads showing disk and ray florets.

The largest family of flowering plants is the orchids (Orchidaceae), estimated by some specialists to include up to 35,000 species, which often have highly specialized flowers that attract particular insects for pollination. The stamens are modified to produce pollen in clusters called pollinia, which become attached to insects that crawl into the flower. The flower shapes may force insects to pass by the pollen, which is "glued" to the insect. Some orchids are even more highly specialized, with flower shapes that mimic the shape of insects to attract them to 'mate' with the flowers, a few even have scents that mimic insect pheromones.

Another large group of flowering plants is the Asteraceae or sunflower family with close to 22,000 species, which also have highly modified inflorescences that are flowers collected together in heads composed of a composite of individual flowers called florets. Heads with florets of one sex, when the flowers are pistillate or functionally staminate, or made up of all bisexual florets, are called homogamous and can include discoid and liguliflorous type heads. Some radiate heads may be homogamous too. Plants with heads that have florets of two or more sexual forms are called heterogamous and include radiate and disciform head forms, though some radiate heads may be heterogamous too.

Ferns

Ferns typically produce large diploid sporophytes with rhizomes, roots and leaves; and on fertile leaves called sporangium, spores are produced. The spores are released and germinate to produce short, thin gametophytes that are typically heart shaped, small and green in color. The gametophytes or thallus, produce both motile sperm in the antheridia and egg cells in separate archegonia. After rains or when dew deposits a film of water, the motile sperm are splashed away from the antheridia, which are normally produced on the top side of the thallus, and swim in the film of water to the antheridia where they fertilize the egg. To promote out crossing or cross fertilization the sperm are released before the eggs are receptive of the sperm, making it more likely that the sperm will fertilize the eggs of different thallus. A zygote is formed after fertilization, which grows into a new sporophytic plant. The condition of having separate sporophyte and gametophyte plants is call alternation of generations. Other plants with similar reproductive means include the *Psilotum*, *Lycopodium*, *Selaginella* and *Equisetum*.

Bryophytes

The bryophytes, which include liverworts, hornworts and mosses, reproduce both sexually and vegetatively. The gametophyte is the most commonly known phase of the plant. All are small plants found growing in moist locations and like ferns, have motile sperm with flagella and need water to facilitate sexual reproduction. These plants start as a haploid spore that grows into the dominate form, which is a multicellular haploid body with leaf-like structures that photosynthesize. Haploid gametes are produced in antherida and archegonia by mitosis. The sperm released from the antheridia respond to chemicals released by ripe archegonia and swim to them in a film of water and fertilize the egg cells, thus producing zygotes that are diploid. The zygote divides by mitotic division and grows into a sporophyte that is diploid. The multicellular diploid sporophyte produces structures called spore capsules. The spore capsules produce spores by meiosis, and when ripe, the capsules burst open and the spores are released. Bryophytes show considerable variation in their breeding structures and the above is a basic outline. In some species each gametophyte is one sex while other species produce both antheridia and archegonia on the same gametophyte which is thus hermaphrodite.

Sexual Morphology

Many plants have evolved complex sexual reproductive systems, which is expressed in different combinations of their reproductive organs. Some species have separate male and female plants, and some have separate male and female flowers on the same plant, but the majority of plants have both male and female parts in the same flower. Some plants change their morphological expression depending on a number of factors like age, time of day, or because of environmental conditions. Plant sexual morphology also varies within different populations of some species.

Asexual Reproduction

Asexual reproduction is a type of reproduction by which offspring arise from a single organism, and inherit the genes of that parent only; it does not involve the fusion of gametes and almost never changes the number of chromosomes. Asexual reproduction is the primary form of reproduction for single-celled organism as the archaea, bacteria, and protists. Many plants and fungi reproduce asexually as well.

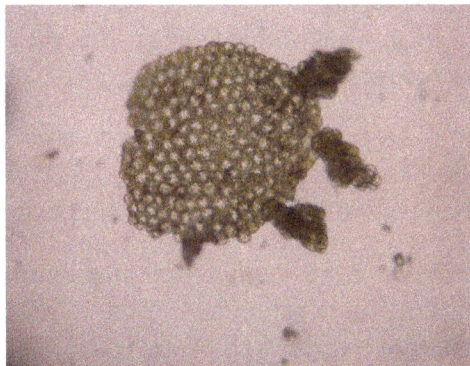

Asexual reproduction in liverworts: a caducous phylloid germinating

While all prokaryotes reproduce asexually (without the formation and fusion of gametes), mechanisms for lateral gene transfer such as conjugation, transformation and transduction are sometimes likened to sexual reproduction (or at least with sex, in the sense of genetic recombination). A complete lack of sexual reproduction is relatively rare among multicellular organisms, particularly animals. It is not entirely understood why the ability to reproduce sexually is so common among them. Current hypotheses suggest that asexual reproduction may have short term benefits when rapid population growth is important or in stable environments, while sexual reproduction offers a net advantage by allowing more rapid generation of genetic diversity, allowing adaptation to changing environments. Developmental constraints may underlie why few animals have relinquished sexual reproduction completely in their life-cycles. Another constraint on switching from sexual to asexual reproduction would be the concomitant loss of meiosis and the protective recombinational repair of DNA damage afforded as one function of meiosis.

Types

Binary Fission

An important form of fission is binary fission. In binary fission, the parent organism is replaced by two daughter organisms, because it literally divides in two. Only prokaryotes (the archaea and the bacteria) reproduce asexually through binary fission. Eukaryotes (such as protists and unicellular fungi) reproduce by mitosis; most of these are also capable of sexual reproduction.

Another type of fission is multiple fission that is advantageous to the plant life cycle. Multiple fission at the cellular level occurs in many protists, e.g. sporozoans and algae. The nucleus of the parent cell divides several times by mitosis, producing several nuclei. The cytoplasm then separates, creating multiple daughter cells.

In apicomplexans, multiple fission, or schizogony, is manifested either as merogony, sporogony or gametogony. Merogony results in merozoites, which are multiple daughter cells, that originate within the same cell membrane, sporogony results in sporozoites, and gametogony results in microgametes.

Budding

Some cells split via budding (for example baker's yeast), resulting in a "mother" and "daughter" cell. The offspring organism is smaller than the parent. Budding is also known on a multicellular level; an animal example is the hydra, which reproduces by budding. The buds grow into fully matured individuals which eventually break away from the parent organism.

Internal budding or Endodyogeny is a process of asexual reproduction, favoured by parasites such as *Toxoplasma gondii*. It involves an unusual process in which two daughter cells are produced inside a mother cell, which is then consumed by the offspring prior to their separation.

Endopolygeny is the division into several organisms at once by internal budding. Also, budding (external or internal) is present in some worm like Taenia or Echinococci; these worm produce cyst and then produce (invaginated or evaginated) protoscolex with budding.

Vegetative Propagation

Closeup of a Bryophyllum daigremontianum

Vegetative propagation is a type of asexual reproduction found in plants where new individuals are formed without the production of seeds or spores by meiosis or syngamy. Examples of vegetative reproduction include the formation of miniaturized plants called plantlets on specialized leaves (for example in kalanchoe) and some produce new plants out of rhizomes or stolon (for example in strawberry). Other plants reproduce by forming bulbs or tubers (for example tulip bulbs and dahlia tubers). Some plants produce adventitious shoots and omay form a clonal colony, where all the individuals are clones, and the clones may cover a large area.

Sporogenesis

Many multicellular organisms form spores during their biological life cycle in a process called *sporogenesis*. Exceptions are animals and some protists, who undergo *meiosis* immediately followed by fertilization. Plants and many algae on the other hand undergo *sporic meiosis* where meiosis leads to the formation of haploid spores rather than gametes. These spores grow into multicellular individuals (called gametophytes in the case of plants) without a fertilization event. These haploid individuals give rise to gametes through mitosis. Meiosis and gamete formation therefore occur in separate generations or "phases" of the life cycle, referred to as alternation of generations. Since sexual reproduction is often more narrowly defined as the fusion of gametes (fertilization), spore formation in plant sporophytes and algae might be considered a form of asexual reproduction (agamogenesis) despite being the result of meiosis and undergoing a reduction in ploidy. However, both events (spore formation and fertilization) are necessary to complete sexual reproduction in the plant life cycle.

Fungi and some algae can also utilize true asexual spore formation, which involves mitosis giving rise to reproductive cells called mitospores that develop into a new organism after dispersal. This method of reproduction is found for example in conidial fungi and the red algae *Polysiphonia*, and involves sporogenesis without meiosis. Thus the chromosome number of the spore cell is the same as that of the parent producing the spores. However, mitotic sporogenesis is an exception and most spores, such as those of plants, most Basidiomycota, and many algae, are produced by meiosis.

Fragmentation

Fragmentation is a form of asexual reproduction where a new organism grows from a fragment of the parent. Each fragment develops into a mature, fully grown individual. Fragmentation is seen in many organisms such as animals (some planarian and annelid worms, turbellarians and sea stars), fungi, and plants. Some plants have specialized structures for reproduction via fragmentation, such as *gemma* in liverworts. Most lichens, which are a symbiotic union of a fungus and photosynthetic algae or bacteria, reproduce through fragmentation to ensure that new individuals contain both symbiont. These fragments can take the form of *soredia*, dust-like particles consisting of fungal hyphen wrapped around photobiont cells.

Clonal Fragmentation in multicellular or colonial organisms is a form of asexual reproduction or cloning where an organism is split into fragments. Each of these fragments develop into mature, fully grown individuals that are clones of the original organism. In echinoderms, this method of reproduction is usually known as fissiparity.Researchers claim today that due to many environmental and epigenetic differences, that clones originated in the same ancestor might actually be genetically and epigenetically different.

Agamogenesis

Agamogenesis is any form of reproduction that does not involve a male gamete. Examples are parthenogenesis and apomixis.

Parthenogenesis

Parthenogenesis is a form of agamogenesis in which an unfertilized egg develops into a new individual. Parthenogenesis occurs naturally in many plants, invertebrates (e.g. water fleas, rotifers, aphids, stick insects, some ants, bees and parasitic wasps), and vertebrates (e.g. some reptiles, amphibians, rarely birds). In plants, apomixis may or may not involve parthenogenesis.

Apomixis and Nucellar Embryony

Apomixis in plants is the formation of a new sporophyte without fertilization. It is important in ferns and in flowering plants, but is very rare in other seed plants. In flowering plants, the term "apomixis" is now most often used for agamospermy, the formation of seeds without fertilization, but was once used to include vegetative reproduction. An example of an apomictic plant would be the triploid European dandelion. Apomixis mainly occurs in two forms: In gametophytic apomixis, the embryo arises from an unfertilized egg within a diploid embryo sac that was formed without completing meiosis. In nucellar embryony, the embryo is formed from the diploid nucellus tissue surrounding the embryo sac. Nucellar embryony occurs in some citrus seeds. Male apomixis can occur in rare cases, such as the Saharan Cypress *Cupressus dupreziana*, where the genetic material of the embryo are derived entirely from pollen. The term "apomixis" is also used for asexual reproduction in some animals, notably water-fleas, *Daphnia*.

Regeneration

Alternation Between Sexual and Asexual Reproduction

Some species alternate between the sexual and asexual strategies, an ability known as heterogamy, depending on conditions. Alternation is observed in several rotifer species (cyclical parthenogenesis e.g. in Brachionus species) and a few types of insects, such as aphids which will, under certain conditions, produce eggs that have not gone through meiosis, thus cloning themselves. The cape bee *Apis mellifera* subsp. *capensis* can reproduce asexually through a process called thelytoky. A few species of amphibians, reptiles, and birds have a similar ability. For example, the freshwater crustacean *Daphnia* reproduces by parthenogenesis in the spring to rapidly populate ponds, then switches to sexual reproduction as the intensity of competition and predation increases. Another example are monogonont rotifers of the genus *Brachionus*, which reproduce via cyclical parthenogenesis: at low population densities females produce asexually and at higher densities a chemical cue accumulates and induces the transition to sexual reproduction. Many protists and fungi alternate between sexual and asexual reproduction.

For example, the slime mold *Dictyostelium* undergoes binary fission (mitosis) as single-celled amoebae under favorable conditions. However, when conditions turn unfavorable, the cells aggregate and follow one of two different developmental pathways, depending on conditions. In the social pathway, they form a multicellular slug which then forms a fruiting body with asexually generated spores. In the sexual pathway, two cells fuse to form a giant cell that develops into a large cyst. When this macrocyst germinates, it releases hundreds of amoebic cells that are the product of meiotic recombination between the original two cells.

The hyphae of the common mold (*Rhizopus*) are capable of producing both mitotic as well as meiotic spores. Many algae similarly switch between sexual and asexual reproduction. A number of plants use both sexual and asexual means to produce new plants, some species alter their primary modes of reproduction from sexual to asexual under varying environmental conditions.

Inheritance in Sexual Species

For example, in the rotifer *Brachionus calyciflorus* asexual reproduction (obligate parthenogenesis) can be inherited by a recessive allele, which leads to loss of sexual reproduction in homozygous offspring. Inheritance of asexual reproduction by a single recessive locus has also been found in the parasitoid wasp *Lysiphlebus fabarum*.

Examples in Animals

There are examples of parthenogenesis in the hammerhead shark and the blacktip shark. In both cases, the sharks had reached sexual maturity in captivity in the absence of males, and in both cases the offspring were shown to be genetically identical to the mothers. The New Mexico whiptail is another example.

Reptiles use the ZW sex-determination system, which produces either males (with ZZ sex chromosomes) or females (with ZW or WW sex chromosomes). Until 2010, it was thought that the ZW chromosome system used by reptiles was incapable of producing viable WW offspring, but a (ZW) female boa constrictor was discovered to have produced viable female offspring with WW chromosomes. The female boa could have chosen any number of male partners (and had successfully in the past) but on these occasions she reproduced asexually, creating 22 female babies with WW sex-chromosomes.

Polyembryony is a widespread form of asexual reproduction in animals, whereby the fertilized egg or a later stage of embryonic development splits to form genetically identical clones. Within animals, this phenomenon has been best studied in the parasitic Hymenoptera. In the 9-banded armadillos, this process is obligatory and usually gives rise to genetically identical quadruplets. In other mammals, monozygotic twinning has no apparent genetic basis, though its occurrence is common. There are at least 10 million identical human twins and triplets in the world today.

Bdelloid rotifers reproduce exclusively asexually, and all individuals in the class Bdelloidea are females. Asexuality evolved in these animals millions of years ago and has persisted since. There is evidence to suggest that asexual reproduction has allowed the animals to evolve new proteins through the Meselson effect that have allowed them to survive better in periods of dehydration.

Molecular evidence strongly suggest that several species of the stick insect genus *Timema* have used only asexual (parthenogenetic) reproduction for millions of years, the longest period known for any insect.

In the grass thrips genus *Aptinothrips* there have been several transitions to asexuality, likely due to different causes.

Types of Asexual Reproduction

Fission (Biology)

In biology, fission is the division of a single entity into two or more parts and the regeneration of those parts into separate entities resembling the original. The object experiencing fission is usually a cell, but the term may also refer to how organisms, bodies, populations, or species split into discrete parts. The fission may be *binary fission*, in which a single entity produces two parts, or *multiple fission*, in which a single entity produces multiple parts.

Binary Fission

Organisms in the domains of Archaea and Bacteria reproduce with binary fission. This form of asexual reproduction and cell division is also used by some organelles within eukaryotic organisms (e.g., mitochondria). Binary fission results in the reproduction of a living prokaryotic cell (or organelle) by dividing the cell into two parts, each with the potential to grow to the size of the original.

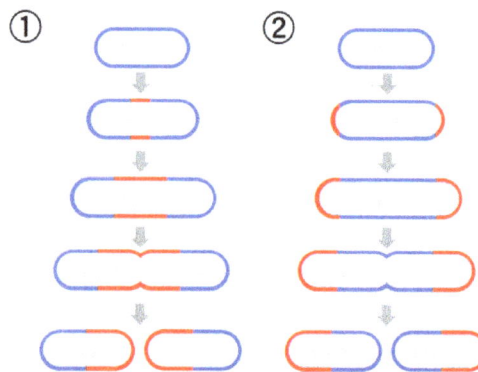

Schematic diagram of cellular growth (longation) and binary fission of bacilli. Blue and red lines indicate old and newly generated bacterial cell wall, respectively. (1) growth at the center of bacterial body. e.g. *Bacillus subtilis, E. coli,* and others. (2) apical growth. e.g. *Corynebacterium diphtheriae.*

Fission of Prokaryotes

Unlike the process of mitosis used by eukaryotic cells, binary fission takes place without the formation of a spindle apparatus in the cell. The single DNA molecule first replicates, then attaches each copy to a different part of the cell membrane. When the cell begins to pull apart, the replicated and original chromosomes are separated. The consequence of this asexual method of reproduction is that all the cells are genetically identical, meaning that they have the same genetic material (barring random mutations).

Process of Bacterial Fission

The process of binary fission in bacteria involves the following steps. First, the cell's DNA is replicated. The replicated DNA copies then move to opposite poles of the cell in an energy-dependent process. The cell lengthens. Then, the equatorial plate of the cell constricts and separates the plasma membrane such that each new cell has exactly the same genetic material.

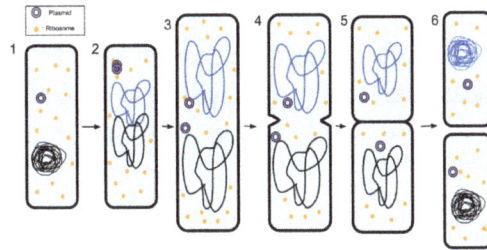

Binary Fission In A Prokaryote

More specifically, the following steps occur, as seen in the figure Binary fission in a prokaryote:

1. The bacterium before binary fission is when the DNA is tightly coiled.

2. The DNA of the bacterium has uncoiled and duplicated.

3. The DNA is pulled to the separate poles of the bacterium as it increases size to prepare for splitting.

4. The growth of a new cell wall begins to separate the bacterium.

5. The new cell wall fully develops, resulting in the complete split of the bacterium.

6. The new daughter cells have tightly coiled DNA rods, ribosomes, and plasmids; these are now brand new organisms.

Speed of Bacterial Fission

Binary fission is generally rapid though its speed varies between species. For *E. coli*, cells typically divide about every 20 minutes at 37 °C. Because the new cells will, in turn, undergo binary fission on their own, the time binary fission requires is also the time the bacterial culture requires to double in the number of cells it contains. This time period can therefore be referred to as the doubling time. Some species other than *E. coli* may have faster or slower doubling times: some strains of *Mycobacterium tuberculosis* may have doubling times of nearly 100 hours. Bacterial growth is limited by factors including nutrient availability and available space, so binary fission occurs at much lower rates in bacterial cultures once they enter the stationary phase of growth.

Fission of Organelles

Some organelles in eukaryotic cells reproduce using binary fission. Mitochondrial fission occurs frequently within the cell, even when the cell is not actively undergoing mitosis, and is necessary to regulate the cell's metabolism.

Multiple Fission

Fission of Protists

Multiple fission at the cellular level occurs in many protists, e.g. sporozoans and algae. The nucleus of the parent cell divides several times by mitosis, producing several nuclei. The cytoplasm then separates, creating multiple daughter cells.

Some parasitic, single-celled organisms undergo a multiple fission-like process to produce numerous daughter cells from a single parent cell. Isolates of the human parasite *Blastocystis hominis* were observed to begin such a process within 4 to 6 days. Cells of the fish parasite *Trypanoplasma borreli* have also been observed participating in both binary and multiple fission.

Fission of Apicomplexans

In the apicomplexans, a phylum of parasitic protists, multiple fission, or schizogony, is manifested either as merogony, sporogony or gametogony. Merogony results in merozoites, which are multiple daughter cells, that originate within the same cell membrane, sporogony results in sporozoites, and gametogony results in microgametes.

Fission of Green Algae

Green algae can divide into more than two daughter cells. The exact number of daughter cells depends on the species of algae and is an effect of temperature and light.

Multiple Fission of Bacteria

Most species of bacteria primarily undergo binary reproduction. Some species and groups of bacteria may undergo multiple fission as well, sometimes beginning or ending with the production of spores. The species *Metabacterium polyspora*, a symbiont of guinea pigs, has been found to produce multiple endospores in each division. Some species of cyanobacteria have also been found to reproduce through multiple fission.

Clonal Fragmentation

Fragmentation in multicellular or colonial organisms is a form of asexual reproduction or cloning where an organism is split into fragments. Each of these fragments develop into mature, fully grown individuals that are clones of the original organism. In echinoderms, this method of reproduction is usually known as fissiparity.

Population Fission

Any splitting of a single population of individuals into discrete parts may be considered fission. A population may undergo fission for a variety of reasons, including migration or geographic isolation. Because the fission leads to genetic variance in the newly isolated, smaller populations, population fission is a precursor to speciation.

Budding

Budding is a form of asexual reproduction in which a new organism develops from an outgrowth or bud due to cell division at one particular site. The new organism remains attached as it grows, separating from the parent organism only when it is mature, leaving behind scar tissue. Since the reproduction is asexual, the newly created organism is a clone and is genetically identical to the parent organism.

Organisms such as hydra use regenerative cells for reproduction in the process of budding. In hydra, a bud develops as an outgrowth due to repeated cell division at one specific site. These buds develop into tiny individuals and, when fully mature, detach from the parent body and become new independent individuals.

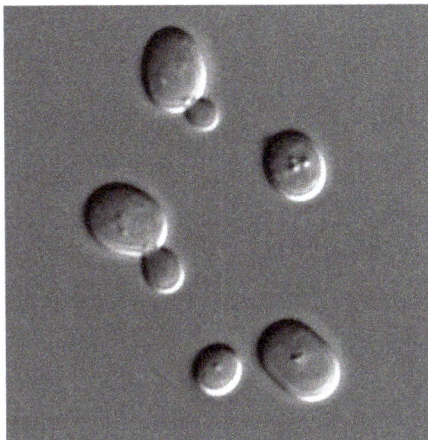

Saccharomyces cerevisiae reproducing by budding

Internal budding or endodyogeny is a process of asexual reproduction, favoured by parasites such as *Toxoplasma gondii*. It involves an unusual process in which two daughter cells are produced inside a mother cell, which is then consumed by the offspring prior to their separation.

Endopolygeny is the division into several organisms at once by internal budding.

Cellular Reproduction

Some cells divide asymmetrically by budding, for example *Saccharomyces cerevisiae*, the yeast species used in baking and brewing. This process results in a 'mother' cell and a smaller 'daughter' cell. Cryo-electron tomography recently revealed that mitochondria in cells divide by budding.

Animal Reproduction

In some multicellular animals (metazoans), offspring may develop as outgrowths of the mother. Animals that reproduce by budding include corals, some sponges, some acoel flatworms (e.g., *Convolutriloba*), and echinoderm larvae.

Hydra with two buds

Colony Division

Colonies of some bee species have also exhibited budding behavior, such as *Apis dorsata*. Although budding behavior is rare in this bee species, it has been observed when a group of workers leave the natal nest and construct a new nest usually near the natal one.

Virology

In virology, *budding* is a form of viral shedding by which enveloped viruses acquire their external envelope from the host cell membrane, which bulges outwards and encloses the virion.

Plant Multiplication

In agriculture and horticulture, budding refers to grafting the bud of one plant onto another.

Vegetative Reproduction

Vegetative reproduction (vegetative propagation, vegetative multiplication, vegetative cloning) is a form of asexual reproduction in plants. It is a process by which new organisms arise without production of seeds or spores. It can occur naturally or be induced by horticulturists.

A bulb of *Muscari* has reproduced vegetatively underground to make two bulbs, each of which produces a flower stem.

Although most plants normally reproduce sexually, many have the ability for vegetative propagation, or can be vegetatively propagated if small pieces are subjected to chemical (hormonal) treatments. This is because meristematic cells capable of cellular differentiation are present in many plant tissues. Horticulturalists are interested in understanding how meristematic cells can be induced to reproduce an entire plant.

Success rates and difficulty of propagation vary greatly. For example willow and coleus can be propagated merely by inserting a stem in water or moist soil. On the other hand, monocotyledons, unlike dicotyledons, typically lack a vascular cambium and therefore are harder to propagate.

Types

In a wide sense, methods of vegetative propagation include cutting, vegetative apomixis, layering, division, budding, grafting and tissue culture. Cutting is the most common artificial vegetative propagation method, where pieces of the "parent" plant are removed and placed in a suitable environment so that they can grow into a whole new plant, the "clone", which is genetically identical to the parent. Cutting exploits the ability of plants to grow adventitious roots (i.e. root material that can generate from a location other than the existing or primary root system, as in from a leaf or cut stem) under certain conditions. Vegetative propagation is usually considered a cloning method. However, there are several cases where vegetatively propagated plants are not genetically identical. Root cuttings of thornless blackberries will revert to thorny type because the adventitious shoot develops from a cell that is genetically thorny. Thornless blackberry is a chimera, with the epidermal layers genetically thornless but the tissue beneath it genetically thorny. Similarly, leaf cutting propagation of certain chimeral variegated plants, such as snake plant (Sansevieria trifasciata), will produce mainly nonvariegated plants. Grafting is often not a complete cloning method because seedlings are used as rootstocks. In that case only the top of the plant is clonal. In some crops, particularly apples, the rootstocks are vegetatively propagated so the entire graft can be clonal if the scion and rootstock are both clones. Apomixis (including apospory and diplospory) is a type of reproduction that does not involve fertilisation. In flowering plants, unfertilized seeds are involved, or plantlets that grow instead of flowers. Hawkweed (*Hieracium*), dandelion (*Taraxacum*), some citrus (*Citrus*) and many grasses such as Kentucky blue grass (*Poa pratensis*) all use this form of asexual reproduction. Bulbils are sometimes formed instead of the flowers of garlic.

Vegetative Structures

Virtually all types of shoots and roots are capable of vegetative propagation, including stems, basal shoots, tubers, rhizomes, stolons, corms, bulbs, and buds. In a few species (such as *Kalanchoë*), leaves are involved in vegetative reproduction.

- The rhizome is a modified underground stem serving as an organ of vegetative reproduction, e. g. *Polypodium* (polypody), Iris, Couch Grass and Nettles.

- Prostrate aerial stems, called runners or stolons are important vegetative reproduction organs in some species, such as the strawberry, numerous grasses, and some ferns.

- Adventitious buds form on roots near the ground surface, on damaged stems (as on the stumps of cut trees), or on old roots. These develop into above-ground stems and leaves.

- A form of budding called suckering is the reproduction or regeneration of a plant by shoots that arise from an existing root system. Species that characteristically produce suckers include Elm (*Ulmus*) and members of the Rose Family (*Rosa*).

- Another type of a vegetative reproduction is the production of bulbs. Plants like onion (*Allium cepa*), hyacinth (*Hyacinth*), narcissus (*Narcissus*) and tulips (*Tulipa*) reproduce by forming bulbs.

- Other plants like potatoes (*Solanum tuberosum*) and dahlia (*Dahlia*) reproduce by a method similar to bulbs: they produce tubers.

- Gladioli and crocuses (*Crocus*) reproduce by forming a bulb-like structure called a corm.

- Some orchids reproduce by the growth of keikis from the stem or cane of the parent plant.

Natural Vegetative Propagation

Natural vegetative propagation is mostly a process found in herbaceous and woody perennial plants, and typically involves structural modifications of the stem, although any horizontal, underground part of a plant (whether stem, leaf, or root) can contribute to vegetative reproduction of a plant. Most plant species that survive and significantly expand by vegetative reproduction would be perennial almost by definition, since specialized organs of vegetative reproduction, like seeds of annuals, serve to survive seasonally harsh conditions. A plant that persists in a location through vegetative reproduction of individuals over a long period of time constitutes a clonal colony.

In a sense, this process is not one of reproduction but one of survival and expansion of biomass of the individual. When an individual organism increases in size via cell multiplication and remains intact, the process is called "vegetative growth". However, in vegetative reproduction, the new plants that result are new individuals in almost every respect except genetic. Of considerable interest is how this process appears to reset the aging clock.

Artificial Vegetative Propagation

Mass Propagation of Eucalyptus Seedlings

Vegetative propagation of particular cultivars that have desirable characteristics is very common practice. Reasons for preferring vegetative rather than sexual means of reproduction vary, but commonly include greater ease and speed of propagation of certain plants, such as many perennial root crops and vines. Another major attraction is that the resulting plant amounts to a clone of the parent plant and accordingly is of a more predictable quality than most seedlings. However, as can be seen in many variegated plants, this does not always apply, because many plants actually are chimeras and cuttings might reflect the attributes of only one or some of the parent cell lines. Man-made methods of vegetative reproduction are usually enhancements of natural processes, but they range from rooting cuttings to grafting and artificial propagation by laboratory tissue culture.

In horticulture, a "cutting" is a piece that has been cut off from a mother plant and then caused to grow into a whole plant. Often this involves a piece of stem that is treated with rooting liquid or powder containing hormones. In some species root cuttings can produce shoot growth. When the cutting has become a self-sufficient plant, it is genetically identical to the mother plant except when chimeric tissues or similar complications affect the outcome.

A related form of regeneration is that of grafting. A stem piece or a single bud (the scion) is joined onto the stem of a plant that has roots (the rootstock), or a stem piece can be joined to a root piece. A popular use of grafting is to produce fruit trees, sometimes with more than one variety of the same fruit species growing from the same stem. Rootstocks for fruit trees are either seedlings or propagated by layering.

Fragmentation (Reproduction)

Fragmentation or clonal fragmentation in multicellular or colonial organisms is a form of asexual reproduction or cloning in which an organism is split into fragments. Each of these fragments develop into mature, fully grown individuals that are clones of the original organism.

The splitting may or may not be *intentional* – it may occur due to man-made or natural damage by the environment or predators. This kind of organism may develop specific organs or zones that may be shed or easily broken off. If the splitting occurs without the prior preparation of the organism, both fragments must be able to regenerate the complete organism for it to function as reproduction.

Fragmentation, also known as splitting, as a method of reproduction is seen in many organisms such as filamentous cyanobacteria, molds, lichens, many plants, and animals such as sponges, acoel flatworms, some annelid worms and sea stars.

Fragmentation in Various Organisms

Moulds, yeasts and mushrooms, all of which are part of the Fungi kingdom, produce tiny filaments called hyphae. These hyphae obtain food and nutrients from the body of other organisms to grow and fertilize. Then a piece of hyphae breaks off and grows into a new indiv and the cycle continues.

Many lichens produce specialized structures that can easily break away and disperse. These struc-tures contain both the hyphae of the mycobiont and the algae(phycobiont). Larger fragments of the thallus may break away when the lichen dries or due to mechanical distur-bances.

Plants

Fragmentation is a very common type of vegetative reproduction in plants. Many trees, shrubs, nonwoody perennials, and ferns form clonal colonies by producing new rooted shoots by rhizomes or stolons, which increases the diameter of the colony. If a rooted shoot becomes detached from the colony, then fragmentation has occurred. There are several other mechanisms of natural frag-mentation in plants.

- Production of specialized reproductive structures: A few plants produce adventitious plant-lets on their leaves, which drop off and form independent plants, e.g. *Tolmiea menziesii* and *Kalanchoe daigremontiana*. Others produce organs like bulbils and turions.

- Easily lost parts that have high potential to grow into a complete plant: Some woody plants like the willow naturally shed twigs. This is termed cladoptosis. The lost twigs may form roots in a suitable environment to establish a new plant. River currents often tear off branch fragments from certain cottonwood species growing on riverbanks. Fragments reaching suitable environments can root and establish new plants. Some cacti and other plants have jointed stems. When a stem segment, called a pad, falls off, it can root and form a new plant. Leaves of some plants readily root when they fall off, e.g. *Sedum* and *Echeveria*.

- Fragmentation is observed in nonvascular plants as well, for example, in liverworts and mosses. Small pieces of moss "stems" or "leaves" are often scattered by wind, water or animals. If a moss fragment reaches a suitable environment, it can establish a new plant. They also produce gemma that are easily broken off and distributed.

People use fragmentation to artificially propagate many plants via division, layering, cuttings, grafting, micropropagation and storage organs, such as bulbs, corms, tubers and rhizomes.

Animals

Animals like sponges and coral colonies naturally fragment and reproduce. Many species of annelids and flat worms reproduce by this method.

When the splitting occurs due to specific developmental changes, the terms architomy, paratomy and budding are used. In architomy the animal splits at a particular point and the two fragments regenerate the missing organs and tissues. The splitting is not preceded by the development of the tissues to be lost. Prior to splitting, the animal may develop furrows at the zone of splitting. The headless fragment has to regenerate a complete head.

In paratomy, the split occurs perpendicular to the antero-posterior axis and the split is preceded by the "pregeneration" of the anterior structures in the posterior portion. The two organisms have their body axis aligned i.e. they develop in a head to tail fashion. Budding can be considered to be similar to paratomy except that the body axes need not be aligned: the new head may grow toward the side or even point backward (e.g. *Convolutriloba retrogemma* an acoel flat worm).

Coral

Corals can be multiplied in aquaria by attaching "frags" from a mother colony to a suitable substrate, such as a ceramic plug or a piece of live rock. This aquarium is designed specifically for growing coral colonies from frags.

Many types of coral colonies can increase in number by fragmentation that occurs naturally or artificially. Within the reef aquarium hobby, enthusiasts regularly fragment corals for a multitude of purposes including shape control; selling to, trading with, or sharing with others; regrowth experiments; and minimizing damage to natural coral reefs. Both hard and soft corals can be fragmented, with the level of success depending on the skill of the aquarist, method used, tolerance of the specific species, and conditions of care. Genera that have shown to be highly tolerant of fragmentation include *Acropora*, *Montipora*, *Pocillopora*, *Euphyllia*, and *Caulastraea* among many others.

Echinoderms

In echinoderms, the process is usually known as fissiparity (a term also used infrequently for fission in general). Some species can intentionally reproduce in this manner through autotomy. This method is more common during the larval stages.

Disadvantage of This Process of Reproduction

As this process is a form of asexual reproduction, it does not produce genetic diversity in the offspring. Therefore, these are more vulnerable to changing environments.

Parthenogenesis

Parthenogenesis is a natural form of asexual reproduction in which growth and development of embryos occur without fertilization. In animals, parthenogenesis means development of an embryo from an unfertilized egg cell and is a component process of apomixis.

The asexual, all-female whiptail species *Cnemidophorus neomexicanus* (center), which reproduces via parthenogenesis, is shown flanked by two sexual species having males *C. inornatus* (left) and *C. tigris* (right), which hybridized naturally to form the *C. neomexicanus* species.

Gynogenesis and pseudogamy are closely related phenomena in which a sperm or pollen triggers the development of the egg cell into an embryo but makes no genetic contribution to the embryo. The rest of the cytology and genetics of these phenomena are mostly identical to that of parthenogenesis.

The term is sometimes used inaccurately to describe reproduction modes in hermaphroditic species that can reproduce by themselves because they contain reproductive organs of both sexes in a single individual's body. However, these species still use fertilization.

Parthenogenesis occurs naturally in many plants, some invertebrate animal species (including nematodes, water fleas, some scorpions, aphids, some mites, some bees, some Phasmida and parasitic wasps) and a few vertebrates (such as some fish, amphibians, reptiles and very rarely birds). This type of reproduction has been induced artificially in a few species including fish and amphibians.

Normal egg cells form after meiosis and are haploid, with half as many chromosomes as their mother's body cells. Haploid individuals, however, are usually non-viable, and parthenogenetic offspring usually have the diploid chromosome number. Depending on the mechanism involved in restoring the diploid number of chromosomes, parthenogenetic offspring may have anywhere between all and half of the mother's alleles. The offspring having all of the mother's genetic material are called full clones and those having only half are called half clones. Full clones are usually formed without meiosis. If meiosis occurs, the offspring will get only a fraction of the mother's alleles.

Parthenogenetic offspring in species that use either the XY or the Xo sex-determination system have two X chromosomes and are female. In species that use the ZW sex-determination system, they have either two Z chromosomes (male) or two W chromosomes (mostly non-viable but rarely a female), or they could have one Z and one W chromosome (female).

Life History Types

Some species reproduce exclusively by parthenogenesis (such as the Bdelloid rotifers), while others can switch between sexual reproduction and parthenogenesis. This is called facultative parthenogenesis (other terms are cyclical parthenogenesis, heterogamy or heterogony). The switch between sexuality and parthenogenesis in such species may be triggered by the season (aphid, some gall wasps), or by a lack of males or by conditions that favour rapid population growth (rotifers and cladocerans like daphnia). In these species asexual reproduction occurs either in summer (aphids) or as long as conditions are favourable. This is because in asexual reproduction a successful genotype can spread quickly without being modified by sex or wasting resources on male offspring who won't give birth. In times of stress, offspring produced by sexual reproduction may be fitter as they have new, possibly beneficial gene combinations. In addition, sexual reproduction provides the benefit of meiotic recombination between non-sister chromosomes, a process associated with repair of DNA double-strand breaks and other DNA damages that may be induced by stressful conditions.

Many taxa with heterogony have within them species that have lost the sexual phase and are now completely asexual. Many other cases of obligate parthenogenesis (or gynogenesis) are found among polyploids and hybrids where the chromosomes cannot pair for meiosis.

The production of female offspring by parthenogenesis is referred to as thelytoky (e.g., aphids) while the production of males by parthenogenesis is referred to as arrhenotoky (e.g., bees). When unfertilized eggs develop into both males and females, the phenomenon is called deuterotoky.

Types and Mechanisms

Parthenogenesis can occur without meiosis through mitotic oogenesis. This is called apomictic parthenogenesis. Mature egg cells are produced by mitotic divisions, and these cells directly develop into embryos. In flowering plants, cells of the gametophyte can undergo this process. The offspring produced by apomictic parthenogenesis are *full clones* of their mother. Examples include aphids.

Parthenogenesis involving meiosis is more complicated. In some cases, the offspring are haploid (e.g., male ants). In other cases, collectively called automictic parthenogenesis, the ploidy is restored to diploidy by various means. This is because haploid individuals are not viable in most species. In automictic parthenogenesis the offspring differ from one another and from their mother. They are called *half clones* of their mother.

Automictic Parthenogenesis

Automixis is a term that covers several reproductive mechanisms, some of which are parthenogenetic.

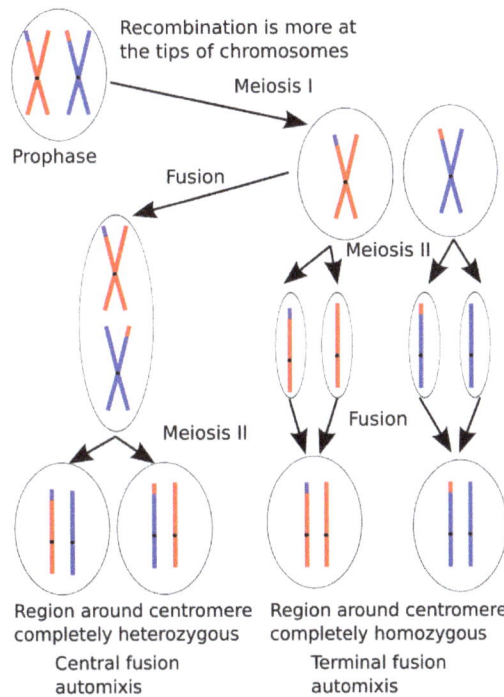

The effects of central fusion and terminal fusion on heterozygosity

Diploidy might be restored by the doubling of the chromosomes without cell division before meiosis begins or after meiosis is completed. This is referred to as an *endomitotic* cycle. This may also happen by the fusion of the first two blastomeres. Other species restore their ploidy by the fusion of the meiotic products. The chromosomes may not separate at one of the two anaphases (called restitutional meiosis), or the nuclei produced may fuse or one of the polar bodies may fuse with the egg cell at some stage during its maturation.

Some authors consider all forms of automixis sexual as they involve recombination. Many others classify the endomitotic variants as asexual, and consider the resulting embryos parthenogenetic. Among these authors the threshold for classifying automixis as a sexual process depends on when the products of anaphase I or of anaphase II are joined together. The criterion for "sexuality" varies from all cases of restitutional meiosis, to those where the nuclei fuse or to only those where gametes are mature at the time of fusion. Those cases of automixis that are classified as sexual reproduction are compared to self-fertilization in their mechanism and consequences.

The genetic composition of the offspring depends on what type of apomixis takes place. When endomitosis occurs before meiosis or when *central fusion* occurs (restitutional meiosis of anaphase I or the fusion of its products), the offspring get all to more than half of the mother's genetic material and heterozygosity is mostly preserved (if the mother has two alleles for a locus, it is likely that the offspring will get both). This is because in anaphase I the homologous chromosomes are separated. Heterozygosity is not completely preserved when crossing over occurs in central fusion. In the case of pre-meiotic doubling, recombination -if it happens- occurs between identical sister chromatids.

If *terminal fusion* (restitutional meiosis of anaphase II or the fusion of its products) occurs, a little over half the mother's genetic material is present in the offspring and the offspring are mostly homozygous. This is because at anaphase II the sister chromatids are separated and whatever heterozygosity is present is due to crossing over. In the case of endomitosis after meiosis the offspring is completely homozygous and has only half the mother's genetic material.

This can result in parthenogenetic offspring being unique from each other and from their mother.

Sex of The Offspring

In apomictic parthenogenesis, the offspring are clones of the mother and hence are usually (except for aphids) female. In the case of aphids, parthenogenetically produced males and females are clones of their mother except that the males lack one of the X chromosomes (XO).

When meiosis is involved, the sex of the offspring will depend on the type of sex determination system and the type of apomixis. In species that use the XY sex-determination system, parthenogenetic offspring will have two X chromosomes and are female. In species that use the ZW sex-determination system the offspring genotype may be one of ZW (female), ZZ (male), or WW (non-viable in most species but a fertile, viable female in a few (e.g., boas)). ZW offspring are produced by endoreplication before meiosis or by central fusion. ZZ and WW offspring occur either by terminal fusion or by endomitosis in the egg cell.

In polyploid obligate parthenogens like the whiptail lizard, all the offspring are female.

In many hymenopteran insects such as honeybees, female eggs are produced sexually, using sperm from a drone father, while the production of further drones (males) depends on the queen (and occasionally workers) producing unfertilised eggs. This means that females (workers and queens) are always diploid, while males (drones) are always haploid, and produced parthenogenetically.

Facultative Parthenogenesis

Facultative parthenogenesis is the term for when a female can produce offspring either sexually or via asexual reproduction. Facultative parthenogenesis is extremely rare in nature, with only a few examples of animal taxa capable of facultative parthenogenesis. One of the best known examples of taxa exhibiting facultative parthenogenesis are mayflies; presumably this is the default reproductive mode of all species in this insect order. Facultative parthenogenesis is believed to be a response to a lack of a viable male. A female may undergo facultative parthenogenesis if a male is absent from the habitat or if it is unable to produce viable offspring.

Facultative parthenogenesis is often incorrectly used to describe cases of accidental or spontaneous parthenogenesis in normally sexual animals. For example, many cases of accidental parthenogenesis in sharks, some snakes, Komodo dragons and a variety of domesticated birds were widely perpetuated as facultative parthenogenesis. These cases are, however, examples of accidental parthenogenesis, given the frequency of asexually produced eggs and their hatching rates are extremely low, in contrast to true facultative parthenogenesis where the majority of asexually produced eggs hatch. In addition, asexually produced offspring in vertebrates are virtually always sterile, highlighting that this mode of reproduction is not adaptive, because the ability to reproduce asexually is not inherited to the next generation. The occurrence of such asexually produced eggs in sexual animals can be explained by a meiotic error, leading to eggs produced via automixis.

Obligate Parthenogenesis

Obligate parthenogenesis is the process in which organisms exclusively reproduce through asexual means. Many species have been shown to transition to obligate parthenogenesis over evolutionary time. Among these species, one of the most well documented transitions to obligate parthenogenesis was found in almost all metazoan taxa, albeit through highly diverse mechanisms. These transitions often occur as a result of inbreeding or mutation within large populations. There are a number of documented species, specifically salamanders and geckos, that rely on obligate parthenogenesis as their major method of reproduction. As such, there are over 80 species of unisex reptiles (mostly lizards but including a single snake species), amphibians and fishes in nature for which males are no longer a part of the reproductive process. A female will produce an ovum with a full set (two sets of genes) provided solely by the mother. Thus, a male is not needed to provide sperm to fertilize the egg. This form of asexual reproduction is thought in some cases to be a serious threat to biodiversity for the subsequent lack of gene variation and potentially decreased fitness of the offspring.

Natural Occurrence

Parthenogenesis is seen to occur naturally in aphids, *Daphnia*, rotifers, nematodes and some other invertebrates, as well as in many plants. Among vertebrates, strict parthenogenesis is only known to occur in lizards, snakes, birds and sharks, with fish, amphibians and reptiles exhibiting various forms of gynogenesis and hybridogenesis (an incomplete form of parthenogenesis). The first all-female (unisexual) reproduction in vertebrates was described in the fish *Poecilia formosa* in 1932. Since then at least 50 species of unisexual vertebrate have been described, including at least 20 fish, 25 lizards, a single snake species, frogs, and salamanders. Other, usually sexual, species may occasionally reproduce parthenogenetically and Komodo dragons; the hammerhead and blacktip sharks are recent additions to the known list of spontaneous parthenogenetic vertebrates. As with all types of asexual reproduction, there are both costs (low genetic diversity and therefore susceptibility to adverse mutations that might occur) and benefits (reproduction without the need for a male) associated with parthenogenesis.

Parthenogenesis is distinct from artificial animal cloning, a process where the new organism is necessarily genetically identical to the cell donor. In cloning, the nucleus of a diploid cell from a donor organism is inserted into an enucleated egg cell and the cell is then stimulated to undergo continued mitosis, resulting in an organism that is genetically identical to the donor. Parthenogenesis is different, in that it originates from the genetic material contained within an egg cell and the new organism is not necessarily genetically identical to the parent.

Parthenogenesis may be achieved through an artificial process as described below under the discussion of mammals.

Insects

Parthenogenesis in insects can cover a wide range of mechanisms. The offspring produced by parthenogenesis may be of both sexes, only female (thelytoky, e.g. aphids and some hymenopterans) or only male (arrhenotoky, e.g. most hymenopterans). Both true parthenogenesis and pseudogamy (gynogenesis or sperm-dependent parthenogenesis) are known to occur. The egg cells, depending on the species may be produced without meiosis (apomictically) or by one of the several automictic mechanisms.

A related phenomenon, polyembryony is a process that produces multiple clonal offspring from a single egg cell. This is known in some hymenopteran parasitoids and in Strepsiptera.

In automictic species the offspring can be haploid or diploid. Diploids are produced by doubling or fusion of gametes after meiosis. Fusion is seen in the Phasmatodea, Hemiptera (Aleurodids and Coccidae), Diptera, and some Hymenoptera.

In addition to these forms is hermaphroditism, where both the eggs and sperm are produced by the same individual, but is not a type of parthenogenesis. This is seen in three species of *Icerya* scale insects.

Parasitic bacteria like *Wolbachia* have been noted to induce automictic thelytoky in many insect species with haplodiploid systems. They also cause gamete duplication in unfertilized eggs causing them to develop into female offspring.

Honey Bee on a plum blossom

Among species with the haplo-diploid sex-determination system, such as hymenopterans (ants, bees and wasps) and thysanopterans (thrips), haploid males are produced from unfertilized eggs. Usually eggs are laid only by the queen, but the unmated workers may also lay haploid, male eggs either regularly (e.g. stingless bees) or under special circumstances. An example of non-viable parthenogenesis is common among domesticated honey bees. The queen bee is the only fertile female in the hive; if she dies without the possibility for a viable replacement queen, it is not uncommon for the worker bees to lay eggs. This is a result of the lack of the queen's pheromones and the pheromones secreted by uncapped brood, which normally suppress ovarian development in workers. Worker bees are unable to mate, and the unfertilized eggs produce only drones (males), which can mate only with a queen. Thus, in a relatively short period, all the worker bees die off, and the new drones follow if they have not been able to mate before the collapse of the colony. This behaviour is believed to have evolved to allow a doomed colony to produce drones which may mate with a virgin queen and thus preserve the colony's genetic progeny.

A few ants and bees are capable of producing diploid female offspring parthenogenetically. These include a honey bee subspecies from South Africa, *Apis mellifera capensis*, where workers are capable of producing diploid eggs parthenogenetically, and replacing the queen if she dies; other examples include some species of small carpenter bee, (genus *Ceratina*). Many parasitic wasps are known to be parthenogenetic, sometimes due to infections by *Wolbachia*.

The workers in five ant species and the queens in some ants are known to reproduce by parthenogenesis. In *Cataglyphis cursor*, a European formicine ant, the queens and workers can produce new queens by parthenogenesis. The workers are produced sexually.

In Central and South American electric ants, *Wasmannia auropunctata*, queens produce more queens through automictic parthenogenesis with central fusion. Sterile workers usually are produced from eggs fertilized by males. In some of the eggs fertilized by males, however, the fertilization can cause the female genetic material to be ablated from the zygote. In this way, males pass on only their genes to become fertile male offspring. This is the first recognized example of an animal species where both females and males can reproduce clonally resulting in a complete separation of male and female gene pools. As a consequence, the males will only have fathers and the queens only mothers, while the sterile workers are the only ones with both parents of both genders.

These ants get both the benefits of both asexual and sexual reproduction—the daughters who can reproduce (the queens) have all of the mother's genes, while the sterile workers whose physical strength and disease resistance are important are produced sexually.

Other examples of insect parthenogenesis can be found in gall-forming aphids (e.g., *Pemphigus betae*), where females reproduce parthenogenetically during the gall-forming phase of their life cycle and in grass thrips. In the grass thrips genus *Aptinothrips* there have been, despite the very limited number of species in the genus, several transitions to asexuality.

Crustaceans

Crustacean reproduction varies both across and within species. The water flea *Daphnia pulex* alternates between sexual and parthenogenetic reproduction. Among the better-known large decapod crustaceans, some crayfish reproduce by parthenogensis. "Marmorkrebs" are parthenogenetic crayfish that were discovered in the pet trade in the 1990s. Offspring are genetically identical to the parent, indicating it reproduces by apomixis, i.e. parthenogenesis in which the eggs did not undergo meiosis. Spinycheek crayfish (*Orconectes limosus*) can reproduce both sexually and by parthenogenesis. The Louisiana red swamp crayfish (*Procambarus clarkii*), which normally reproduces sexually, has also been suggested to reproduce by parthenogenesis, although no individuals of this species have been reared this way in the lab. *Artemia parthenogenetica* is a species or series of populations of parthenogenetic brine shrimps.

Spiders

At least two species of spiders in the family Oonopidae (goblin spiders), *Heteroonops spinimanus* and *Triaeris stenaspis*, are thought to be parthenogenetic, as no males have ever been collected. Parthenogenetic reproduction has been demonstrated in the laboratory for *T. stenaspis*.

Rotifers

In bdelloid rotifers, females reproduce exclusively by parthenogenesis (obligate parthenogenesis), while in monogonont rotifers, females can alternate between sexual and asexual reproduction (cyclical parthenogenesis). At least in one normally cyclical parthenogenetic species obligate parthenogenesis can be inherited: a recessive allele leads to loss of sexual reproduction in homozygous offspring.

Flatworms

At least two species in the genus *Dugesia*, flatworms in the Turbellaria sub-division of the phylum Platyhelminthes, include polyploid individuals that reproduce by parthenogenesis. This type of parthenogenesis requires mating, but the sperm does not contribute to the genetics of the off-spring (the parthenogenesis is pseudogamous, alternatively referred to as gynogenetic). A complex cycle of matings between diploid sexual and polyploid parthenogenetic individuals produces new parthenogenetic lines.

Snails

Several species of parthenogenetic gastropods have been studied, especially with respect to their status as invasive species. Such species include the New Zealand mud snail (*Potamopyrgus antipodarum*), the red-rimmed melania (*Melanoides tuberculata*), and the Quilted melania (*Tarebia granifera*).

Squamata

Komodo dragon, *Varanus komodoensis*, rarely reproduces offspring via parthenogenesis.

Most reptiles of the squamata order (lizards and snakes) reproduce sexually, but parthenogenesis has been observed to occur naturally in certain species of whiptails, some geckos, rock lizards,- Komodo dragons and snakes. Some of these like the mourning gecko *Lepidodactylus lugubris*,

Indo-Pacific house gecko *Hemidactylus garnotii*, the hybrid whiptails *Cnemidophorus*, Caucasian rock lizards *Darevskia*, and the brahminy blindsnake, *Indotyphlops braminus* are unisexual and obligately parthenogenetic. Others reptiles, such as the Komodo dragon, other monitor lizards, and some species of boas, pythons, filesnakes, gartersnakes and rattlesnakes were previously considered as cases of facultative parthenogenesis, but are in fact cases of accidental parthenogenesis.

In 2012, facultative parthenogenesis was reported in wild vertebrates for the first time by US researchers amongst captured pregnant copperhead and cottonmouth female pit-vipers. The Komodo dragon, which normally reproduces sexually, has also been found able to reproduce asexually by parthenogenesis. A case has been documented of a Komodo dragon reproducing via sexual reproduction after a known parthenogenetic event, highlighting that these cases of parthenogenesis are reproductive accidents, rather than adaptive, facultative parthenogenesis.

Some reptile species use a ZW chromosome system, which produces either males (ZZ) or females (ZW). Until 2010, it was thought that the ZW chromosome system used by reptiles was incapable of producing viable WW offspring, but a (ZW) female boa constrictor was discovered to have produced viable female offspring with WW chromosomes.

Parthenogenesis has been studied extensively in the New Mexico whiptail in the genus *Cnemidophorus* (also known as *Aspidoscelis*) of which 15 species reproduce exclusively by parthenogenesis. These lizards live in the dry and sometimes harsh climate of the southwestern United States and northern Mexico. All these asexual species appear to have arisen through the hybridization of two or three of the sexual species in the genus leading to polyploid individuals. The mechanism by which the mixing of chromosomes from two or three species can lead to parthenogenetic reproduction is unknown. Recently, a hybrid parthenogenetic whiptail lizard was bred in the laboratory from a cross between an asexual and a sexual whiptail. Because multiple hybridization events can occur, individual parthenogenetic whiptail species can consist of multiple independent asexual lineages. Within lineages, there is very little genetic diversity, but different lineages may have quite different genotypes.

An interesting aspect to reproduction in these asexual lizards is that mating behaviors are still seen, although the populations are all female. One female plays the role played by the male in closely related species, and mounts the female that is about to lay eggs. This behaviour is due to the hormonal cycles of the females, which cause them to behave like males shortly after laying eggs, when levels of progesterone are high, and to take the female role in mating before laying eggs, when estrogen dominates. Lizards who act out the courtship ritual have greater fecundity than those kept in isolation, due to the increase in hormones that accompanies the mounting. So, although the populations lack males, they still require sexual behavioral stimuli for *maximum* reproductive success.

Some lizard parthenogens show a pattern of geographic parthenogenesis, occupying high mountain areas where their ancestral forms have an inferior competition ability. In Caucasian rock lizards of genus *Darevskia*, which have six parthenogenetic forms of hybrid origin hybrid parthenogenetic form *D. "dahli"* has a broader niche than either of its bisexual ancestors and its expansion throughout the Central Lesser Caucasus caused decline of the ranges of both its maternal and paternal species.

Amphibians

Sharks

Parthenogenesis in sharks has been confirmed in at least three species, the bonnethead, the black-tip shark, and the zebra shark, and reported in others.

A bonnethead, a type of small hammerhead shark, was found to have produced a pup, born live on 14 December 2001 at Henry Doorly Zoo in Nebraska, in a tank containing three female hammerheads, but no males. The pup was thought to have been conceived through parthenogenic means. The shark pup was apparently killed by a stingray within days of birth. The investigation of the birth was conducted by the research team from Queen's University Belfast, Southeastern University in Florida, and Henry Doorly Zoo itself, and it was concluded after DNA testing that the reproduction was parthenogenic. The testing showed the female pup's DNA matched only one female who lived in the tank, and that no male DNA was present in the pup. The pup was not a twin or clone of her mother, but rather, contained only half of her mother's DNA ("automictic parthenogenesis"). This type of reproduction had been seen before in bony fish, but never in cartilaginous fish such as sharks, until this documentation.

In the same year, a female Atlantic blacktip shark in Virginia reproduced via parthenogenesis. On 10 October 2008 scientists confirmed the second case of a virgin birth in a shark. The Journal of Fish Biology reported a study in which scientists said DNA testing proved that a pup carried by a female Atlantic blacktip shark in the Virginia Aquarium & Marine Science Center contained no genetic material from a male.

In 2002, two white-spotted bamboo sharks were born at the Belle Isle Aquarium in Detroit. They hatched 15 weeks after being laid. The births baffled experts as the mother shared an aquarium with only one other shark, which was female. The female bamboo sharks had laid eggs in the past. This is not unexpected, as many animals will lay eggs even if there is not a male to fertilize them. Normally, the eggs are assumed to be inviable and are discarded. This batch of eggs was left undisturbed by the curator as he had heard about the previous birth in 2001 in Nebraska and wanted to observe whether they would hatch. Other possibilities had been considered for the birth of the Detroit bamboo sharks including thoughts that the sharks had been fertilized by a male and stored the sperm for a period of time, as well as the possibility that the Belle Isle bamboo shark is a hermaphrodite, harboring both male and female sex organs, and capable of fertilizing its own eggs, but that is not confirmed.

In 2008, a Hungarian aquarium had another case of parthenogenesis after its lone female shark produced a pup without ever having come into contact with a male shark.

The repercussions of parthenogenesis in sharks, which fails to increase the genetic diversity of the offspring, is a matter of concern for shark experts, taking into consideration conservation management strategies for this species, particularly in areas where there may be a shortage of males due to fishing or environmental pressures. Although parthenogenesis may help females who cannot find mates, it does reduce genetic diversity.

In 2011, recurring shark parthenogenesis over several years was demonstrated in a captive zebra shark, a type of carpet shark.

Birds

Parthenogenesis in birds is known mainly from studies of domesticated turkeys and chickens, although it has also been noted in the domestic pigeon. In most cases the egg fails to develop normally or completely to hatching. The first description of parthenogenetic development in a passerine was demonstrated in captive zebra finches, although the dividing cells exhibited irregular nuclei and the eggs did not hatch.

Parthenogenesis in turkeys appears to result from a conversion of haploid cells to diploid; most embryos produced in this way die early in development. Rarely, viable birds result from this process, and the rate at which this occurs in turkeys can be increased by selective breeding, however male turkeys produced from parthenogenesis exhibit smaller testes and reduced fertility.

Mammals

There are no known cases of naturally occurring mammalian parthenogenesis in the wild. Parthenogenetic progeny of mammals would have two X chromosomes, and would therefore be female.

In 1936, Gregory Goodwin Pincus reported successfully inducing parthenogenesis in a rabbit. In April 2004, scientists at Tokyo University of Agriculture used parthenogenesis successfully to create a fatherless mouse. Using gene targeting, they were able to manipulate two imprinted loci H19/IGF2 and DLK1/MEG3 to produce bi-maternal mice at high frequency and subsequently show that fatherless mice have enhanced longevity.

Induced parthenogenesis in mice and monkeys often results in abnormal development. This is because mammals have imprinted genetic regions, where either the maternal or the paternal chromosome is inactivated in the offspring in order for development to proceed normally. A mammal created by parthenogenesis would have double doses of maternally imprinted genes and lack paternally imprinted genes, leading to developmental abnormalities. It has been suggested that defects in placental folding or interdigitation are one cause of swine parthenote abortive development. As a consequence, research on human parthenogenesis is focused on the production of embryonic stem cells for use in medical treatment, not as a reproductive strategy.

Use of an electrical or chemical stimulus can produce the beginning of the process of parthenogenesis in the asexual development of viable offspring.

Induction of parthenogenesis in swine. Parthenogenetic development of swine oocytes. High metaphase promoting factor (MPF) activity causes mammalian oocytes to arrest at the metaphase II stage until fertilization by a sperm. The fertilization event causes intracellular calcium oscillations, and targeted degradation of cyclin B, a regulatory subunit of MPF, thus permitting the MII-arrested oocyte to proceed through meiosis. To initiate parthenogenesis of swine oocytes, various methods exist to induce an artificial activation that mimics sperm entry, such as calcium ionophore treatment, microinjection of calcium ions, or electrical stimulation. Treatment with cycloheximide, a non-specific protein synthesis inhibitor, enhances parthenote development in swine presumably by continual inhibition of MPF/cyclin B. As meiosis proceeds, extrusion of the second polar is blocked by exposure to cytochalasin B. This treatment results in a diploid (2 maternal genomes) parthenote. Parthenotes can be surgically transferred to a recipient oviduct for further development, but will succumb by developmental failure after $\approx$30 days of gestation. The swine parthenote placentae often appears hypo-vascular and is approximately 50% smaller than biparental offspring placentae.

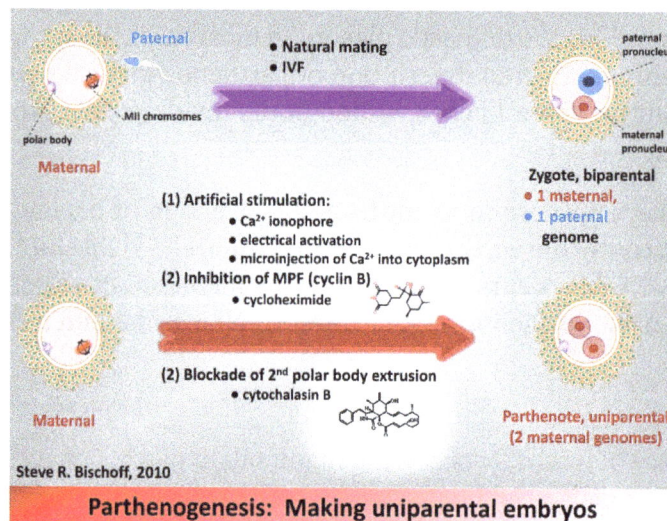

Parthenogenesis: Making uniparental embryos

During oocyte development, high metaphase promoting factor (MPF) activity causes mammalian oocytes to arrest at the metaphase II stage until fertilization by a sperm. The fertilization event causes intracellular calcium oscillations, and targeted degradation of cyclin B, a regulatory subunit of MPF, thus permitting the MII-arrested oocyte to proceed through meiosis.

To initiate parthenogenesis of swine oocytes, various methods exist to induce an artificial activation that mimics sperm entry, such as calcium ionophore treatment, microinjection of calcium ions, or electrical stimulation. Treatment with cycloheximide, a non-specific protein synthesis inhibitor, enhances parthenote development in swine presumably by continual inhibition of MPF/cyclin B. As meiosis proceeds, extrusion of the second polar is blocked by exposure to cytochalasin B. This treatment results in a diploid (2 maternal genomes) parthenote Parthenotes can be surgically transferred to a recipient oviduct for further development, but will succumb to developmental failure after ≈30 days of gestation. The swine parthenote placentae often appears hypo-vascular.

Humans

On June 26, 2007, International Stem Cell Corporation (ISCC), a California-based stem cell research company, announced that their lead scientist, Dr. Elena Revazova, and her research team were the first to intentionally create human stem cells from unfertilized human eggs using parthenogenesis. The process may offer a way for creating stem cells that are genetically matched to a particular female for the treatment of degenerative diseases that might affect her. In December 2007, Dr. Revazova and ISCC published an article illustrating a breakthrough in the use of parthenogenesis to produce human stem cells that are homozygous in the HLA region of DNA. These stem cells are called HLA homozygous parthenogenetic human stem cells (hpSC-Hhom) and have unique characteristics that would allow derivatives of these cells to be implanted into millions of people without immune rejection. With proper selection of oocyte donors according to HLA haplotype, it is possible to generate a bank of cell lines whose tissue derivatives, collectively, could be MHC-matched with a significant number of individuals within the human population.

On August 2, 2007, after much independent investigation, it was revealed that discredited South Korean scientist Hwang Woo-Suk unknowingly produced the first human embryos resulting from parthenogenesis. Initially, Hwang claimed he and his team had extracted stem cells from cloned human embryos, a result later found to be fabricated. Further examination of the chromosomes of these cells show indicators of parthenogenesis in those extracted stem cells, similar to those found in the mice created by Tokyo scientists in 2004. Although Hwang deceived the world about being the first to create artificially cloned human embryos, he did contribute a major breakthrough to stem cell research by creating human embryos using parthenogenesis. The truth was discovered in 2007, long after the embryos were created by him and his team in February 2004. This made Hwang the first, unknowingly, to successfully perform the process of parthenogenesis to create a human embryon and, ultimately, a human parthenogenetic stem cell line.

Oomycetes

Apomixis can apparently occur in *Phytophthora*, an Oomycete. Oospores derived after an experimental cross were germinated, and some of the progeny were genetically identical to one or other parent, which would imply that meiosis did not occur and the oospores developed by parthenogenesis.

Similar Phenomena

Gynogenesis

A form of asexual reproduction related to parthenogenesis is gynogenesis. Here, offspring are produced by the same mechanism as in parthenogenesis, but with the requirement that the egg merely be stimulated by the *presence* of sperm in order to develop. However, the sperm cell does not contribute any genetic material to the offspring. Since gynogenetic species are all female, activation of their eggs requires mating with males of a closely related species for the needed stimulus. Some salamanders of the genus *Ambystoma* are gynogenetic and appear to have been so for over a million years. It is believed that the success of those salamanders may be due to rare fertilization of eggs by males, introducing new material to the gene pool, which may result from perhaps only one mating out of a million. In addition, the amazon molly is known to reproduce by gynogenesis.

Hybridogenesis

Hybridogenesis is a mode of reproduction of hybrids. Hybridogenetic hybrids (for example AB genome), usually females, during gametogenesis exclude one of parental genomes (A) and produce gametes with unrecombined genome of second parental species (B), instead of containing mixed recombined parental genomes. First genome (A) is restored by fertilization of these gametes with gametes from the first species (AA, sexual host, usually male).

So hybridogenesis is not completely asexual, but instead hemiclonal: half of genome is passed to the next generation clonally, unrecombined, intact (B), other half sexually, recombined (A).

This process continues, so that each generation is half (or hemi-) clonal on the mother's side and has half new genetic material from the father's side.

This form of reproduction is seen in some live-bearing fish of the genus *Poeciliopsis* as well as in some of the *Pelophylax* spp. ("green frogs" or "waterfrogs"):

- *P. kl. esculentus* (edible frog): *P. lessonae* × *P. ridibundus*,

- *P. kl. grafi* (Graf's hybrid frog): *P. perezi* × *P. ridibundus*

- *P. kl. hispanicus* (Italian edible frog) – unknown origin: *P. bergeri* × *P. ridibundus* or *P. kl. esculentus*

and perhaps in *P. demarchii*.

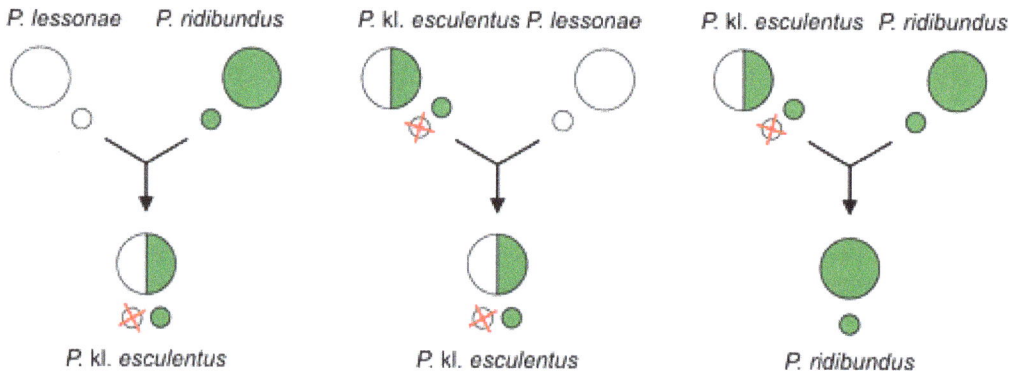

Example crosses between pool frog (*Pelophylax lessonae*), marsh frog (*P. ridibundus*) and their hybrid – edible frog (*P.* kl. *esculentus*). First one is the primary hybridisation generating hybrid, second one is most widespread type of hybridogenesis.

Other examples where hybridogenesis is at least one of modes of reproduction include i.e.

- Iberian minnow *Tropidophoxinellus alburnoides* (*Squalius pyrenaicus* × hypothetical ancestor related with Anaecypris hispanica)

- spined loaches *Cobitis hankugensis* × *C. longicorpus*

- *Bacillus* stick insects B. rossius × Bacillus grandii benazzii

Apomixis

Vegetative apomixis in *Poa bulbosa*; bulbils form instead of flowers

In botany, apomixis was defined by Hans Winkler as replacement of the normal sexual reproduction by asexual reproduction, without fertilization. Its etymology is Greek for "away from" + "mixing". This definition notably does not mention meiosis. Thus "normal asexual reproduction" of plants, such as propagation from cuttings or leaves, has never been considered to be apomixis, but replacement of the seed by a plantlet or replacement of the flower by bulbils are types of apomixis. Apomictically produced offspring are genetically identical to the parent plant.

Some authors included all forms of asexual reproduction within apomixis, but that generalization of the term has since died out.

In flowering plants, the term "apomixis" is commonly used in a restricted sense to mean agamospermy, i.e. clonal reproduction through seeds. Although agamospermy could theoretically occur in gymnosperms, it appears to be absent in that group.

Apogamy is a related term that has had various meanings over time. In plants with independent gametophytes (notably ferns), the term is still used interchangeably with "apomixis", and both refer to the formation of sporophytes by parthenogenesis of gametophyte cells.

Male apomixis (paternal apomixis) involves replacement of the genetic material of the egg cell by that from the pollen.

Apomixis and Evolution

Because apomictic plants are genetically identical from one generation to the next, each lineage has some of the characters of a true species, maintaining distinctions from other apomictic lineages within the same genus, while having much smaller differences than is normal between species of most genera. They are therefore often called microspecies. In some genera, it is possible to identify and name hundreds or even thousands of microspecies, which may be grouped together as species aggregates, typically listed in floras with the convention "*Genus species* agg." (such as the bramble, *Rubus fruticosus* agg.). In some plant families, genera with apomixis are quite common, for example in Asteraceae, Poaceae, and Rosaceae. Examples of apomixis can be found in the genera *Crataegus* (hawthorns), *Amelanchier* (shadbush), *Sorbus* (rowans and whitebeams), *Rubus* (brambles or blackberries), *Poa* (meadow grasses), *Nardus stricta* (Matgrass), *Hieracium* (hawkweeds) and *Taraxacum* (dandelions).

Although the evolutionary advantages of sexual reproduction are lost, apomixis can pass along traits fortuitous for evolutionary fitness. As Jens Clausen put it.

The apomicts actually have discovered the effectiveness of mass production long before Mr Henry Ford applied it to the production of the automobile. ... Facultative apomixis ... does not prevent variation; rather, it multiplies certain varietal products.

Facultative apomixis means that apomixis does not always occur, i.e. sexual reproduction can also happen. It appears likely that all apomixis in plants is facultative; in other words, that "obligate apomixis" is an artifact of insufficient observation (missing uncommon sexual reproduction).

Apogamy and Apospory in Non-flowering Plants

The gametophytes of bryophytes, and less commonly ferns and lycopods can develop a group of cells that grow to look like a sporophyte of the species but with the ploidy level of the gametophyte, a phenomenon known as apogamy. The sporophytes of plants of these groups may also have the ability to form a plant that looks like a gametophyte but with the ploidy level of the sporophyte, a phenomenon known as apospory.

Apomixis in Flowering Plants (Angiosperms)

Agamospermy, asexual reproduction through seeds, occurs in flowering plants through many different mechanisms and a simple hierarchical classification of the different types is not possible. Consequently, there are almost as many different usages of terminology for apomixis in angiosperms as there are authors on the subject. For English speakers, Maheshwari 1950 is very influential. German speakers might prefer to consult Rutishauser 1967. Some older text books on the basis of misinformation (that the egg cell in a meiotically unreduced gametophyte can never be fertilized) attempted to reform the terminology to match parthenogenesis as it is used in zoology, and this continues to cause much confusion.

Agamospermy occurs mainly in two forms: In *gametophytic apomixis*, the embryo arises from an unfertilized egg cell (i.e. by parthenogenesis) in a gametophyte that was produced from a cell that did not complete meiosis. In *adventitious embryony* (sporophytic apomixis), an embryo is formed directly (not from a gametophyte) from nucellus or integument tissue.

Types of Apomixis in Flowering Plants

Maheshwari used the following simple classification of types of apomixis in flowering plants:

Caribbean Agave producing plantlets on the old flower stem.

- Nonrecurrent apomixis: In this type "the megaspore mother cell undergoes the usual meiotic divisions and a haploid embryo sac [megagametophyte] is formed. The new embryo may then arise either from the egg (haploid parthenogenesis) or from some other cell of the gametophyte (haploid apogamy)." The haploid plants have half as many chromosomes as the mother plant, and "the process is not repeated from one generation to another" (which is why it is called nonrecurrent).

- Recurrent apomixis, is now more often called gametophytic apomixis: In this type, the megagametophyte has the same number of chromosomes as the mother plant because meiosis was not completed. It generally arises either from an archesporial cell or from some other part of the nucellus.

- Adventive embryony, also called sporophytic apomixis, sporophytic budding, or nucellar embryony: Here there may be a megagametophyte in the ovule, but the embryos do not arise from the cells of the gametophyte; they arise from cells of nucellus or the integument. Adventive embryony is important in several species of *Citrus*, in *Garcinia, Euphorbia dulcis, Mangifera indica* etc.

- Vegetative apomixis: In this type "the flowers are replaced by bulbils or other vegetative propagules which frequently germinate while still on the plant". Vegetative apomixis is important in *Allium, Fragaria, Agave*, and some grasses, among others.

Types of Gametophytic Apomixis

Gametophytic apomixis in flowering plants develops in several different ways. A megagametophyte develops with an egg cell within it that develops into an embryo through parthenogenesis. The central cell of the megagametophyte may require fertilization to form the endosperm, pseudogamous gametophytic apomixis, or in autonomous gametophytic apomixis fertilization is not required.

- In diplospory (also called generative apospory), the megagametophyte arises from a cell of the archesporium.

- In apospory (also called somatic apospory), the megagametophyte arises from some other nucellus cell.

Considerable confusion has resulted because diplospory is often defined to involve the megaspore mother cell only, but a number of plant families have a multicellular archesporium and the megagametophyte could originate from another archesporium cell.

Diplospory is further subdivided according to how the megagametophyte forms:

- *Allium odorum–A. nutans* type. The chromosomes double (endomitosis) and then meiosis proceeds in an unusual way, with the chromosome copies pairing up (rather than the original maternal and paternal copies pairing up).

- *Taraxacum* type: Meiosis I fails to complete, meiosis II creates two cells, one of which degenerates; three mitotic divisions form the megagametophyte.

- *Ixeris* type: Meiosis I fails to complete; three rounds of nuclear division occur without cell-wall formation; wall formation then occurs.

- *Blumea–Elymus* types: A mitotic division is followed by degeneration of one cell; three mitotic divisions form the megagametophyte.

- *Antennaria–Hieracium* types: three mitotic divisions form the megagametophyte.

- *Eragrostis–Panicum* types: Two mitotic division give a 4-nucleate megagametophyte, with cell walls to form either three or four cells.

Incidence of Apomixis in Flowering Plants

Apomixis occurs in at least 33 families of flowering plants, and has evolved multiple times from sexual relatives. Apomictic species or individual plants often have a hybrid origin, and are usually polyploid.

In plants with both apomictic and meiotic embryology, the proportion of the different types can differ at different times of year, and photoperiod can also change the proportion. It appears unlikely that there are any truly completely apomictic plants, as low rates of sexual reproduction have been found in several species that were previously thought to be entirely apomictic.

The genetic control of apomixis can involve a single genetic change that affects all the major developmental components, formation of the megagametophyte, parthenogenesis of the egg cell, and endosperm development. However, the timing of the various developmental processes is critical to successful development of an apomictic seed, and the timing can be affected by multiple genetic factors.

Some Terms Related to Apomixis

- Apomeiosis: "Without meiosis"; usually meaning the production of a meiotically unreduced gametophyte.

- Parthenogenesis: Development of an embryo directly from an egg cell without fertilization is called parthenogenesis. It is of two types:

  o Haploid parthenogenesis: Parthenogenesis of a normal haploid egg (a meiotically reduced egg) into an embryo is termed haploid parthenogenesis. If the mother plant was diploid, then the haploid embryo that results is monoploid, and the plant that grows from the embryo is sterile. If they are not sterile, they are sometimes useful to plant breeders (especially in potato breeding, see dihaploidy). This type of apomixis has been recorded in *Solanum nigrum*, *Lilium* spp., *Orchis maculata*, *Nicotiana tabacum*, etc.

  o Diploid parthenogenesis: When the megagametophyte develops without completing meiosis, so that the megagametophyte and all cells within it are meiotically unreduced (a.k.a. diploid, but diploid is an ambiguous term), this is called diploid parthenogenesis, and the plant that develops from the embryo will have the same number of chromosomes as the mother plant. Diploid parthenogenesis is a component process of gametophytic apomixis.

- Androgenesis and androclinesis are synonyms. These terms are used for two different processes that both have the effect of producing an embryo that has "male inheritance".

  The first process is a natural one. It may also be referred to as male apomixis or paternal apomixis. It involves fusion of the male and female gametes and replacement of the female nucleus by the male nucleus. This has been noted as a rare phenomenon in many plants (e.g. *Nicotiana* and *Crepis*), and occurs as the regular reproductive method in the Saharan Cypress, *Cupressus dupreziana*.

The second process that is referred to as androgenesis or androclinesis involves (artificial) culture of haploid plants from anther tissue or microspores.

- Apogamy: Although this term was (before 1908) used for other types of apomixis, and then discarded as too confusing, it is still sometimes used when an embryo develops from a cell of the megagametophyte other than the egg cell. In flowering plants, the cells involved in apogamy would be synergids or antipodal cells.

- Addition hybrids, called B_{III} hybrids by Rutishauser: An embryo is formed after a meiotically unreduced egg cell is fertilized. The ploidy level of the embryo is therefore higher than that of the mother plant. This process occurs in some plants that are otherwise apomictic, and may play a significant role in producing tetraploid plants from triploid apomictic mother plants (if they receive pollen from diploids). Because fertilization is involved, this process does not fit the definition of apomixis.

- Pseudogamy refers to any reproductive process that requires pollination but does not involve male inheritance. It is sometimes used in a restrictive sense to refer to types of apomixis in which the endosperm is fertilized but the embryo is not. A better term for the restrictive sense is centrogamy.

- Agamospecies, the concept introduced by Göte Turesson: "an apomict population the constituents of which, for morphological, cytological or other reasons, are to be considered as having a common origin," i.e., basically synonymous with "microspecies.

Sexual Reproduction

In the first stage of sexual reproduction, "meiosis", the number of chromosomes is reduced from a diploid number (2n) to a haploid number (n). During "fertilization", haploid gametes come together to form a diploid zygote and the original number of chromosomes is restored.

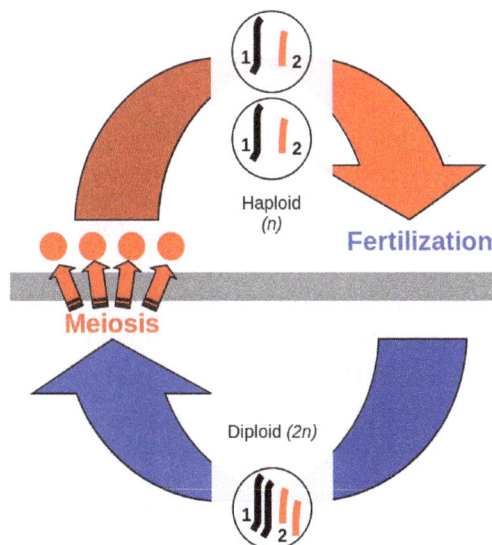

Sexual reproduction is a form of reproduction where two morphologically distinct types of specialized reproductive cells called gametes fuse together, involving a female's large ovum (or egg) and a male's smaller sperm. Each gamete contains half the number of chromosomes of normal cells. They are created by a specialized type of cell division, which only occurs in eukaryotic cells, known as meiosis. The two gametes fuse during fertilization to produce DNA replication and the creation of a single-celled zygote which includes genetic material from both gametes. In a process called genetic recombination, genetic material (DNA) joins up so that homologous chromosome sequences are aligned with each other, and this is followed by exchange of genetic information. Two rounds of cell division then produce four daughter cells with half the number of chromosomes from each original parent cell, and the same number of chromosomes as both parents, though self-fertilization can occur. For instance, in human reproduction each human cell contains 46 chromosomes, 23 pairs, except gamete cells, which only contain 23 chromosomes, so the child will have 23 chromosomes from each parent genetically recombined into 23 pairs. Cell division initiates the development of a new individual organism in multicellular organisms, including animals and plants, for the vast majority of whom this is the primary method of reproduction.

The evolution of sexual reproduction is a major puzzle because asexual reproduction should be able to outcompete it as every young organism created can bear its own young. This implies that an asexual population has an intrinsic capacity to grow more rapidly with each generation. This 50% cost is a fitness disadvantage of sexual reproduction. The two-fold cost of sex includes this cost and the fact that any organism can only pass on 50% of its own genes to its offspring. One definite advantage of sexual reproduction is that it prevents the accumulation of genetic mutations.

Sexual selection is a mode of natural selection in which some individuals out-reproduce others of a population because they are better at securing mates for sexual reproduction. It has been described as "a powerful evolutionary force that does not exist in asexual populations."

Prokaryotes, whose initial cell has additional or transformed genetic material, reproduce through asexual reproduction but may, in lateral gene transfer, display processes such as bacterial conjugation, transformation and transduction, which are similar to sexual reproduction although they do not lead to reproduction.

Evolution

The first fossilized evidence of sexual reproduction in eukaryotes is from the Stenian period, about 1 to 1.2 billion years ago.

Biologists studying evolution propose several explanations for why sexual reproduction developed and why it is maintained. These reasons include fighting the accumulation of deleterious mutations, increasing rate of adaptation to changing environments, dealing with competition, or masking deleterious mutations. All of these ideas about why sexual reproduction has been maintained are generally supported, but ultimately the size of the population determines if sexual reproduction is entirely beneficial. Larger populations appear to respond more quickly to benefits obtained through sexual reproduction than do smaller population sizes.

Maintenance of sexual reproduction has been explained by theories that work at several levels of selection, though some of these models remain controversial.

New models presented in recent years suggest a basic advantage for sexual reproduction in slowly reproducing complex organisms. Sexual reproduction allows these species to exhibit characteristics that depend on the specific environment that they inhabit, and the particular survival strategies that they employ.

Sexual Selection

In order to sexually reproduce, both males and females need to find a mate. Generally in animals mate choice is made by females while males compete to be chosen. This can lead organisms to extreme efforts in order to reproduce, such as combat and display, or produce extreme features caused by a positive feedback known as a Fisherian runaway. Thus sexual reproduction, as a form of natural selection, has an effect on evolution. Sexual dimorphism is where the basic phenotypic traits vary between males and females of the same species. Dimorphism is found in both sex organs and in secondary sex characteristics, body size, physical strength and morphology, biological ornamentation, behavior and other bodily traits. However, sexual selection is only implied over an extended period of time leading to sexual dimorphism.

Sex Ratio

Apart from some eusocial wasps, organisms which reproduce sexually have a 1:1 sex ratio of male and female births. The English statistician and biologist Ronald Fisher outlined why this is so in what has come to be known as Fisher's principle. This essentially says the following:

1. Suppose male births are less common than female.

2. A newborn male then has better mating prospects than a newborn female, and therefore can expect to have more offspring.

3. Therefore parents genetically disposed to produce males tend to have more than average numbers of grandchildren born to them.

4. Therefore the genes for male-producing tendencies spread, and male births become more common.

5. As the 1:1 sex ratio is approached, the advantage associated with producing males dies away.

6. The same reasoning holds if females are substituted for males throughout. Therefore 1:1 is the equilibrium ratio.

Animals

Insects

Insect species make up more than two-thirds of all extant animal species. Most insect species reproduce sexually, though some species are facultatively parthenogenetic. Many insects species have sexual dimorphism, while in others the sexes look nearly identical. Typically they have two sexes with males producing spermatozoa and females ova. The ova develop into eggs that have a covering called the chorion, which forms before internal fertilization. Insects have very diverse mating and reproductive strategies most often resulting in the male depositing spermatophore within the female, which she stores until she is ready for egg fertilization. After fertilization, and the formation of a zygote, and varying degrees of development, in many species the eggs are deposited outside the female; while in others, they develop further within the female and are born live.

Australian emperor laying egg, guarded by the male

Mammals

There are three extant kinds of mammals: monotremes, placentals and marsupials, all with internal fertilization. In placental mammals, offspring are born as juveniles: complete animals with the sex organs present although not reproductively functional. After several months or years, depending on the species, the sex organs develop further to maturity and the animal becomes sexually mature. Most female mammals are only fertile during certain periods during their estrous cycle, at which point they are ready to mate. Individual male and female mammals meet and carry out copulation. For most mammals, males and females exchange sexual partners throughout their adult lives.

Fish

The vast majority of fish species lay eggs that are then fertilized by the male, some species lay their eggs on a substrate like a rock or on plants, while others scatter their eggs and the eggs are fertilized as they drift or sink in the water column.

Some fish species use internal fertilization and then disperse the developing eggs or give birth to live offspring. Fish that have live-bearing offspring include the guppy and mollies or *Poecilia*. Fishes that give birth to live young can be ovoviviparous, where the eggs are fertilized within the female and the eggs simply hatch within the female body, or in seahorses, the male carries the developing young within a pouch, and gives birth to live young. Fishes can also be viviparous, where the female supplies nourishment to the internally growing offspring. Some fish are hermaphrodites, where a single fish is both male and female and can produce eggs and sperm. In hermaphroditic fish, some are male and female at the same time while in other fish they are serially hermaphroditic; starting as one sex and changing to the other. In at least one hermaphroditic species, self-fertilization occurs when the eggs and sperm are released together. Internal self-fertilization may occur in some other species. One fish species does not reproduce by sexual reproduction but uses sex to produce offspring; *Poecilia formosa* is a unisex species that uses a form of parthenogenesis called gynogenesis, where unfertilized eggs develop into embryos that produce female offspring. *Poecilia formosa* mate with males of other fish species that use internal fertilization, the sperm does not fertilize the eggs but stimulates the growth of the eggs which develops into embryos.

Plants

Animals typically produce gametes directly by meiosis. Male gametes are called sperm, and female gametes are called eggs or ova. In animals, fertilization follows immediately after meiosis. Plants on the other hand have mitosis occurring in spores, which are produced by meiosis. The spores germinate into the gametophyte phase. The gametophytes of different groups of plants vary in size; angiosperms have as few as three cells in pollen, and mosses and other so called primitive plants may have several million cells. Plants have an alternation of generations where the sporophyte phase is succeeded by the gametophyte phase. The sporophyte phase produces spores within the sporangium by meiosis.

Flowering Plants

Flowers are the sexual organs of flowering plants.

Flowering plants are the dominant plant form on land and they reproduce either sexually or asexually. Often their most distinguishing feature is their reproductive organs, commonly called flowers. The anther produces pollen grains which contain the male gametophytes (sperm). For pollination to occur, pollen grains must attach to the stigma of the female reproductive structure (carpel), where the female gametophytes (ovules) are located inside the ovary. After the pollen tube grows through the carpel's style, the sex cell nuclei from the pollen grain migrate into the ovule to fertilize the egg cell and endosperm nuclei within the female gametophyte in a process termed double fertilization. The resulting zygote develops into an embryo, while the triploid endosperm (one sperm cell plus two female cells) and female tissues of the ovule give rise to the surrounding tissues in the developing seed. The ovary, which produced the female gametophyte(s), then grows into a fruit, which surrounds the seed(s). Plants may either self-pollinate or cross-pollinate.

Nonflowering plants like ferns, moss and liverworts use other means of sexual reproduction.

In 2013, flowers dating from the Cretaceous (100 million years before present) were found encased in amber, the oldest evidence of sexual reproduction in a flowering plant. Microscopic images showed tubes growing out of pollen and penetrating the flower's stigma. The pollen was sticky, suggesting it was carried by insects.

Ferns

Ferns mostly produce large diploid sporophytes with rhizomes, roots and leaves; and on fertile leaves called sporangium, spores are produced. The spores are released and germinate to produce short, thin gametophytes that are typically heart shaped, small and green in color. The gametophytes or thallus, produce both motile sperm in the antheridia and egg cells in separate archegonia. After rains or when dew deposits a film of water, the motile sperm are splashed away from the antheridia, which are normally produced on the top side of the thallus, and swim in the film of water to the archegonia where they fertilize the egg. To promote out crossing or cross fertilization the sperm are released before the eggs are receptive of the sperm, making it more likely that the sperm will fertilize the eggs of different thallus. A zygote is formed after fertilization, which grows into a new sporophytic plant. The condition of having separate sporephyte and gametophyte plants is called alternation of generations. Other plants with similar reproductive means include the *Psilotum*, *Lycopodium*, *Selaginella* and *Equisetum*.

Bryophytes

The bryophytes, which include liverworts, hornworts and mosses, reproduce both sexually and vegetatively. They are small plants found growing in moist locations and like ferns, have motile sperm with flagella and need water to facilitate sexual reproduction. These plants start as a haploid spore that grows into the dominate form, which is a multicellular haploid body with leaf-like structures that photosynthesize. Haploid gametes are produced in antherida and archegonia by mitosis. The sperm released from the antherida respond to chemicals released by ripe archegonia and swim to them in a film of water and fertilize the egg cells thus producing a zygote. The zygote divides by mitotic division and grows into a sporophyte that is diploid. The multicellular diploid sporophyte produces structures called spore capsules, which are connected by seta to the archegonia. The spore capsules produce spores by meiosis, when ripe the capsules burst open and the spores are released. Bryophytes show considerable variation in their breeding structures and the above is a basic outline. Also in some species each plant is one sex while other species produce both sexes on the same plant.

Fungi

Fungi are classified by the methods of sexual reproduction they employ. The outcome of sexual reproduction most often is the production of resting spores that are used to survive inclement times and to spread. There are typically three phases in the sexual reproduction of fungi: plasmogamy, karyogamy and meiosis. The cytoplasm of two parent cells fuse during plasmogamy and the nuclei fuse during karyogamy. New haploid gametes are formed during meiosis and develop into spores.

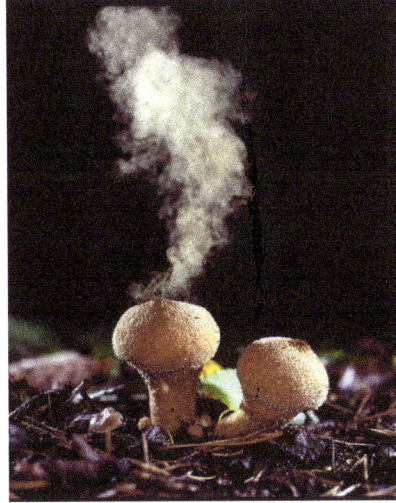

Puffballs emitting spores

Bacteria and Archaea

Three distinct processes in prokaryotes are regarded as similar to eukaryotic sex: bacterial transformation, which involves the incorporation of foreign DNA into the bacterial chromosome; bacterial conjugation, which is a transfer of plasmid DNA between bacteria, but the plasmids are rarely incorporated into the bacterial chromosome; and gene transfer and genetic exchange in archaea.

Bacterial transformation involves the recombination of genetic material and its function is mainly associated with DNA repair. Bacterial transformation is a complex process encoded by numerous bacterial genes, and is a bacterial adaptation for DNA transfer. This process occurs naturally in at least 40 bacterial species. For a bacterium to bind, take up, and recombine exogenous DNA into its chromosome, it must enter a special physiological state referred to as competence. Sexual reproduction in early single-celled eukaryotes may have evolved from bacte-rial transformation, or from a similar process in archaea.

On the other hand, bacterial conjugation is a type of direct transfer of DNA between two bacteria through an external appendage called the conjugation pilus. Bacterial conjugation is controlled by plasmid genes that are adapted for spreading copies of the plasmid between bacteria. The infrequent integration of a plasmid into a host bacterial chromosome, and the subsequent transfer of a part of the host chromosome to another cell do not appear to be bacterial adaptations.

Exposure of hyperthermophilic archaeal Sulfolobus species to DNA damaging conditions induces cellular aggregation accompanied by high frequency genetic marker exchange. Ajon et al. hypothesized that this cellular aggregation enhances species-specific DNA repair by homologous recombination. DNA transfer in Sulfolobus may be an early form of sexual interaction similar to the more well-studied bacterial transformation systems that also involve species-specific DNA transfer leading to homologous recombinational repair of DNA damage.

Plant Reproductive Morphology

Plant reproductive morphology is the study of the physical form and structure (the morphology) of those parts of plants directly or indirectly concerned with sexual reproduction.

Close-up of a flower of *Schlumbergera* (Christmas or Holiday Cactus), showing part of the gynoecium (the stigma and part of the style is visible) and the stamens that surround it

Among all living organisms, flowers, which are the reproductive structures of angiosperms, are the most varied physically and show a correspondingly great diversity in methods of reproduction. Plants that are not flowering plants (green algae, mosses, liverworts, hornworts, ferns and gymnosperms such as conifers) also have complex interplays between morphological adaptation and environmental factors in their sexual reproduction. The breeding system, or how the sperm from one plant fertilizes the ovum of another, depends on the reproductive morphology, and is the single most important determinant of the genetic structure of nonclonal plant populations. Christian Konrad Sprengel (1793) studied the reproduction of flowering plants and for the first time it was understood that the pollination process involved both biotic and abiotic interactions. Charles Darwin's theories of natural selection utilized this work to build his theory of evolution, which includes analysis of the coevolution of flowers and their insect pollinators.

Use of Sexual Terminology

Dioicous gametophytes of the liverwort *Marchantia polymorpha*. In this species, gametes are produced on different plants on umbrella-shaped gametophores with different morphologies. The radiating arms of female gameteophores (left) protect archegonia that produce eggs. Male gametophores (right) are topped with antheridia that produce sperm.

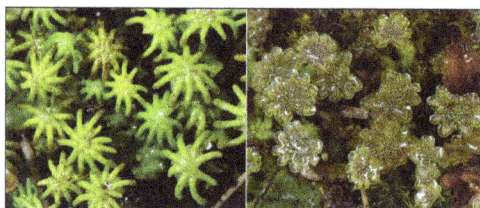

Plants have a complex lifecycle involving alternation of generations. One generation, the sporophyte, gives rise to the next generation via spores. Spores may come in different sizes (microspores and megaspores), but strictly speaking, spores and sporophytes are neither male nor female. The alternate generation, the gametophyte, produces eggs and sperm. A gametophyte can be either female (producing eggs), male (producing sperm) or hermaphrodite (monoicous, producing both eggs and sperm).

In groups like liverworts, mosses and hornworts, the dominant generation is the sexual gametophyte. In ferns and seed plants (including cycads, conifers, flowering plants, etc.) the sporophyte is by far the most dominant generation. The obvious visible plant, whether a small herb or a large tree, is the sporophyte, and the gametophyte is very small. In seed plants, the female gametophyte, and the spores that give rise to it, are hidden within the sporophyte and are entirely dependent on it for nutrition. The male gametophyte consists of a few cells within a pollen grain.

The sporophyte of a flowering plant is often described using sexual terms (e.g. "female" or "male") *based on the sexuality of the gametophyte it gives rise to.* For example, a sporophyte that produces spores that give rise only to male gametophytes may be described as "male", even though the sporophyte itself is asexual, producing only spores. Similarly, flowers produced by the sporophyte may be described as "unisexual" or "bisexual", meaning that they give rise to either one sex of gametophyte or both sexes of gametophyte.

Flowering plants

Basic flower morphology

The flower is the characteristic structure concerned with sexual reproduction in flowering plants (angiosperms). Flowers vary enormously in their construction (morphology). A "complete" flower, like that of *Ranunculus glaberrimus* shown in the figure, has a calyx of outer sepals and a corolla of inner petals. (The sepals and petals together form the perianth.) Next inwards there are numerous stamens, which produce pollen grains, each containing a microscopic male gametophyte. Finally in the middle there are partially joined carpels, which at maturity contain ovules, and within the ovules are tiny female gametophytes. Each carpel in *Ranunculus* species produces one ovule, which when fertilized becomes a seed.

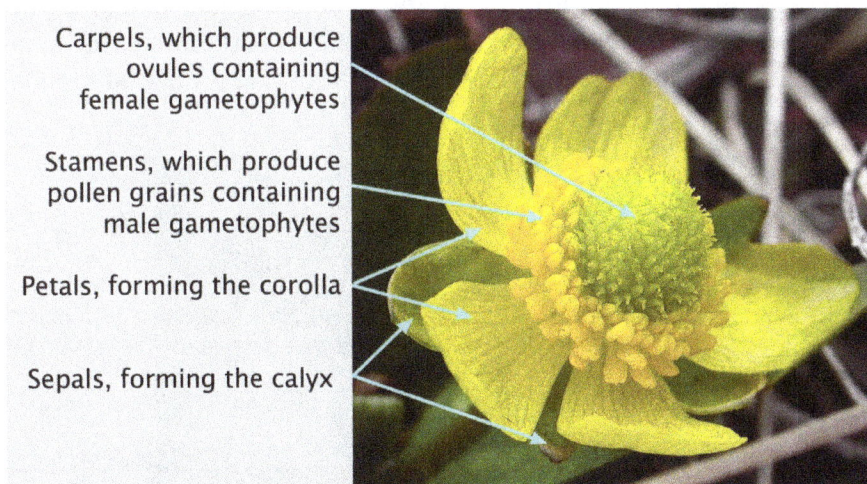

Carpels, which produce ovules containing female gametophytes

Stamens, which produce pollen grains containing male gametophytes

Petals, forming the corolla

Sepals, forming the calyx

Flower of *Ranunculus glaberrimus*

Stamens may be called the "male" parts of a flower; collectively they form the androecium. Carpels may be called the "female" parts of a flower; collectively they form the gynoecium. The carpels are often fused together to varying degrees; the entire structure may be called a pistil. The lower part of the carpel or of the fused pistil, where the ovules are produced, is called the ovary; it may be divided into chambers (locules) corresponding to the separate carpels.

Variations

A "perfect" flower has both stamens and carpels, and may be described as "bisexual" or "hermaph-roditic". A "unisexual" flower is one in which either the stamens or the carpels are missing, vestigial or otherwise non-functional. Each flower is either "staminate" (having only functional stamens) and thus "male", or "carpellate" (or "pistillate") (having only functional carpels) and thus "female". If separate staminate and carpellate flowers are always found on the same plant, the species is called monoecious. If separate staminate and carpellate flowers are always found on different plants, the species is called dioecious. A 1995 study found that about 6% of angiosperm species are dioecious, and that 7% of genera contain some dioecious species.

Alnus serrulata has unisexual flowers and is monoecious. Shown here: maturing male flower catkins on the right, last year's female catkins on the left.

Members of the birch family (Betulaceae) are examples of monoecious plants with unisexual flowers. A mature alder tree (*Alnus* species) produces long catkins containing only male flowers, each with four stamens and a minute perianth, and separate stalked groups of female flowers, each without a perianth.

Ilex aquifolium is dioecious: (above) shoot with flowers from male plant; (top right) male flower enlarged, showing stamens with pollen and reduced, sterile stigma; (below) shoot with flowers from female plant; (lower right) female flower enlarged, showing stigma and reduced, sterile stamens (staminodes) with no pollen

Most hollies (members of the genus *Ilex*) are dioecious. Each plant produces either functionally male flowers or functionally female flowers. In *Ilex aquifolium* (see the illustration), the common

European holly, both kinds of flower have four sepals and four white petals; male flowers have four stamens, female flowers usually have four non-functional reduced stamens and a four-celled ovary. Since only female plants are able to set fruit and produce berries, this has consequences for gardeners. *Amborella* represents the first known group of flowering plants to separate from their common ancestor. It too is dioecious; at any one time, each plant produces either flowers with functional stamens but no carpels, or flowers with a few non-functional stamens and a number of fully functional carpels. However, *Amborella* plants may change their "gender" over time. In one study, five cuttings from a male plant produced only male flowers when they first flowered, but at their second flowering three switched to producing female flowers.

In extreme cases, all of the parts present in a complete flower may be missing, so long as at least one carpel or one stamen is present. This situation is reached in the female flowers of duckweeds (*Lemna*), which comprise a single carpel, and in the male flowers of spurges (*Euphorbia*) which comprise a single stamen.

A species such as *Fraxinus excelsior*, the common ash of Europe, demonstrates one possible kind of variation. Ash flowers are wind-pollinated and lack petals and sepals. Structurally, the flowers may be bisexual, consisting of two stamens and an ovary, or may be male (staminate), lacking a functional ovary, or female (carpellate), lacking functional stamens. Different forms may occur on the same tree, or on different trees. The Asteraceae (sunflower family), with close to 22,000 species worldwide, have highly modified inflorescences made up of flowers (florets) collected together into tightly packed heads. Heads may have florets of one sexual morphology – all bisexual, all carpellate or all staminate (when they are called homogamous), or may have mixtures of two or more sexual forms (heterogamous). Thus goatsbeards (*Tragopogon* species) have heads of bisexual florets, like other members of the tribe Cichorieae, whereas marigolds (*Calendula* species) generally have heads with the outer florets bisexual and the inner florets staminate (male).

Like *Amborella*, some plants undergo sex-switching. For example, *Arisaema triphyllum* (Jack-in-the-pulpit) expresses sexual differences at different stages of growth: smaller plants produce all or mostly male flowers; as plants grow larger over the years the male flowers are replaced by more female flowers on the same plant. *Arisaema triphyllum* thus covers a multitude of sexual conditions in its lifetime: nonsexual juvenile plants, young plants that are all male, larger plants with a mix of both male and female flowers, and large plants that have mostly female flowers. Other plant populations have plants that produce more male flowers early in the year and as plants bloom later in the growing season they produce more female flowers.

Terminology

The complexity of the morphology of flowers and its variation within populations has led to a rich terminology.

- Androdioecious: having male flowers on some plants, bisexual ones on others.

- Androecious: having only male flowers (the male of a dioecious population); producing pollen but no seed.

- Androgynous.

- Androgynomonoecious: having male, female, and bisexual flowers on the same plant, also called trimonoecious.

- Andromonoecious: having both bisexual and male flowers on the same plant.

- Bisexual: each flower of each individual has both male and female structures, i.e. it combines both sexes in one structure. Flowers of this kind are called perfect, having both stamens and carpels. Other terms used for this condition are androgynous, hermaphroditic, monoclinous and synoecious.

- Dichogamous: having sexes developing at different times; producing pollen when the stigmas are not receptive, either protandrous or protogynous. This promotes outcrossing by limiting self-pollination. Some dichogamous plants have bisexual flowers, others have unisexual flowers.

- Diclinous

- Dioecious: having either only male or only female flowers. No individual plant of the population produces both pollen and ovules.

- Gynodioecious: having hermaphrodite flowers and female flowers on separate plants.

- Gynoecious: having only female flowers (the female of a dioecious population); producing seed but not pollen.

- Gynomonoecious: having both bisexual and female flowers on the same plant.

- Hermaphroditic.

- Imperfect: (of flowers) having some parts that are normally present not developed, e.g. lacking stamens.

- Monoclinous.

- Monoecious: In the commoner narrow sense of the term, it refers to plants with unisexual flowers which occur on the same individual. In the broad sense of the term, it also includes plants with bisexual flowers. Individuals bearing separate flowers of both sexes at the same time are called simultaneously or synchronously monoecious. Individuals that bear flowers of one sex at one time are called consecutively monoecious.

- Perfect: (of flowers) .

- Polygamodioecious: mostly dioecious, but with either a few flowers of the opposite sex or a few bisexual flowers on the same plant.

- Polygamomonoecious.

- Polygamous: having male, female, and bisexual flowers on the same plant. Also called polygamomonoecious or trimonoecious. Or, with bisexual and at least one of male and female flowers on the same plant.

- Protandrous: (of dichogamous plants) having male parts of flowers developed before female parts, e.g. having flowers that function first as male and then change to female or producing pollen before the stigmas of the same plant are receptive. (Protoandrous is also used.)

- Protogynous: (of dichogamous plants) having female parts of flowers developed before male parts, e.g. having flowers that function first as female and then change to male or producing pollen after the stigmas of the same plant are receptive.

- Subandroecious: having mostly male flowers, with a few female or bisexual flowers.

- Subdioecious: having some individuals in otherwise dioecious populations with flowers that are not clearly male or female. The population produces normally male or female plants with unisexual flowers, but some plants may have bisexual flowers, some both male and female flowers, and others some combination thereof, such as female and bisexual flowers. The condition is thought to represent a transition between bisexuality and dioecy.

- Subgynoecious: having mostly female flowers, with a few male or bisexual flowers.

- Synoecious.

- Trimonoecious.

- Trioecious.

- Unisexual: having either functionally male or functionally female flowers. This condition is also called diclinous, incomplete or imperfect.

Outcrossing

Outcrossing, cross-fertilization or allogamy, in which offspring are formed by the fusion of the gametes of two different plants, is the most common mode of reproduction among higher plants. About 55% of higher plant species reproduce in this way. An additional 7% are partially cross-fertilizing and partially self-fertilizing (autogamy). About 15% produce gametes but are principally self-fertilizing with significant out-crossing lacking. Only about 8% of higher plant species reproduce exclusively by non-sexual means. These include plants that reproduce vegetatively by runners or bulbils, or which produce seeds without embryo fertilization (apomixis). The selective advantage of outcrossing appears to be the masking of deleterious recessive mutations.

The primary mechanism used by flowering plants to ensure outcrossing involves a genetic mechanism known as self-incompatibility. Various aspects of floral morphology promote allogamy. In plants with bisexual flowers, the anthers and carpels may mature at different times, plants being protandrous (with the anthers maturing first) or protogynous (with the carpels mature first). Monoecious species, with unisexual flowers on the same plant, may produce male and female flowers at different times.

Dioecy, the condition of having unisexual flowers on different plants, necessarily results in outcrossing, and might thus be thought to have evolved for this purpose. However, "dioecy has proven difficult to explain simply as an outbreeding mechanism in plants that lack self-in-compatibility". Resource-allocation constraints may be important in the evolution of dioecy, for example, with wind-pollination, separate male flowers arranged in a catkin that vibrates in the wind may provide better pollen dispersal. In climbing plants, rapid upward growth may be essential, and resource allocation to fruit production may be incompatible with rapid growth, thus giving an advantage to delayed production of female flowers. Dioecy has evolved separately in many different lineages, and monoecy in the plant lineage correlates with the evolution of dioecy, suggesting that dioecy can evolve more readily from plants that already produce separate male and female flowers.

Evolution of Sexual Reproduction

The evolution of sexual reproduction describes how sexually reproducing animals, plants, fungi and protists evolved from a common ancestor that was a single celled eukaryotic species. There are a few species which have secondarily lost the ability to reproduce sexually, such as Bdelloidea and some parthenocarpic plants. The evolution of sex contains two related, yet distinct, themes: its *origin* and its *maintenance*.

Ladybirds mating

Moths mating *Laothoe populi*

The maintenance of sexual reproduction in a highly competitive world has long been one of the major mysteries of biology given that asexual reproduction can reproduce much more quickly as

50% of offspring in sexual reproduction are males, unable to produce offspring themselves. However, research published in 2015 indicates that sexual selection can explain the persistence of sexual reproduction in animals.

Since hypotheses for the origins of sex are difficult to test experimentally (outside of Evolutionary computation), most current work has focused on the maintenance of sexual reproduction. Sexual reproduction must offer significant fitness advantages to a species because despite the two-fold cost of sex, it dominates among multicellular forms of life, implying that the fitness of offspring produced outweighs the costs. Sexual reproduction derives from recombination, where parent genotypes are reorganized and shared with the offspring. This stands in contrast to single-parent asexual replication, where the offspring is identical to the parents. Recombination supplies two fault-tolerance mechanisms at the molecular level: *recombinational DNA repair* (promoted during meiosis because homologous chromosomes pair at that time) and *complementation* (also known as heterosis, hybrid vigor or masking of mutations).

Sexual reproduction has probably contributed to the evolution of sexual dimorphism, where organisms within a species adopted different strategies of parental investment. Males adopt strategies with lower investment in individual gametes and may present a higher mutation rate, while females may invest more resources and serve to conserve better-adapted solutions.

Historical Perspective

Modern philosophical-scientific thinking on the problem can be traced back to Erasmus Darwin in the 18th century; it also features in Aristotle's writings. The thread was later picked up by August Weismann in 1889, who argued that the purpose of sex was to generate genetic variation, as is detailed in the majority of the explanations below. On the other hand, Charles Darwin concluded that the effects of hybrid vigor (complementation) "is amply sufficient to account for the ... genesis of the two sexes." This is consistent with the repair and complementation hypothesis, given below under "Other explanations."

Several explanations have been suggested by biologists including W. D. Hamilton, Alexey Kondrashov, George C. Williams, Harris Bernstein, Carol Bernstein, Michael M. Cox, Frederic A. Hopf and Richard E. Michod to explain how sexual reproduction is maintained in a vast array of different living organisms.

Questions

Some questions biologists have attempted to answer include:

- Why sexual reproduction exists, if in many organisms it has a 50% cost (fitness disadvantage) in relation to asexual reproduction?

- Did mating types (types of gametes, according to their compatibility) arise as a result of anisogamy (gamete dimorphism), or did mating types evolve before anisogamy?

- Why do most sexual organisms use a binary mating system? Why do some organisms have gamete dimorphism?

Two-fold Cost of Sex

In most multicellular sexual species, the population consists of two sexes, only one of which is capable of bearing young (with the exception of simultaneous hermaphrodites). In an asexual species, each member of the population is capable of bearing young. This implies that an asexual population has an intrinsic capacity to grow more rapidly with each generation. The cost was first described in mathematical terms by John Maynard Smith. He imagined an asexual mutant arising in a sexual population, half of which comprises males that cannot themselves produce offspring. With female-only offspring, the asexual lineage doubles its representation in the population each generation, all else being equal. Technically this is not a problem of sex but a problem of some multicellular sexually reproducing organisms. There are numerous isogamous species which are sexual and do not have the problem of producing individuals which cannot directly replicate themselves. The principal costs of sex is that males and females must search for each other in order to mate, and sexual selection often favours traits that reduce the survival of individuals.

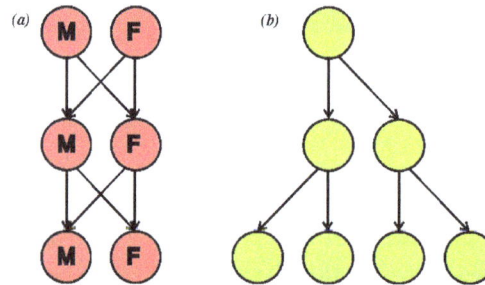

This diagram illustrates the *twofold cost of sex*. If each individual were to contribute to the same number of offspring (two), *(a)* the sexual population remains the same size each generation, where the *(b)* asexual population doubles in size each generation.

Evidence that the cost is surmountable comes from George C. Williams, who noted the existence of species which are capable of both asexual and sexual reproduction, such as certain lizards. These species time their sexual reproduction with periods of environmental uncertainty, and reproduce asexually when conditions are more favourable. The important point is that these species are observed to reproduce sexually when they could choose not to, implying that there is a selective advantage to sexual reproduction.

It is widely believed that a disadvantage of sexual reproduction is that a sexually reproducing organism will only be able to pass on 50% of its genes to each offspring. This is a consequence of the fact that gametes from sexually reproducing species are haploid. This, however, conflates sex and reproduction which are two separate events. The "two-fold cost of sex" may more accurately be described as the cost of anisogamy. Not all sexual organisms are anisogamous. There are numerous species which are sexual and do not have this problem because they do not produce males or females. Yeast, for example, are isogamous sexual organisms which have two mating types which fuse and recombine their haploid genomes. Both sexes reproduce during the haploid and diploid stages of their life cycle and have a 100% chance of passing their genes into their offspring.

The two-fold cost of sex may be compensated for in some species in many ways. Females may eat males after mating, males may be much smaller or rarer, or males may help raise offspring.

Sex Decoupled from Reproduction

Some species avoid the cost of 50% of sexual reproduction, although they have "sex" (in the sense of genetic recombination). In these species (e.g., bacteria, ciliates, dinoflagellates and diatoms), "sex" and reproduction occurs separately.

Promotion of Genetic Variation

August Weismann proposed in 1889 an explanation for the evolution of sex, where the advantage of sex is the creation of variation among siblings. It was then subsequently explained in genetics terms by Fisher and Muller and has been recently summarised by Burt in 2000.

George C. Williams gave an example based around the elm tree. In the forest of this example, empty patches between trees can support one individual each. When a patch becomes available because of the death of a tree, other trees' seeds will compete to fill the patch. Since the chance of a seed's success in occupying the patch depends upon its genotype, and a parent cannot anticipate which genotype is most successful, each parent will send many seeds, creating competition between siblings. Natural selection therefore favours parents which can produce a variety of offspring.

A similar hypothesis is named the *tangled bank hypothesis* after a passage in Charles Darwin's *The Origin of Species*:

> "It is interesting to contemplate an entangled bank, clothed with many plants of many kinds, with birds singing on the bushes, with various insects flitting about, and with worms crawling through the damp earth, and to reflect that these elaborately constructed forms, so different from each other, and dependent on each other in so complex a manner, have all been produced by laws acting around us."

The hypothesis, proposed by Michael Ghiselin in his 1974 book, *The Economy of Nature and the Evolution of Sex*, suggests that a diverse set of siblings may be able to extract more food from its environment than a clone, because each sibling uses a slightly different niche. One of the main proponents of this hypothesis is Graham Bell of McGill University. The hypothesis has been criticised for failing to explain how asexual species developed sexes. In his book, *Evolution and Human Behavior* (MIT Press, 2000), John Cartwright comments:

> "Although once popular, the tangled bank hypothesis now seems to face many problems, and former adherents are falling away. The theory would predict a greater interest in sex among animals that produce lots of small offspring that compete with each other. In fact, sex is invariably associated with organisms that produce a few large offspring, whereas organisms producing small offspring frequently engage in parthenogenesis [asexual reproduction]. In addition, the evidence from fossils suggests that species go for vast periods of [geologic] time without changing much."

In contrast to the view that sex promotes genetic variation, Heng, and Gorelick and Heng reviewed evidence that sex actually acts as a constraint on genetic variation. They consider that sex acts as a coarse filter, weeding out major genetic changes, such as chromosomal rearrangements, but permitting minor variation, such as changes at the nucleotide or gene level (that are often neutral) to pass through the sexual sieve.

Advantages Conferred by Sex

It is important to mention that the concept of sex includes two fundamental phenomena: the sexual process (fusion of genetic information of two individuals) and sexual differentiation (separation of this information into two parts). Depending on the presence or absence of these phenomena, the existing ways of reproduction can be divided into asexual, hermaphrodite and dioecious forms. The sexual process and sexual differentiation are different phenomena, and, in essence, are diametrically opposed. The first creates (increases) diversity of genotypes, and the second decreases it in half.

Reproductive advantages of the asexual forms are in quantity of the progeny and the advantages of the hermaphrodite forms – in maximum diversity. Transition from the hermaphrodite to dioecious state leads to a loss of at least half of the diversity. So, the main question is to explain the advantages given by sexual differentiation, i.e. the benefits of two separate sexes compare to hermaphrodites rather than to explain benefits of sexual forms (hermaphrodite + dioecious) over asexual ones. It has already been understood that since sexual reproduction is not associated with any clear reproductive advantages, as compared with asexual, there should be some important advantages in evolution.

Advantages Due to Genetic Variation

For the advantage due to genetic variation, there are three possible reasons this might happen. First, sexual reproduction can combine the effects of two beneficial mutations in the same individual (i.e. sex aids in the spread of advantageous traits). Also, the necessary mutations do not have to have occurred one after another in a single line of descendants. Second, sex acts to bring together currently deleterious mutations to create severely unfit individuals that are then eliminated from the population (i.e. sex aids in the removal of deleterious genes). However, in organisms containing only one set of chromosomes, deleterious mutations would be eliminated immediately, and therefore removal of harmful mutations is an unlikely benefit for sexual reproduction. Lastly, sex creates new gene combinations that may be more fit than previously existing ones, or may simply lead to reduced competition among relatives.

For the advantage due to DNA repair, there is an immediate large benefit of removing DNA damage by recombinational DNA repair during meiosis, since this removal allows greater survival of progeny with undamaged DNA. The advantage of complementation to each sexual partner is avoidance of the bad effects of their deleterious recessive genes in progeny by the masking effect of normal dominant genes contributed by the other partner.

The classes of hypotheses based on the creation of variation are further broken down below. It is important to realise that any number of these hypotheses may be true in any given species (they are not mutually exclusive), and that different hypotheses may apply in different species. However, a research framework based on creation of variation has yet to be found that allows one to determine whether the reason for sex is universal for all sexual species, and, if not, which mechanisms are acting in each species.

On the other hand, the maintenance of sex based on DNA repair and complementation applies widely to all sexual species.

Novel Genotypes

Sex could be a method by which novel genotypes are created. Because sex combines genes from two individuals, sexually reproducing populations can more easily combine advantageous genes than can asexual populations. If, in a sexual population, two different advantageous alleles arise at different loci on a chromosome in different members of the population, a chromosome containing the two advantageous alleles can be produced within a few generations by recombination. However, should the same two alleles arise in different members of an asexual population, the only way that one chromosome can develop the other allele is to independently gain the same mutation, which would take much longer. Several studies have addressed counterarguments, and the question of whether this model is sufficiently robust to explain the predominance of sexual versus asexual reproduction.

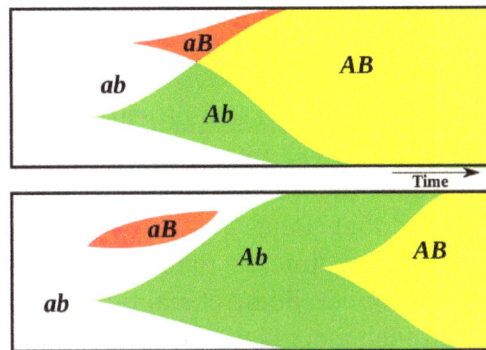

This diagram illustrates how sex might create novel genotypes more rapidly. Two advantageous alleles *A* and *B* occur at random. The two alleles are recombined rapidly in a sexual population (top), but in an asexual population (bottom) the two alleles must independently arise because of clonal interference.

Ronald Fisher also suggested that sex might facilitate the spread of advantageous genes by allowing them to better escape their genetic surroundings, if they should arise on a chromosome with deleterious genes.

Supporters of these theories respond to the balance argument that the individuals produced by sexual and asexual reproduction may differ in other respects too – which may influence the persistence of sexuality. For example, in the heterogamous water fleas of the genus *Cladocera*, sexual offspring form eggs which are better able to survive the winter versus those the fleas produce asexually.

Increased Resistance to Parasites

One of the most widely discussed theories to explain the persistence of sex is that it is maintained to assist sexual individuals in resisting parasites, also known as the Red Queen's Hypothesis.

When an environment changes, previously neutral or deleterious alleles can become favourable. If the environment changed sufficiently rapidly (i.e. between generations), these changes in the environment can make sex advantageous for the individual. Such rapid changes in environment are caused by the co-evolution between hosts and parasites.

Imagine, for example that there is one gene in parasites with two alleles *p* and *P* conferring two types of parasitic ability, and one gene in hosts with two alleles *h* and *H*, conferring two types of parasite resistance, such that parasites with allele p can attach themselves to hosts with the allele h, and P to H. Such a situation will lead to cyclic changes in allele frequency - as p increases in frequency, h will be disfavoured.

In reality, there will be several genes involved in the relationship between hosts and parasites. In an asexual population of hosts, offspring will only have the different parasitic resistance if a mutation arises. In a sexual population of hosts, however, offspring will have a new combination of parasitic resistance alleles.

In other words, like Lewis Carroll's Red Queen, sexual hosts are continually "running" (adapting) to "stay in one place" (resist parasites).

Evidence for this explanation for the evolution of sex is provided by comparison of the rate of molecular evolution of genes for kinases and immunoglobulins in the immune system with genes coding other proteins. The genes coding for immune system proteins evolve considerably faster.

Further evidence for the Red Queen hypothesis was provided by observing long-term dynamics and parasite coevolution in a "mixed" (sexual and asexual) population of snails (*Potamopyrgus antipodarum*). The number of sexuals, the number asexuals, and the rates of parasite infection for both were monitored. It was found that clones that were plentiful at the beginning of the study became more susceptible to parasites over time. As parasite infections increased, the once plentiful clones dwindled dramatically in number. Some clonal types disappeared entirely. Meanwhile, sexual snail populations remained much more stable over time.

However, Hanley et al. studied mite infestations of a parthenogenetic gecko species and its two related sexual ancestral species. Contrary to expectation based on the Red Queen hypothesis, they found that the prevalence, abundance and mean intensity of mites in sexual geckos was significantly higher than in asexuals sharing the same habitat.

In 2011, researchers used the microscopic roundworm *Caenorhabditis elegans* as a host and the pathogenic bacteria *Serratia marcescens* to generate a host-parasite coevolutionary system in a controlled environment, allowing them to conduct more than 70 evolution experiments testing the Red Queen Hypothesis. They genetically manipulated the mating system of *C. elegans*, causing populations to mate either sexually, by self-fertilization, or a mixture of both within the same population. Then they exposed those populations to the *S. marcescens* parasite. It was found that the self-fertilizing populations of *C. elegans* were rapidly driven extinct by the coevolving parasites while sex allowed populations to keep pace with their parasites, a result consistent with the Red Queen Hypothesis.

Critics of the Red Queen hypothesis question whether the constantly changing environment of hosts and parasites is sufficiently common to explain the evolution of sex. In particular, Otto and Nuismer presented results showing that species interactions (e.g. host vs parasite interactions) typically select against sex. They concluded that, although the Red Queen hypothesis favors sex under certain circumstances, it alone does not account for the ubiquity of sex. Otto and Gerstein further stated that "it seems doubtful to us that strong selection per gene is sufficiently commonplace for the Red Queen hypothesis to explain the ubiquity of sex." Parker reviewed numerous genetic studies on plant disease resistance and failed to uncover a single example consistent with the assumptions of the Red Queen hypothesis.

Deleterious Mutation Clearance

Mutations can have many different effects upon an organism. It is generally believed that the majority of non-neutral mutations are deleterious, which means that they will cause a decrease in the organism's overall fitness. If a mutation has a deleterious effect, it will then usually be removed from the population by the process of natural selection. Sexual reproduction is believed to be more efficient than asexual reproduction in removing those mutations from the genome.

There are two main hypotheses which explain how sex may act to remove deleterious genes from the genome.

Evading Harmful Mutation Build-up

While DNA is able to recombine to modify alleles, DNA is also susceptible to mutations within the sequence that can affect an organism in a negative manner. Asexual organisms do not have the ability to recombine their genetic information to form new and differing alleles. Once a mutation occurs in the DNA or other genetic carrying sequence, there is no way for the mutation to be removed from the population until another mutation occurs that ultimately deletes the primary mutation. This is rare among organisms. Hermann Joseph Muller introduced the idea that mutations build up in asexual reproducing organisms. Muller described this occurrence by comparing the mutations that accumulate as a ratchet. Each mutation that arises in asexually reproducing organisms turns the ratchet once. The ratchet is unable to be rotated backwards, only forwards. The next mutation that occurs turns the ratchet once more. Additional mutations in a population continually turn the ratchet and the mutations, mostly deleterious, continually accumulate without recombination. These mutations are passed onto the next generation because the offspring are exact genetic clones of their parent. The genetic load of organisms and their populations will increase due to the addition of multiple deleterious mutations and decrease the overall reproductive success and fitness.

For sexually reproducing populations, mutations in the DNA are more likely to be removed due to recombination in the process of meiosis. The offspring are also not direct genetic clones of a single parent. The alleles from both parents contribute to the offspring. This creates the ability to mask a mutation in the form of heterozygotes. Selection can also work in removing mutations from a sexual population. The lessened amounts of harmful mutations within an organism can lead to increased reproductive success. Natural selection will select for the reduced number of deleterious mutations. Many believe that this ability to evade the accumulation of harmful and possibly lethal mutations produces a substantial advantage for sexually reproducing populations.

Removal of Deleterious Genes

Diagram illustrating different relationships between numbers of mutations and fitness. Kondrashov's model requires *synergistic epistasis*, which is represented by the red line - each subsequent mutation has a disproportionately large effect on the organism's fitness.

This hypothesis was proposed by Alexey Kondrashov, and is sometimes known as the *deterministic mutation hypothesis*. It assumes that the majority of deleterious mutations are only slightly deleterious, and affect the individual such that the introduction of each additional mutation has an increasingly large effect on the fitness of the organism. This relationship between number of mutations and fitness is known as *synergistic epistasis*.

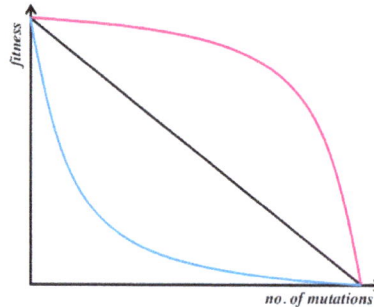

By way of analogy, think of a car with several minor faults. Each is not sufficient alone to prevent the car from running, but in combination, the faults combine to prevent the car from functioning.

Similarly, an organism may be able to cope with a few defects, but the presence of many mutations could overwhelm its backup mechanisms.

Kondrashov argues that the slightly deleterious nature of mutations means that the population will tend to be composed of individuals with a small number of mutations. Sex will act to recombine these genotypes, creating some individuals with fewer deleterious mutations, and some with more. Because there is a major selective disadvantage to individuals with more mutations, these individuals die out. In essence, sex compartmentalises the deleterious mutations.

There has been much criticism of Kondrashov's theory, since it relies on two key restrictive conditions. The first requires that the rate of deleterious mutation should exceed one per genome per generation in order to provide a substantial advantage for sex. While there is some empirical evidence for it (for example in Drosophila and E. coli), there is also strong evidence against it. Thus, for instance, for the sexual species *Saccharomyces cerevisiae* (yeast) and *Neurospora crassa* (fungus), the mutation rate per genome per replication are 0.0027 and 0.0030 respectively. For the nematode worm *Caenorhabditis elegans*, the mutation rate per effective genome per sexual generation is 0.036. Secondly, there should be strong interactions among loci (synergistic epistasis), a mutation-fitness relation for which there is only limited evidence. Conversely, there is also the same amount of evidence that mutations show no epistasis (purely additive model) or antagonistic interactions (each additional mutation has a disproportionally *small* effect).

Other Explanations

Geodakyan's Evolutionary Theory of Sex

Geodakyan suggested that sexual dimorphism provides a partitioning of a species' phenotypes into at least two functional partitions: a female partition that secures beneficial features of the species and a male partition that emerged in species with more variable and unpredictable environments. The male partition is suggested to be an "experimental" part of the species that allows the species to expand their ecological niche, and to have alternative configurations. This theory underlines the higher variability and higher mortality in males, in comparison to females. This functional partitioning also explains the higher suceptibility to disease in males, in comparison to females and therefore includes the idea of "protection against parasites" as another functionality of male sex. Geodakyan's evolutionary theory of sex was developed in Russia in 1960-80 and was not known to the West till the era of the Internet. Trofimova, who analysed psychological sex differences, pointed out that the male sex might also provide a "redundancy pruning" function.

Speed of Evolution

Ilan Eshel suggested that sex prevents rapid evolution. He suggests that recombination breaks up favourable gene combinations more often than it creates them, and sex is maintained because it ensures selection is longer-term than in asexual populations - so the population is less affected by short-term changes. This explanation is not widely accepted, as its assumptions are very restrictive.

It has recently been shown in experiments with *Chlamydomonas* algae that sex can remove the speed limit on evolution.

An information theoretic analysis using a simplified but useful model shows that in asexual reproduction, the information gain per generation of a species is limited to 1 bit per generation, while in sexual reproduction, the information gain is bounded by , where is the size of the genome in bits.

DNA Repair and Complementation

As discussed in the earlier part of this section, sexual reproduction is conventionally explained as an adaptation for producing genetic variation through allelic recombination. As acknowledged above, however, serious problems with this explanation have led many biologists to conclude that the benefit of sex is a major unsolved problem in evolutionary biology.

An alternative "informational" approach to this problem has led to the view that the two fundamental aspects of sex, genetic recombination and outcrossing, are adaptive responses to the two major sources of "noise" in transmitting genetic information. Genetic noise can occur as either physical damage to the genome (e.g. chemically altered bases of DNA or breaks in the chromosome) or replication errors (mutations) This alternative view is referred to as the repair and complementation hypothesis, to distinguish it from the traditional variation hypothesis.

The repair and complementation hypothesis assumes that genetic recombination is fundamentally a DNA repair process, and that when it occurs during meiosis it is an adaptation for repairing the genomic DNA which is passed on to progeny. Recombinational repair is the only repair process known which can accurately remove double-strand damages in DNA, and such damages are both common in nature and ordinarily lethal if not repaired. For instance, double-strand breaks in DNA occur about 50 times per cell cycle in human cells. Recombinational repair is prevalent from the simplest viruses to the most complex multicellular eukaryotes. It is effective against many different types of genomic damage, and in particular is highly efficient at overcoming double-strand damages. Studies of the mechanism of meiotic recombination indicate that meiosis is an adaptation for repairing DNA. These considerations form the basis for the first part of the repair and complementation hypothesis.

In some lines of descent from the earliest organisms, the diploid stage of the sexual cycle, which was at first transient, became the predominant stage, because it allowed complementation — the masking of deleterious recessive mutations (i.e. hybrid vigor or heterosis). Outcrossing, the second fundamental aspect of sex, is maintained by the advantage of masking mutations and the disadvantage of inbreeding (mating with a close relative) which allows expression of recessive mutations (commonly observed as inbreeding depression). This is in accord with Charles Darwin, who concluded that the adaptive advantage of sex is hybrid vigor; or as he put it, "the offspring of two individuals, especially if their progenitors have been subjected to very different conditions, have a great advantage in height, weight, constitutional vigor and fertility over the self fertilised offspring from either one of the same parents."

However, outcrossing may be abandoned in favor of parthenogenesis or selfing (which retain the advantage of meiotic recombinational repair) under conditions in which the costs of mating are very high. For instance, costs of mating are high when individuals are rare in a geographic area, such as when there has been a forest fire and the individuals entering the burned area are the initial ones to arrive. At such times mates are hard to find, and this favors parthenogenic species.

In the view of the repair and complementation hypothesis, the removal of DNA damage by recombinational repair produces a new, less deleterious form of informational noise, allelic recombination, as a by-product. This lesser informational noise generates genetic variation, viewed by some as the major effect of sex.

Libertine Bubble Theory

The evolution of sex can alternatively be described as a kind of gene exchange that is independent from reproduction. According to the Thierry Lodé's "libertine bubble theory", sex originated from an archaic gene transfer process among prebiotic bubbles. The contact among the pre-biotic bubbles could, through simple food or parasitic reactions, promote the transfer of genetic material from one bubble to another. That interactions between two organisms be in balance appear to be a sufficient condition to make these interactions evolutionarily efficient, i.e. to select bubbles that tolerate these interactions ("libertine" bubbles) through a blind evolutionary process of self-reinforcing gene correlations and compatibility.

The "libertine bubble theory" proposes that meiotic sex evolved in proto-eukaryotes to solve a problem that bacteria did not have, namely a large amount of DNA material, occurring in an archaic step of proto-cell formation and genetic exchanges. So that, rather than providing selective advantages through reproduction, sex could be thought of as a series of separate events which combines step-by-step some very weak benefits of recombination, meiosis, gametogenesis and syngamy. Therefore, current sexual species could be descendants of primitive organisms that practiced more stable exchanges in the long term, while asexual species have emerged, much more recently in evolutionary history, from the conflict of interest resulting from anisogamy.

Origin of Sexual Reproduction

Many protists reproduce sexually, as do the multicellular plants, animals, and fungi. In the eukaryotic fossil record, sexual reproduction first appeared by 1.2 billion years ago in the Proterozoic Eon. All sexually reproducing eukaryotic organisms derive from a single-celled common ancestor. There are a few species which have secondarily lost this feature, such as Bdelloidea and some parthenocarpic plants.

Organisms need to replicate their genetic material in an efficient and reliable manner. The necessity to repair genetic damage is one of the leading theories explaining the origin of sexual reproduction. Diploid individuals can repair a damaged section of their DNA via homologous recombination, since there are two copies of the gene in the cell and one copy is presumed to be undamaged. A mutation in a haploid individual, on the other hand, is more likely to become resident, as the DNA repair machinery has no way of knowing what the original undamaged sequence was. The most primitive form of sex may have been one organism with damaged DNA replicating an undamaged strand from a similar organism in order to repair itself.

If, as evidence indicates, sexual reproduction arose very early in eukaryotic evolution, the essential features of meiosis may have already been present in the prokaryotic ancestors of eukaryotes. In extant organisms, proteins with central functions in meiosis are similar to key proteins in bacterial transformation. For example, recA recombinase, that catalyses the key functions of DNA homology search and strand exchange in the bacterial sexual process of transformation, has orthologs in eukaryotes that perform similar functions in meiotic recombination. Both bacterial transformation and meiosis in eukaryotic microorganisms are induced by stressful circumstances such as overcrowding, resource depletion and DNA damaging conditions. This suggests that these sexual processes are adaptations for dealing with stress, particularly stress that causes DNA damage. In bacteria, these stresses induce an altered physiologic state, termed competence, that allows active take-up of DNA from a donor bacterium and the integration of this DNA into the recipient genome allowing recombinational repair of the recipients' damaged DNA. If environmental stresses leading to DNA damage were a persistent challenge to the survival of early microorganisms, then selection would likely have been continuous through the prokaryote to eukaryote transition, and adaptative adjustments would have followed a course in which bacterial transformation naturally gave rise to sexual reproduction in eukaryotes.

Sex may also have been present even earlier, in the RNA world that is considered to have preceded DNA cellular life forms. A proposal for the origin of sex in the RNA world was based on the type of sexual interaction that is known to occur in extant single-stranded segmented RNA viruses such as influenza virus, and in extant double-stranded segmented RNA viruses such as reovirus. Exposure to conditions that cause RNA damage could have led to blockage of replication and death of these early RNA life forms. Sex would have allowed re-assortment of segments between two individuals with damaged RNA, permitting undamaged combinations of RNA segments to come together, thus allowing survival. Such a regeneration phenomenon, known as multiplicity reactivation, occurs in influenza virus and reovirus

Another theory is that sexual reproduction originated from selfish parasitic genetic elements that exchange genetic material (that is: copies of their own genome) for their transmission and propagation. In some organisms, sexual reproduction has been shown to enhance the spread of parasitic genetic elements (e.g.: yeast, filamentous fungi). Bacterial conjugation, a form of genetic exchange that some sources describe as sex, is not a form of reproduction, but rather an example of horizontal gene transfer. However, it does support the selfish genetic element theory, as it is propagated through such a "selfish gene", the F-plasmid. Similarly, it has been proposed that sexual reproduction evolved from ancient haloarchaea through a combination of jumping genes, and swapping plasmids.

A third theory is that sex evolved as a form of cannibalism. One primitive organism ate another one, but rather than completely digesting it, some of the 'eaten' organism's DNA was incorporated into the 'eater' organism.

Sex may also be derived from another prokaryotic process. A comprehensive 'origin of sex as vaccination' theory proposes that eukaryan sex-as-syngamy (fusion sex) arose from prokaryan unilateral sex-as-infection when infected hosts began swapping nuclearised genomes containing coevolved, vertically transmitted symbionts that provided protection against horizontal superinfection by more virulent symbionts. Sex-as-meiosis (fission sex) then evolved as a host strategy to uncouple (and thereby emasculate) the acquired symbiont genomes.

Mechanistic Origin of Sexual Reproduction

While theories positing fitness benefits that led to the origin of sex are often problematic, several theories addressing the emergence of the mechanisms of sexual reproduction have been proposed.

Viral Eukaryogenesis

The viral eukaryogenesis (VE) theory proposes that eukaryotic cells arose from a combination of a lysogenic virus, an archaeon and a bacterium. This model suggests that the nucleus originated when the lysogenic virus incorporated genetic material from the archaeon and the bacterium and took over the role of information storage for the amalgam. The archaeal host transferred much of its functional genome to the virus during the evolution of cytoplasm but retained the function of gene translation and general metabolism. The bacterium transferred most of its functional genome to the virus as it transitioned into a mitochondrion.

For these transformations to lead to the eukaryotic cell cycle, the VE hypothesis specifies a pox-like virus as the lysogenic virus. A pox-like virus is a likely ancestor because of its fundamental similarities with eukaryotic nuclei. These include a double stranded DNA genome, a linear chromosome with short telomeric repeats, a complex membrane bound capsid, the ability to produce capped mRNA, and the ability to export the capped mRNA across the viral membrane into the cytoplasm. The presence of a lysogenic pox-like virus ancestor explains the development of meiotic division, an essential component of sexual reproduction.

Meiotic division in the VE hypothesis arose because of the evolutionary pressures placed on the lysogenic virus as a result of its inability to enter into the lytic cycle. This selective pressure resulted in the development of processes allowing the viruses to spread horizontally throughout the population. The outcome of this selection was cell-to-cell fusion. (This is distinct from the conjugation methods used by bacterial plasmids under evolutionary pressure, with important consequences.) The possibility of this kind of fusion is supported by the presence of fusion proteins in the envelopes of the pox viruses that allow them to fuse with host membranes. These proteins could have been transferred to the cell membrane during viral reproduction, enabling cell-to-cell fusion between the virus host and an uninfected cell. The theory proposes meiosis originated from the fusion between two cells infected with related but different viruses which recognised each other as uninfected. After the fusion of the two cells, incompatibilities between the two viruses result in a meiotic-like cell division.

The two viruses established in the cell would initiate replication in response to signals from the host cell. A mitosis-like cell cycle would proceed until the viral membranes dissolved, at which point linear chromosomes would be bound together with centromeres. The homologous nature of the two viral centromeres would incite the grouping of both sets into tetrads. It is speculated that this grouping may be the origin of crossing over, characteristic of the first division in modern meiosis. The partitioning apparatus of the mitotic-like cell cycle the cells used to replicate independently would then pull each set of chromosomes to one side of the cell, still bound by centromeres. These centromeres would prevent their replication in subsequent division, resulting in four daughter cells with one copy of one of the two original pox-like viruses. The process resulting from combination of two similar pox viruses within the same host closely mimics meiosis.

Neomuran Revolution

An alternative theory, proposed by Thomas Cavalier-Smith, was labeled the Neomuran revolution. The designation "Neomuran revolution" refers to the appearances of the common ancestors of eukaryotes and archaea. Cavalier-Smith proposes that the first neomurans emerged 850 million years ago. Other molecular biologists assume that this group appeared much earlier, but Cavalier-Smith dismisses these claims because they are based on the "theoretically and empirically" unsound model of molecular clocks. Cavalier-Smith's theory of the Neomuran revolution has implications for the evolutionary history of the cellular machinery for recombination and sex. It suggests that this machinery evolved in two distinct bouts separated by a long period of stasis; first the appearance of recombination machinery in a bacterial ancestor which was maintained for 3 Gy, until the neomuran revolution when the mechanics were adapted to the presence of nucleosomes. The archaeal products of the revolution maintained recombination machinery that was essentially bacterial, whereas the eukaryotic products broke with this bacterial continuity. They introduced cell fusion and ploidy cycles into cell life histories. Cavalier-Smith argues that both bouts of mechanical evolution were motivated by similar selective forces: the need for accurate DNA replication without loss of viability.

References

- Fritz, Robert E.; Simms, Ellen Louise (1992). Plant resistance to herbivores and pathogens: ecology, evolution, and genetics. Chicago: University of Chicago Press. p. 359. ISBN 978-0-226-26554-4.

- Rooting cuttings of tropical trees. London: Commonwealth Science Council. 1993. p. 9. ISBN 978-0-85092-394-0.

- Reiley, H. Edward; Shry, Carroll L. (2004). Introductory horticulture. Albany, NY: Delmar/Thomson Learning. p. 54. ISBN 978-0-7668-1567-4.

- Bernstein H, Hopf FA, Michod RE (1987). "The molecular basis of the evolution of sex". Adv. Genet. Advances in Genetics. 24: 323–70. doi:10.1016/s0065-2660(08)60012-7. ISBN 9780120176243. PMID 3324702.

- Avise, J. (2008) Clonality: The Genetics, Ecology and Evolution of Sexual Abstinence in Vertebrate Animals. See pp. 22-25. Oxford University Press. ISBN 019536967X ISBN 978-0195369670

- Britannica Educational Publishing (2011). Fungi, Algae, and Protists. The Rosen Publishing Group. ISBN 978-1-61530-463-9.

- James Desmond Smyth; Derek Wakelin (1994). Introduction to animal parasitology (3 ed.). Cambridge University Press. pp. 101–102. ISBN 0-521-42811-4.

- R. S. Mehrotra; K. R. Aneja (December 1990). An Introduction to Mycology. New Age International. pp. 83 ff. ISBN 978-81-224-0089-2. Retrieved 4 August 2010.

- Kathleen M. Cole; Robert G. Sheath (1990). Biology of the red algae. Cambridge University Press. pp. 469–. ISBN 978-0-521-34301-5. Retrieved 4 August 2010.

- Edward G. Reekie; Fakhri A. Bazzaz (28 October 2005). Reproductive allocation in plants. Academic Press. pp. 99–. ISBN 978-0-12-088386-8. Retrieved 4 August 2010.

- Britannica Educational Publishing (2011). Fungi, Algae, and Protists. The Rosen Publishing Group. ISBN 978-1-61530-463-9.

- James Desmond Smyth, Derek Wakelin (1994). Introduction to animal parasitology (3 ed.). Cambridge University Press. pp. 101–102. ISBN 0-521-42811-4.

- Bernstein, H; Hopf, FA; Michod, RE (1987). "The molecular basis of the evolution of sex". Adv Genet. Advances in Genetics. 24: 323–370. doi:10.1016/s0065-2660(08)60012-7. ISBN 978-0-12-017624-3. PMID 3324702.

- Bell, G. (1982). The Masterpiece of Nature: The Evolution and Genetics of Sexuality, University of California Press, Berkeley, pp. 1- 635 (see page 295). ISBN 0-520-04583-1 ISBN 978-0-520-04583-5

- Vrijenhoek, Robert C. (1998). "Parthenogenesis and Natural Clones" (PDF). In Knobil, Ernst; Neill, Jimmy D. Encyclopedia of Reproduction. 3. Academic Press. pp. 695–702. ISBN 978-0-12-227020-8.

- Michod, RE; Levin, BE, eds. (1987). The Evolution of sex: An examination of current ideas. Sunderland, Massachusetts: Sinauer Associates. ISBN 978-0878934584.

- Michod, RE (1994). Eros and Evolution: A Natural Philosophy of Sex. Perseus Books. ISBN 978-0201407549.

- Beentje, Henk (2010). The Kew Plant Glossary. Richmond, Surrey: Royal Botanic Gardens, Kew. ISBN 978-1-84246-422-9.

- Bernstein, C. & Bernstein, H. (1991). Aging, Sex, and DNA Repair. San Diego: Academic Press. ISBN 978-0-12-092860-6.

- Bernstein H; Hopf FA; Michod RE (1987). "The Molecular Basis of the Evolution of Sex". Adv. Genet. Advances in Genetics. 24: 323–70. doi:10.1016/S0065-2660(08)60012-7. ISBN 978-0-12-017624-3. PMID 3324702.

- Olivia Judson (2002). Dr. Tatiana's sex advice to all creation. New York: Metropolitan Books. pp. 233–4. ISBN 0-8050-6331-5.

Cell Biology: An Overview

Cells have different structures and different functions, the study of these is known as cell biology. Some of the topics explained in this chapter are cell membrane, cytoskeleton, organelle, cell growth, cell cycle, mitosis and meiosis. This section is an overview of the subject matter incorporating all the major aspects of cell biology.

Cell Biology

Cell biology is a branch of biology that studies the different structures and functions of the cell and focuses mainly on the idea of the cell as the basic unit of life. Cell biology explains the structure, organization of the organelles they contain, their physiological properties, metabolic processes, signaling pathways, life cycle, and interactions with their environment. This is done both on a microscopic and molecular level as it encompasses prokaryotic cells and eukaryotic cells. Knowing the components of cells and how cells work is fundamental to all biological sciences; it is also essential for research in bio-medical fields such as cancer, and other diseases. Research in cell biology is closely related to genetics, biochemistry, molecular biology, immunology, and developmental biology.

Internal Cellular Structures

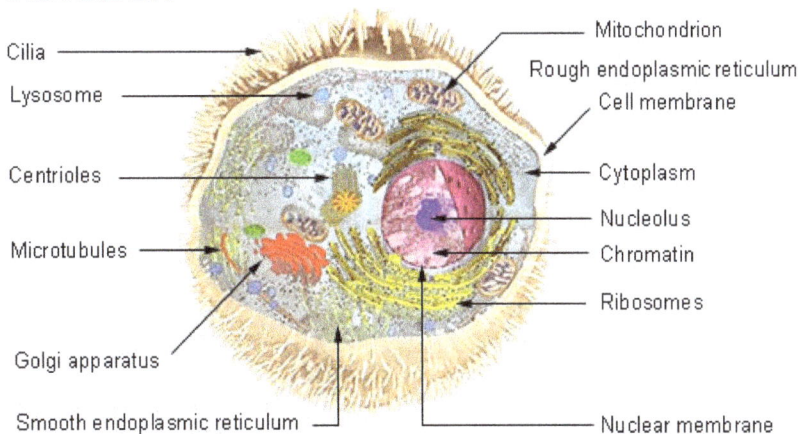

The generalized structure and molecular components of a cell

Chemical and Molecular Environment

The study of the cell is done on a molecular level; however, most of the processes within the cell are made up of a mixture of small organic molecules, inorganic ions, hormones, and water. Approximately 75-85% of the cell's volume is due to water making it an indispensable solvent as a result of its polarity and structure. These molecules within the cell, which operate as substrates, provide a suitable environment for the cell to carry out metabolic reactions and signalling. The cell shape varies among the different types of organisms, and are thus then classified into two categories: eukaryotes and prokaryotes. In the case of eukaryotic cells - which are made up of animal, plant, fungi, and protozoa cells - the shapes are generally round and spherical, while for prokaryotic cells – which are composed of bacteria and archaea - the shapes are: spherical (cocci), rods (bacillus), curved (vibrio), and spirals (*spirochetes)*.

Cell biology focuses more on the study of eukaryotic cells, and their signalling pathways, rather than on prokaryotes which is covered under microbiology. The main constituents of the general molecular composition of the cell includes: proteins and lipids which are either free flowing or membrane bound, along with different internal compartments known as organelles. This environment of the cell is made up of hydrophilic and hydrophobic regions which allows for the exchange of the above-mentioned molecules and ions. The hydrophilic regions of the cell are mainly on the inside and outside of the cell, while the hydrophobic regions are within the phospholipid bilayer of the cell membrane. The cell membrane consists of lipids and proteins which accounts for its hydrophobicity as a result of being non-polar substances. Therefore, in order for these molecules to participate in reactions, within the cell, they need to be able to cross this membrane layer to get into the cell. They accomplish this process of gaining access to the cell via: osmotic pressure, diffusion, concentration gradients, and membrane channels. Inside of the cell are extensive internal sub-cellular membrane-bounded compartments called organelles.

Organelles

- Centrosome - an associated pair of cylindrical shaped protein structures (centrioles) that organize microtubules and aid in forming the mitotic spindle during cell division in eukaryotes

- Cell membrane (plasma membrane) - the part of the cell which separates the cells from the outside environment and protects the cell, as well as regulating what goes in and out of the cell

- Cell wall - extra layer of protection and gives structural support (only found in plant cells)

- Chloroplast - key organelle for photosynthesis (only found in plant cells)

- Cilium - motile structure of eukaryotes having a cytoskeleton, the axoneme.

- Cytoplasm - contents of the main fluid-filled space inside cells, chemical reactions also happen in this jelly-like substance.

- Cytoskeleton - protein filaments inside cells (microfilaments, microtubules, and intermediate filaments)

- Endoplasmic reticulum (rough) - major site of membrane protein synthesis

- Endoplasmic reticulum (smooth) - major site of lipid synthesis

- Endosomes - vesicles that traffic membrane and intra and extra cellular contents for recycling or degradation by lysosomes

- Flagellum - motile structure of bacteria, archaea and eukaryotes

- Golgi apparatus - site of protein glycosylation in the endomembrane system

- Lipid bilayer - fundamental organizational structure of cell membranes

- Lysosome - acidic organelle that breaks down cellular waste products and debris into simple compounds (only found in animal cells)

- Microvilli - increases surface area for absorption of nutrients from surrounding medium

- Mitochondrion - major energy-producing organelle by releasing energy in the form of ATP

- Nucleus - contains chromosomes composed of DNA, the building block of life. Nuclear Architecture is important for dictating nuclear function.

- Organelle - term used for major subcellular structures

- Peroxisomes - a very small organelle that uses oxygen to breakdown and detoxify long fatty acids and other molecules

- Pili - also called fimbria is used for conjugation and sometimes movement

- Ribosome - RNA and protein complex required for protein synthesis in cells

- Starch grain - found in the cytoplasm of a typical plant cell, it stores chemical energy of the plant.

- Vacuole - contain cell sap or other storage material.

- Vesicle - small membrane-bounded spheres inside cells which transport substances.

cell surface membrane protects the cell

Processes

Growth and Development

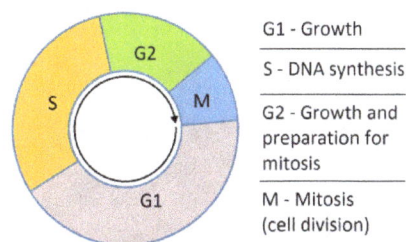

General concept of the cell cycle.

The growth process of the cell does not refer to the size of the cell, but instead the density of the number of cells present in the organism at a given time. Cell growth pertains to the increase in the number of cells present in an organism as it grows and develops; as the organism gets larger so too does the number of cells present. Cells are the foundation of all organisms, they are the fundamental unit of life. The growth and development of the cell are essential for the maintenance of the host, and survival of the organisms. For this process the cell goes through the steps of the cell cycle and development which involves cell growth, DNA replication, cell division, regeneration, specialization, and cell death. The cell cycle is divided into four distinct phases, G1, S, G2, and M. The G phases – which is the cell growth phase - makes up approximately 95% of the cycle. The proliferation of cells is instigated by progenitors, the cells then differentiate to become specialized, where specialized cells of the same type aggregate to form tissues, then organs and ultimately systems. The G phases along with the S phase – DNA replication, damage and repair - are considered to be the interphase portion of the cycle. While the M phase (mitosis and cytokinesis) is the cell division portion of the cycle. The cell cycle is regulated by a series of signalling factors and complexes such as CDK's, kinases, and p53. to name a few. When the cell has completed its growth process, and if it is found to be damaged or altered it undergoes cell death, either by apoptosis or necrosis, to eliminate the threat it cause to the organism's survival.

Other Cellular Processes

- Active transport and Passive transport - Movement of molecules into and out of cells.

- Autophagy - The process whereby cells "eat" their own internal components or microbial invaders.

- Adhesion - Holding together cells and tissues.

- Cell movement - Chemotaxis, contraction, cilia and flagella.

- Cell signaling - Regulation of cell behavior by signals from outside.

- Division - By which cells reproduce either by mitosis (to produce clones of the parent cell) or Meiosis (to produce haploid gametes)

- DNA repair - Cell death and cell senescence.

- Metabolism - Glycolysis, respiration, photosynthesis, and chemosynthesis.

- Signalling - The process by which the activities in the cell are regulated

- Transcription and mRNA splicing - Gene expression.

Techniques Used to Study Cells

Electron micrograph of blood cells clotting.

Cells may be observed under the microscope, using several different techniques; these include optical microscopy, transmission electron microscopy, scanning electron microscopy, fluorescence microscopy, and confocal microscopy.

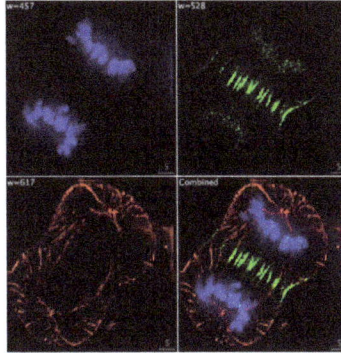

Cell division studied using fluorescence to stain specific structures

There are several different methods used in the study of cells:

- Cell culture is the basic technique of growing cells in a laboratory independent of an organism.

- Immunostaining, also known as immunohistochemistry, is a specialized histological method used to localize proteins in cells or tissue slices. Unlike regular histology, which uses stains to identify cells, cellular components or protein classes, immunostaining requires the reaction of an antibody directed against the protein of interest within the tissue or cell. Through the use of proper controls and published protocols (need to add reference links here), specificity of the antibody-antigen reaction can be achieved. Once this complex is formed, it is identified via either a "tag" attached directly to the antibody, or added in an additional technical step. Commonly used "tags" include fluorophores or enzymes. In the case of the former, detection of the location of the "immuno-stained" protein occurs via fluorescence microscopy. With an enzymatic tag, such as horse radish peroxidase, a chemical reaction is carried out that results in a dark color in the location of the protein of interest. This darkened pattern is then detected using light microscopy.

- Computational genomics is used to find patterns in genomic information

- DNA microarrays identify changes in transcript levels between different experimental conditions.

- Gene knockdown mutates a selected gene.

- In situ hybridization shows which cells are expressing a particular RNA transcript.

- PCR can be used to determine how many copies of a gene are present in a cell.

- Transfection introduces a new gene into a cell, usually an expression construct

Purification of cells and their parts Purification may be performed using the following methods:

- Cell fractionation

    o   Release of cellular organelles by disruption of cells.

    o   Separation of different organelles by centrifugation.

- Flow cytometry

- Immunoprecipitation

    o   The binding of an antibody to a target protein

    o   Collection of the target protein through elution

- Proteins extracted from cell membranes by detergents and salts or other kinds of chemicals.

Cell (Biology)

The cell is the basic structural, functional, and biological unit of all known living organisms. A cell is the smallest unit of life that can replicate independently, and cells are often called the "building blocks of life". The study of cells is called cell biology.

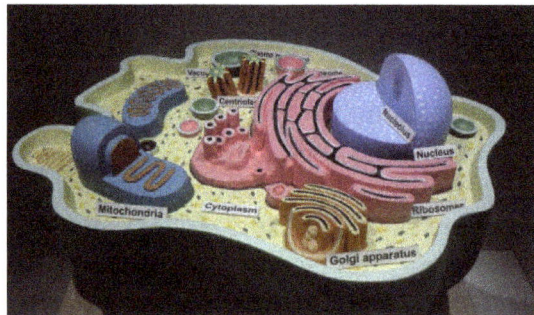

Structure of an animal cell

Cells consist of cytoplasm enclosed within a membrane, which contains many biomolecules such as proteins and nucleic acids. Organisms can be classified as unicellular (consisting of a single cell; including bacteria) or multicellular (including plants and animals). While the number of cells in plants and animals varies from species to species, humans contain more than 10 trillion (10^{12}) cells. Most plant and animal cells are visible only under a microscope, with dimensions between 1 and 100 micrometres.

The cell was discovered by Robert Hooke in 1665, who named the biological unit for its resemblance to cells inhabited by Christian monks in a monastery. Cell theory, first developed in 1839 by Matthias Jakob Schleiden and Theodor Schwann, states that all organisms are composed of one or more cells, that cells are the fundamental unit of structure and function in all living organisms, that all cells come from preexisting cells, and that all cells contain the hereditary information necessary for regulating cell functions and for transmitting information to the next generation of cells. Cells emerged on Earth at least 3.5 billion years ago.

Anatomy

Comparison of features of prokaryotic and eukaryotic cells		
Prokaryotes **Eukaryotes**		
Typical organisms	bacteria, archaea	protists, fungi, plants, animals
Typical size	~ 1–5 μm	~ 10–100 μm
Type of nucleus	nucleoid region; no true nucleus	true nucleus with double membrane
DNA	circular (usually)	linear molecules (chromosomes) with histone proteins
RNA/protein synthesis	coupled in the cytoplasm	RNA synthesis in the nucleus protein synthesis in the cytoplasm
Ribosomes	50S and 30S	60S and 40S
Cytoplasmic structure	very few structures	highly structured by endomembranes and a cytoskeleton
Cell movement	flagella made of flagellin	flagella and cilia containing microtubules; lamellipodia and filopodia containing actin
Mitochondria	none	one to several thousand
Chloroplasts	none	in algae and plants
Organization	usually single cells	single cells, colonies, higher multicellular organisms with specialized cells
Cell division	binary fission (simple division)	mitosis (fission or budding) meiosis
Chromosomes	single chromosome	more than one chromosome
Membranes	cell membrane	Cell membrane and membrane-bound organelles

Cells are of two types, eukaryotic, which contain a nucleus, and prokaryotic, which do not. Prokaryotes are single-celled organisms, while eukaryotes can be either single-celled or multicellular.

Prokaryotic Cells

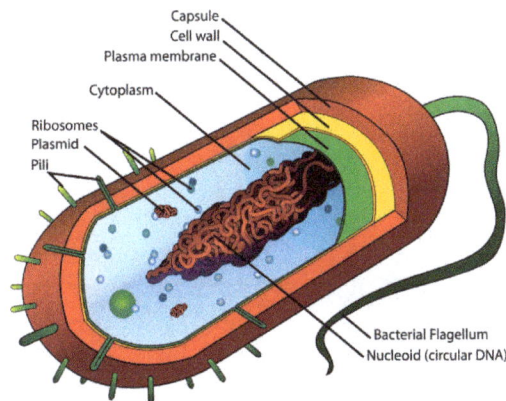

Structure of a typical prokaryotic cell

Prokaryotic cells were the first form of life on Earth, characterised by having vital biological processes including cell signaling and being self-sustaining. They are simpler and smaller than eukaryotic cells, and lack membrane-bound organelles such as the nucleus. Prokaryotes include two of the domains of life, bacteria and archaea. The DNA of a prokaryotic cell consists of a single chromosome that is in direct contact with the cytoplasm. The nuclear region in the cytoplasm is called the nucleoid. Most prokaryotes are the smallest of all organisms ranging from 0.5 to 2.0 μm in diameter.

A Prokaryotic Cell has Three Architectural Regions:

- Enclosing the cell is the cell envelope – generally consisting of a plasma membrane covered by a cell wall which, for some bacteria, may be further covered by a third layer called a capsule. Though most prokaryotes have both a cell membrane and a cell wall, there are exceptions such as *Mycoplasma* (bacteria) and *Thermoplasma* (archaea) which only possess the cell membrane layer. The envelope gives rigidity to the cell and separates the interior of the cell from its environment, serving as a protective filter. The cell wall consists of peptidoglycan in bacteria, and acts as an additional barrier against exterior forces. It also prevents the cell from expanding and bursting (cytolysis) from osmotic pressure due to a hypotonic environment. Some eukaryotic cells (plant cells and fungal cells) also have a cell wall.

- Inside the cell is the cytoplasmic region that contains the genome (DNA), ribosomes and various sorts of inclusions. The genetic material is freely found in the cytoplasm. Prokaryotes can carry extrachromosomal DNA elements called plasmids, which are usually circular. Linear bacterial plasmids have been identified in several species of spirochete bacteria, including members of the genus Borrelia notably *Borrelia burgdorferi*, which causes Lyme disease. Though not forming a *nucleus*, the DNA is condensed in a *nucleoid*. Plasmids encode additional genes, such as antibiotic resistance genes.

- On the outside, flagella and pili project from the cell's surface. These are structures (not present in all prokaryotes) made of proteins that facilitate movement and communication between cells.

Eukaryotic Cells

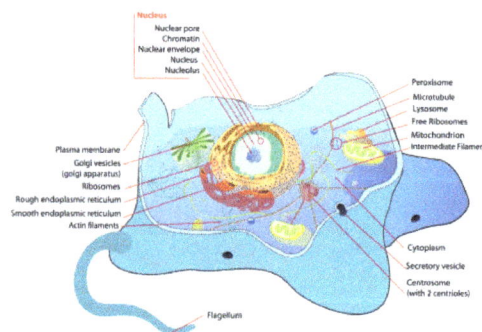

Structure of a typical animal cell

Plants, animals, fungi, slime moulds, protozoa, and algae are all eukaryotic. These cells are about fifteen times wider than a typical prokaryote and can be as much as a thousand times greater in volume. The main distinguishing feature of eukaryotes as compared to prokaryotes is compartmentalization: the presence of membrane-bound organelles (compartments) in which specific metabolic activities take place. Most important among these is a cell nucleus, an organelle that houses the cell's DNA. This nucleus gives the eukaryote its name, which means "true kernel (nucleus)". Other differences include:

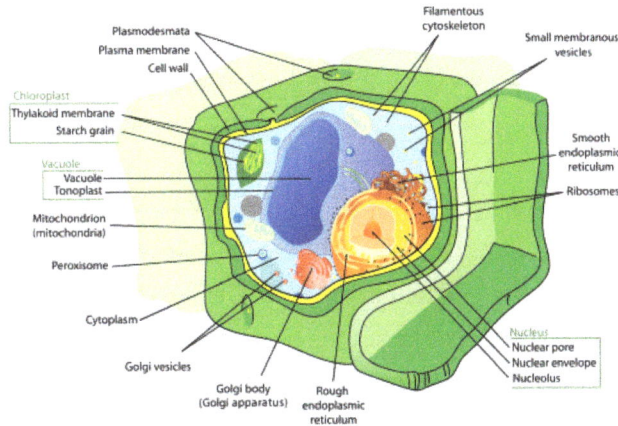

Structure of a typical plant cell

- The plasma membrane resembles that of prokaryotes in function, with minor differences in the setup. Cell walls may or may not be present.

- The eukaryotic DNA is organized in one or more linear molecules, called chromosomes, which are associated with histone proteins. All chromosomal DNA is stored in the *cell nucleus*, separated from the cytoplasm by a membrane. Some eukaryotic organelles such as mitochondria also contain some DNA.

- Many eukaryotic cells are ciliated with *primary cilia*. Primary cilia play important roles in chemosensation, mechanosensation, and thermosensation. Cilia may thus be "viewed as a sensory cellular antennae that coordinates a large number of cellular signaling pathways, sometimes coupling the signaling to ciliary motility or alternatively to cell division and differentiation."

- Motile cells of eukaryotes can move using *motile cilia* or *flagella*. Motile cells are absent in conifers and flowering plants. Eukaryotic flagella are less complex than those of prokaryotes.

Subcellular Components

All cells, whether prokaryotic or eukaryotic, have a membrane that envelops the cell, regulates what moves in and out (selectively permeable), and maintains the electric potential of the cell. Inside the membrane, the cytoplasm takes up most of the cell's volume. All cells (except red blood cells which lack a cell nucleus and most organelles to accommodate maximum space for hemoglobin) possess DNA, the hereditary material of genes, and RNA, containing the information necessary to build various proteins such as enzymes, the cell's primary machinery. There are also other kinds of biomolecules in cells.

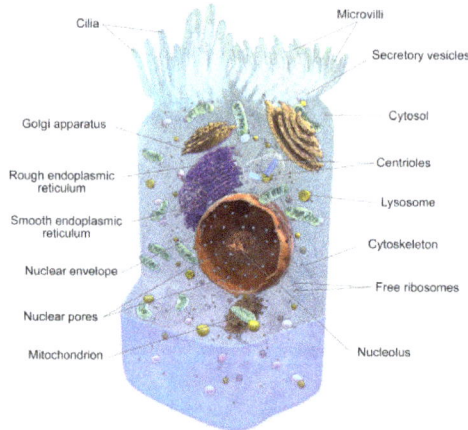

Anatomy of a Cell

Illustration depicting major structures inside a eukaryotic animal cell

Membrane

The cell membrane, or plasma membrane, is a biological membrane that surrounds the cytoplasm of a cell. In animals, the plasma membrane is the outer boundary of the cell, while in plants and prokaryotes it is usually covered by a cell wall. This membrane serves to separate and protect a cell from its surrounding environment and is made mostly from a double layer of phospholipids, which are amphiphilic (partly hydrophobic and partly hydrophilic). Hence, the layer is called a phospholipid bilayer, or sometimes a fluid mosaic membrane. Embedded within this membrane is a variety of protein molecules that act as channels and pumps that move different molecules into and out of the cell. The membrane is said to be 'semi-permeable', in that it can either let a substance (molecule or ion) pass through freely, pass through to a limited extent or not pass through at all. Cell surface membranes also contain receptor proteins that allow cells to detect external signaling molecules such as hormones.

Cytoskeleton

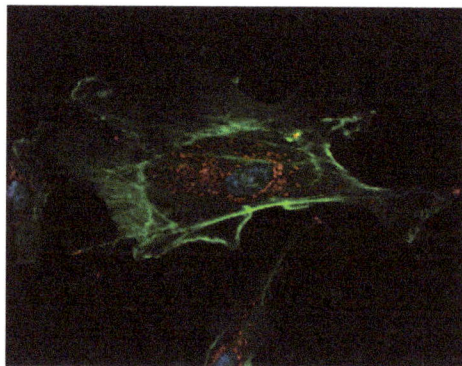

A fluorescent image of an endothelial cell. Nuclei are stained blue, mitochondria are stained red, and microfilaments are stained green.

The cytoskeleton acts to organize and maintain the cell's shape; anchors organelles in place; helps during endocytosis, the uptake of external materials by a cell, and cytokinesis, the separation of daughter cells after cell division; and moves parts of the cell in processes of growth and mobility. The eukaryotic cytoskeleton is composed of microfilaments, intermediate filaments and microtubules. There are a great number of proteins associated with them, each controlling a cell's structure by directing, bundling, and aligning filaments. The prokaryotic cytoskeleton is less well-studied but is involved in the maintenance of cell shape, polarity and cytokinesis. The subunit protein of microfilaments is a small, monomeric protein called actin. The subunit of microtubules is a dimeric molecule called tubulin. Intermediate filaments are heteropolymers whose subunits vary among the cell types in different tissues. But some of the subunit protein of intermediate filaments include vimentin, desmin, lamin (lamins A, B and C), keratin (multiple acidic and basic keratins), neurofilament proteins (NF - L, NF - M).

Genetic Material

Two different kinds of genetic material exist: deoxyribonucleic acid (DNA) and ribonucleic acid (RNA). Cells use DNA for their long-term information storage. The biological information contained in an organism is encoded in its DNA sequence. RNA is used for information transport (e.g., mRNA) and enzymatic functions (e.g., ribosomal RNA). Transfer RNA (tRNA) molecules are used to add amino acids during protein translation.

Prokaryotic genetic material is organized in a simple circular DNA molecule (the bacterial chromosome) in the nucleoid region of the cytoplasm. Eukaryotic genetic material is divided into different, linear molecules called chromosomes inside a discrete nucleus, usually with additional ge-netic material in some organelles like mitochondria and chloroplasts.

A human cell has genetic material contained in the cell nucleus (the nuclear genome) and in the mitochondria (the mitochondrial genome). In humans the nuclear genome is divided into 46 linear DNA molecules called chromosomes, including 22 homologous chromosome pairs and a pair of sex chromosomes. The mitochondrial genome is a circular DNA molecule distinct from the nuclear DNA. Although the mitochondrial DNA is very small compared to nuclear chromosomes, it codes for 13 proteins involved in mitochondrial energy production and specific tRNAs.

Foreign genetic material (most commonly DNA) can also be artificially introduced into the cell by a process called transfection. This can be transient, if the DNA is not inserted into the cell's genome, or stable, if it is. Certain viruses also insert their genetic material into the genome.

Organelles

Organelles are parts of the cell which are adapted and/or specialized for carrying out one or more vital functions, analogous to the organs of the human body (such as the heart, lung, and kidney, with each organ performing a different function). Both eukaryotic and prokaryotic cells have organelles, but prokaryotic organelles are generally simpler and are not membrane-bound.

There are several types of organelles in a cell. Some (such as the nucleus and golgi apparatus) are typically solitary, while others (such as mitochondria, chloroplasts, peroxisomes and lysosomes) can be numerous (hundreds to thousands). The cytosol is the gelatinous fluid that fills the cell and surrounds the organelles.

Eukaryotic

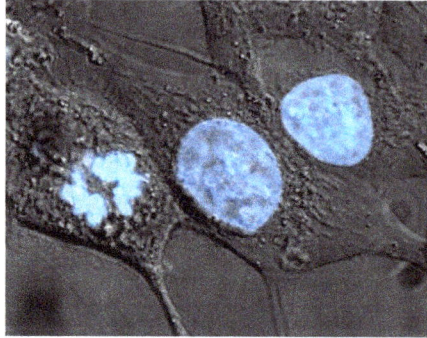

Human cancer cells with nuclei (specifically the DNA) stained blue. The central and rightmost cell are in interphase, so the entire nuclei are labeled. The cell on the left is going through mitosis and its DNA has condensed.

- Cell nucleus: A cell's information center, the cell nucleus is the most conspicuous organelle found in a eukaryotic cell. It houses the cell's chromosomes, and is the place where almost all DNA replication and RNA synthesis (transcription) occur. The nucleus is spherical and separated from the cytoplasm by a double membrane called the nuclear envelope. The nuclear envelope isolates and protects a cell's DNA from various molecules that could accidentally damage its structure or interfere with its processing. During processing, DNA is transcribed, or copied into a special RNA, called messenger RNA (mRNA). This mRNA is then transported out of the nucleus, where it is translated into a specific protein molecule. The nucleolus is a specialized region within the nucleus where ribosome subunits are assembled. In prokaryotes, DNA processing takes place in the cytoplasm.

- Mitochondria and Chloroplasts: generate energy for the cell. Mitochondria are self-replicating organelles that occur in various numbers, shapes, and sizes in the cytoplasm of all eukaryotic cells. Respiration occurs in the cell mitochondria, which generate the cell's energy by oxidative phosphorylation, using oxygen to release energy stored in cellular nutrients (typically pertaining to glucose) to generate ATP. Mitochondria multiply by binary fission, like prokaryotes. Chloroplasts can only be found in plants and algae, and they capture the sun's energy to make carbohydrates through photosynthesis.

Diagram of an endomembrane system

- Endoplasmic reticulum: The endoplasmic reticulum (ER) is a transport network for molecules targeted for certain modifications and specific destinations, as compared to molecules that float freely in the cytoplasm. The ER has two forms: the rough ER, which has ribosomes on its surface that secrete proteins into the ER, and the smooth ER, which lacks ribosomes. The smooth ER plays a role in calcium sequestration and release.

- Golgi apparatus: The primary function of the Golgi apparatus is to process and package the macromolecules such as proteins and lipids that are synthesized by the cell.

- Lysosomes and Peroxisomes: Lysosomes contain digestive enzymes (acid hydrolases). They digest excess or worn-out organelles, food particles, and engulfed viruses or bacteria. Peroxisomes have enzymes that rid the cell of toxic peroxides. The cell could not house these destructive enzymes if they were not contained in a membrane-bound system.

- Centrosome: the cytoskeleton organiser: The centrosome produces the microtubules of a cell – a key component of the cytoskeleton. It directs the transport through the ER and the Golgi apparatus. Centrosomes are composed of two centrioles, which separate during cell division and help in the formation of the mitotic spindle. A single centrosome is present in the animal cells. They are also found in some fungi and algae cells.

- Vacuoles: Vacuoles sequester waste products and in plant cells store water. They are often described as liquid filled space and are surrounded by a membrane. Some cells, most notably *Amoeba*, have contractile vacuoles, which can pump water out of the cell if there is too much water. The vacuoles of plant cells and fungal cells are usually larger than those of animal cells.

Eukaryotic and Prokaryotic

- Ribosomes: The ribosome is a large complex of RNA and protein molecules. They each consist of two subunits, and act as an assembly line where RNA from the nucleus is used to synthesise proteins from amino acids. Ribosomes can be found either floating freely or bound to a membrane (the rough endoplasmatic reticulum in eukaryotes, or the cell membrane in prokaryotes).

Structures Outside The Cell Membrane

Many cells also have structures which exist wholly or partially outside the cell membrane. These structures are notable because they are not protected from the external environment by the semipermeable cell membrane. In order to assemble these structures, their components must be carried across the cell membrane by export processes.

Cell Wall

Many types of prokaryotic and eukaryotic cells have a cell wall. The cell wall acts to protect the cell mechanically and chemically from its environment, and is an additional layer of protection to the cell membrane. Different types of cell have cell walls made up of different materials; plant cell walls are primarily made up of cellulose, fungi cell walls are made up of chitin and bacteria cell walls are made up of peptidoglycan.

Prokaryotic

Capsule

A gelatinous capsule is present in some bacteria outside the cell membrane and cell wall. The capsule may be polysaccharide as in pneumococci, meningococci or polypeptide as *Bacillus anthracis* or hyaluronic acid as in streptococci. Capsules are not marked by normal staining protocols and can be detected by India ink or methyl blue; which allows for higher contrast between the cells for observation.

Flagella

Flagella are organelles for cellular mobility. The bacterial flagellum stretches from cytoplasm through the cell membrane(s) and extrudes through the cell wall. They are long and thick thread-like appendages, protein in nature. A different type of flagellum is found in archaea and a different type is found in eukaryotes.

Fimbria

A fimbria also known as a pilus is a short, thin, hair-like filament found on the surface of bacteria. Fimbriae, or pili are formed of a protein called pilin (antigenic) and are responsible for attachment of bacteria to specific receptors of human cell (cell adhesion). There are special types of specific pili involved in bacterial conjugation.

Cellular Processes

Growth and Metabolism

Between successive cell divisions, cells grow through the functioning of cellular metabolism. Cell metabolism is the process by which individual cells process nutrient molecules. Metabolism has two distinct divisions: catabolism, in which the cell breaks down complex molecules to produce energy and reducing power, and anabolism, in which the cell uses energy and reducing power to construct complex molecules and perform other biological functions. Complex sugars consumed by the organism can be broken down into simpler sugar molecules called monosaccharides such as glucose. Once inside the cell, glucose is broken down to make adenosine triphosphate (ATP), a molecule that possesses readily available energy, through two different pathways.

Replication

Cell division involves a single cell (called a *mother cell*) dividing into two daughter cells. This leads to growth in multicellular organisms (the growth of tissue) and to procreation (vegetative reproduction) in unicellular organisms. Prokaryotic cells divide by binary fission, while eukaryotic cells usually undergo a process of nuclear division, called mitosis, followed by division of the cell, called cytokinesis. A diploid cell may also undergo meiosis to produce haploid cells, usually four. Haploid cells serve as gametes in multicellular organisms, fusing to form new diploid cells.

DNA replication, or the process of duplicating a cell's genome, always happens when a cell divides through mitosis or binary fission. This occurs during the S phase of the cell cycle.

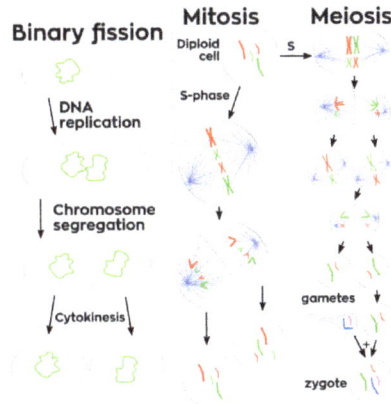

Bacteria divide by binary fission, while eukaryotes divide by mitosis or meiosis.

In meiosis, the DNA is replicated only once, while the cell divides twice. DNA replication only occurs before meiosis I. DNA replication does not occur when the cells divide the second time, in meiosis II. Replication, like all cellular activities, requires specialized proteins for carrying out the job.

Protein Synthesis

An overview of protein synthesis. Within the nucleus of the cell (*light blue*), genes (DNA, *dark blue*) are transcribed into RNA. This RNA is then subject to post-transcriptional modification and control, resulting in a mature mRNA (*red*) that is then transported out of the nucleus and into the cytoplasm (*peach*), where it undergoes translation into a protein. mRNA is translated by ribosomes (*purple*) that match the three-base codons of the mRNA to the three-base anti-codons of the appropriate tRNA. Newly synthesized proteins (*black*) are often further modified, such as by binding to an effector molecule (*orange*), to become fully active.

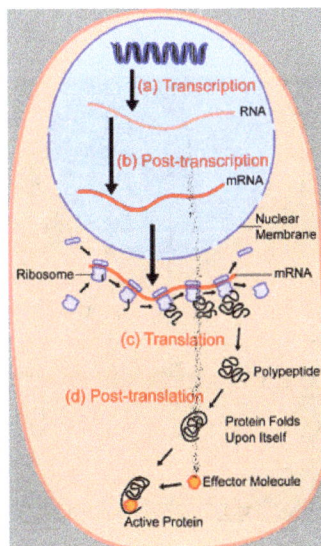

Cells are capable of synthesizing new proteins, which are essential for the modulation and maintenance of cellular activities. This process involves the formation of new protein molecules from amino acid building blocks based on information encoded in DNA/RNA. Protein synthesis generally consists of two major steps: transcription and translation.

Transcription is the process where genetic information in DNA is used to produce a complementary RNA strand. This RNA strand is then processed to give messenger RNA (mRNA), which is free to migrate through the cell. mRNA molecules bind to protein-RNA complexes called ribosomes located in the cytosol, where they are translated into polypeptide sequences. The ribosome mediates the formation of a polypeptide sequence based on the mRNA sequence. The mRNA sequence directly relates to the polypeptide sequence by binding to transfer RNA (tRNA) adapter molecules in binding pockets within the ribosome. The new polypeptide then folds into a functional three-dimensional protein molecule.

Movement or Motility

Unicellular organisms can move in order to find food or escape predators. Common mechanisms of motion include flagella and cilia.

In multicellular organisms, cells can move during processes such as wound healing, the immune response and cancer metastasis. For example, in wound healing in animals, white blood cells move to the wound site to kill the microorganisms that cause infection. Cell motility involves many receptors, crosslinking, bundling, binding, adhesion, motor and other proteins. The process is divided into three steps – protrusion of the leading edge of the cell, adhesion of the leading edge and de-adhesion at the cell body and rear, and cytoskeletal contraction to pull the cell forward. Each step is driven by physical forces generated by unique segments of the cytoskeleton.

Multicellularity

Cell Specialization

Multicellular organisms are organisms that consist of more than one cell, in contrast to single-celled organisms.

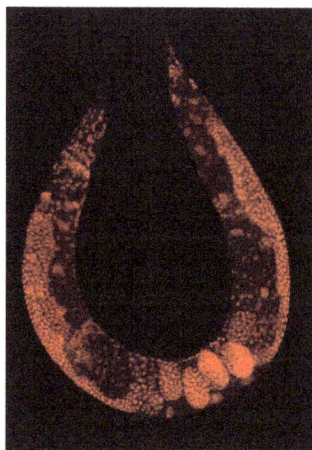

Staining of a *Caenorhabditis elegans* which highlights the nuclei of its cells.

In complex multicellular organisms, cells specialize into different cell types that are adapted to particular functions. In mammals, major cell types include skin cells, muscle cells, neurons, blood cells, fibroblasts, stem cells, and others. Cell types differ both in appearance and function, yet are genetically identical. Cells are able to be of the same genotype but of different cell type due to the differential expression of the genes they contain.

Most distinct cell types arise from a single totipotent cell, called a zygote, that differentiates into hundreds of different cell types during the course of development. Differentiation of cells is driven by different environmental cues (such as cell–cell interaction) and intrinsic differences (such as those caused by the uneven distribution of molecules during division).

Origin of Multicellularity

Multicellularity has evolved independently at least 25 times, including in some prokaryotes, like cyanobacteria, myxobacteria, actinomycetes, *Magnetoglobus multicellularis* or *Methanosarcina*. However, complex multicellular organisms evolved only in six eukaryotic groups: animals, fungi, brown algae, red algae, green algae, and plants. It evolved repeatedly for plants (Chloroplastida), once or twice for animals, once for brown algae, and perhaps several times for fungi, slime molds, and red algae. Multicellularity may have evolved from colonies of interdependent organisms, from cellularization, or from organisms in symbiotic relationships.

The first evidence of multicellularity is from cyanobacteria-like organisms that lived between 3 and 3.5 billion years ago. Other early fossils of multicellular organisms include the contested Grypania spiralis and the fossils of the black shales of the Palaeoproterozoic Francevillian Group Fossil B Formation in Gabon.

The evolution of multicellularity from unicellular ancestors has been replicated in the laboratory, in evolution experiments using predation as the selective pressure.

Origins

The origin of cells has to do with the origin of life, which began the history of life on Earth.

Origin of The First Cell

Stromatolites are left behind by cyanobacteria, also called blue-green algae. They are the oldest known fossils of life on Earth. This one-billion-year-old fossil is from Glacier National Park in the United States.

There are several theories about the origin of small molecules that led to life on the early Earth. They may have been carried to Earth on meteorites, created at deep-sea vents, or synthesized by lightning in a reducing atmosphere. There is little experimental data defining what the first self-replicating forms were.

RNA is thought to be the earliest self-replicating molecule, as it is capable of both storing genetic information and cata-lyzing chemical reactions, but some other entity with the potential to self-replicate could have preceded RNA, such as clay or peptide nucleic acid.

Cells emerged at least 3.5 billion years ago. The current belief is that these cells were heterotrophs. The early cell membranes were probably more simple and permeable than modern ones, with only a single fatty acid chain per lipid. Lipids are known to spontaneously form bilayered vesicles in water, and could have preceded RNA, but the first cell membranes could also have been produced by catalytic RNA, or even have required structural proteins before they could form.

Origin of Eukaryotic Cells

The eukaryotic cell seems to have evolved from a symbiotic community of prokaryotic cells. DNA-bearing organelles like the mitochondria and the chloroplasts are descended from ancient symbiotic oxygen-breathing proteobacteria and cyanobacteria, respectively, which were endosymbiosed by an ancestral archaean prokaryote.

There is still considerable debate about whether organelles like the hydrogenosome predated the origin of mitochondria, or vice versa.

History of Research

- 1632–1723: Antonie van Leeuwenhoek teaches himself to make lenses, constructs basic optical microscopes and draws protozoa, such as *Vorticella* from rain water, and bacteria from his own mouth.

- 1665: Robert Hooke discovers cells in cork, then in living plant tissue using an early compound microscope. He coins the term *cell* (from Latin *cella*, meaning "small room") in his book *Micrographia* (1665).

- 1839: Theodor Schwann and Matthias Jakob Schleiden elucidate the principle that plants and animals are made of cells, concluding that cells are a common unit of structure and development, and thus founding the cell theory.

- 1855: Rudolf Virchow states that new cells come from pre-existing cells by cell division (*omnis cellula ex cellula*).

- 1859: The belief that life forms can occur spontaneously (*generatio spontanea*) is contradicted by Louis Pasteur (1822–1895) (although Francesco Redi had performed an experiment in 1668 that suggested the same conclusion).

- 1931: Ernst Ruska builds the first transmission electron microscope (TEM) at the University of Berlin. By 1935, he has built an EM with twice the resolution of a light microscope, revealing previously unresolvable organelles.

- 1953: Watson and Crick made their first announcement on the double helix structure of DNA on February 28.

- 1981: Lynn Margulis published *Symbiosis in Cell Evolution* detailing the endosymbiotic theory.

Cell Membrane

The cell membrane (also known as the plasma membrane or cytoplasmic membrane) is a biological membrane that separates the interior of all cells from the outside environment. The cell membrane is selectively permeable to ions and organic molecules and controls the movement of substances in and out of cells. The basic function of the cell membrane is to protect the cell from its surroundings.

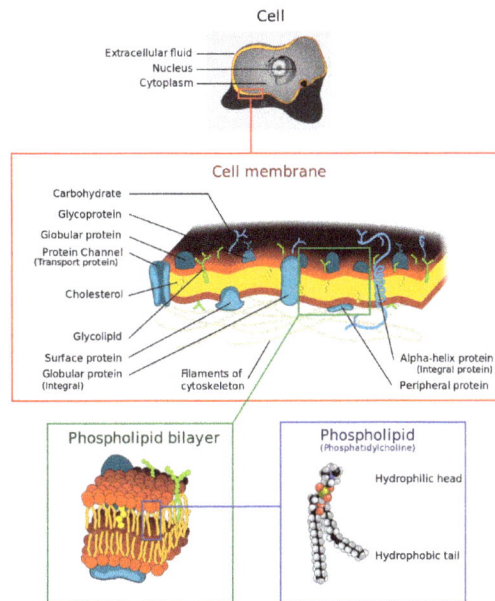

Illustration of a Eukaryotic cell membrane

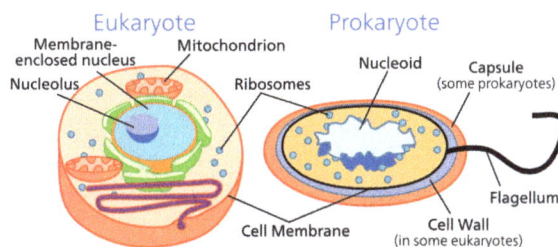

Comparison of Eukaryotes vs. Prokaryotes

It consists of the phospholipid bilayer with embedded proteins. Cell membranes are involved in a variety of cellular processes such as cell adhesion, ion conductivity and cell signalling and serve as the attachment surface for several extracellular structures, including the cell wall, glycocalyx, and intracellular cytoskeleton. Cell membranes can be artificially reassembled.

History

The structure has been variously referred to by different writers as the ectoplast (de Vries, 1885), *Plasmahaut* (plasma skin, Pfeffer, 1877, 1891), *Hautschicht* (skin layer, Pfeffer, 1886; used with a different meaning by Hofmeister, 1867), plasmatic membrane (Pfeffer, 1900), plasma membrane, cytoplasmic membrane, cell envelope and cell membrane.

Some authors that did not believe that there was a functional permeable boundary at the surface of the cell preferred to use the term plasmalemma (coined by Mast, 1924) to the extern region of the cell.

Function

The cell membrane (or plasma membrane or plasmalemma) surrounds the cytoplasm of living cells, physically separating the intracellular components from the extracellular environment. Fungi, bacteria and plants have a cell wall in addition, which provides a mechanical support to the cell and precludes the passage of larger molecules. The cell membrane also plays a role in anchoring the cytoskeleton to provide shape to the cell, and in attaching to the extracellular matrix and other cells to hold them together to form tissues.

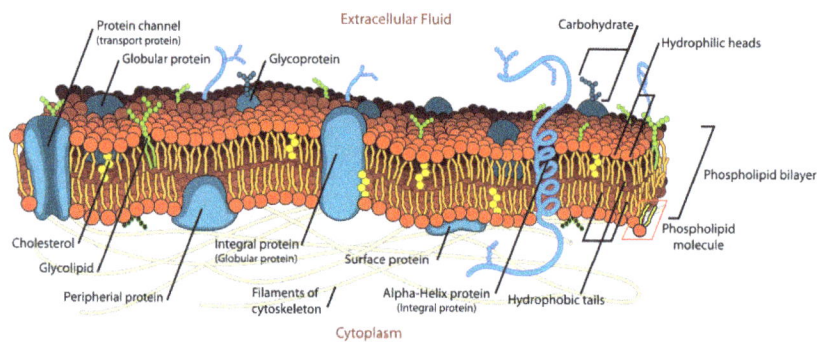

A detailed diagram of the cell membrane

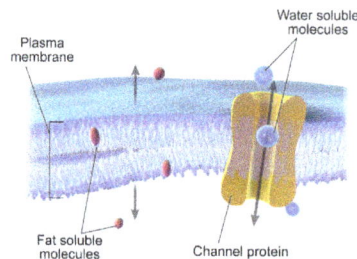

Diffusion Across the Plasma Membrane

Illustration depicting cellular diffusion

The cell membrane is selectively permeable and able to regulate what enters and exits the cell, thus facilitating the transport of materials needed for survival. The movement of substances across the membrane can be either "passive", occurring without the input of cellular energy, or "active", requiring the cell to expend energy in transporting it. The membrane also maintains the cell potential. The cell membrane thus works as a selective filter that allows only certain things to come inside or go outside the cell. The cell employs a number of transport mechanisms that involve biological membranes:

1. Passive osmosis and diffusion: Some substances (small molecules, ions) such as carbon dioxide (CO_2) and oxygen (O_2), can move across the plasma membrane by diffusion, which is a passive transport process. Because the membrane acts as a barrier for certain molecules and ions, they can

occur in different concentrations on the two sides of the membrane. Such a concentration gradient across a semipermeable membrane sets up an osmotic flow for the water.

2. Transmembrane protein channels and transporters: Nutrients, such as sugars or amino acids, must enter the cell, and certain products of metabolism must leave the cell. Such molecules diffuse passively through protein channels such as aquaporins (in the case of water (H_2O)) in facilitated diffusion or are pumped across the membrane by transmembrane transporters. Protein channel proteins, also called *permeases*, are usually quite specific, recognizing and transporting only a limited food group of chemical substances, often even only a single substance.

3. Endocytosis: Endocytosis is the process in which cells absorb molecules by engulfing them. The plasma membrane creates a small deformation inward, called an invagination, in which the substance to be transported is captured. The deformation then pinches off from the membrane on the inside of the cell, creating a vesicle containing the captured substance. Endocytosis is a pathway for internalizing solid particles ("cell eating" or phagocytosis), small molecules and ions ("cell drinking" or pinocytosis), and macromolecules. Endocytosis requires energy and is thus a form of active transport.

4. Exocytosis: Just as material can be brought into the cell by invagination and formation of a vesicle, the membrane of a vesicle can be fused with the plasma membrane, extruding its contents to the surrounding medium. This is the process of exocytosis. Exocytosis occurs in various cells to remove undigested residues of substances brought in by endocytosis, to secrete substances such as hormones and enzymes, and to transport a substance completely across a cellular barrier. In the process of exocytosis, the undigested waste-containing food vacuole or the secretory vesicle budded from Golgi apparatus, is first moved by cytoskeleton from the interior of the cell to the surface. The vesicle membrane comes in contact with the plasma membrane. The lipid molecules of the two bilayers rearrange themselves and the two membranes are, thus, fused. A passage is formed in the fused membrane and the vesicles discharges its contents outside the cell.

Prokaryotes

Gram-negative bacteria have both a plasma membrane and an outer membrane separated by periplasm. Other prokaryotes have only a plasma membrane. Prokaryotic cells are also surrounded by a cell wall composed of peptidoglycan (amino acids and sugars). Some eukaryotic cells also have cell walls, but none that are made of peptidoglycan. The outer membrane of gram negative microbes is rich in lipopolysaccharide and thus is different from cell membrane of the microbes. The outer membrane can bleb out into periplasmic protrusions under stess conditions or upon virulence requirements while encountering a host target cell, and thus such blebs may work as virulence organelles.

Structures

Fluid Mosaic Model

According to the fluid mosaic model of S. J. Singer and G. L. Nicolson (1972), which replaced the earlier model of Davson and Danielli, biological membranes can be considered as a two-dimensional liquid in which lipid and protein molecules diffuse more or less easily. Although the lipid bilayers that form the basis of the membranes do indeed form two-dimensional liquids by themselves, the plasma membrane also contains a large quantity of proteins, which provide more structure. Examples of such structures are protein-protein complexes, pickets and fences formed by the actin-based cytoskeleton, and potentially lipid rafts.

Lipid Bilayer

Lipid bilayers form through the process of self-assembly. The cell membrane consists primarily of a thin layer of amphipathic phospholipids that spontaneously arrange so that the hydrophobic "tail" regions are isolated from the surrounding water while the hydrophilic "head" regions interact with the intracellular (cytosolic) and extracellular faces of the resulting bilayer. This forms a continuous, spherical lipid bilayer. Hydrophobic interactions (also known as the hydrophobic effect) are the major driving forces in the formation of lipid bilayers. An increase in interactions between hydrophobic molecules (causing clustering of hydrophobic regions) allows water molecules to bond more freely with each other, increasing the entropy of the system. This complex interaction can include noncovalent interactions such as van der Waals, electrostatic and hydrogen bonds.

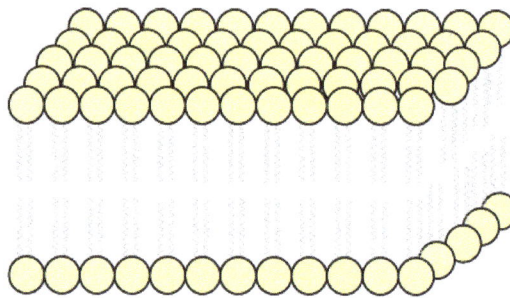

Diagram of the arrangement of amphipathic lipid molecules to form a lipid bilayer. The yellow polar head groups separate the grey hydrophobic tails from the aqueous cytosolic and extracellular environments.

Lipid bilayers are generally impermeable to ions and polar molecules. The arrangement of hydrophilic heads and hydrophobic tails of the lipid bilayer prevent polar solutes (ex. amino acids, nucleic acids, carbohydrates, proteins, and ions) from diffusing across the membrane, but generally allows for the passive diffusion of hydrophobic molecules. This affords the cell the ability to control the movement of these substances via transmembrane protein complexes such as pores, channels and gates.

Flippases and scramblases concentrate phosphatidyl serine, which carries a negative charge, on the inner membrane. Along with NANA, this creates an extra barrier to charged moieties moving through the membrane.

Membranes serve diverse functions in eukaryotic and prokaryotic cells. One important role is to regulate the movement of materials into and out of cells. The phospholipid bilayer structure (fluid mosaic model) with specific membrane proteins accounts for the selective permeability of the membrane and passive and active transport mechanisms. In addition, membranes in prokaryotes and in the mitochondria and chloroplasts of eukaryotes facilitate the synthesis of ATP through chemiosmosis.

Membrane Polarity

The apical membrane of a polarized cell is the surface of the plasma membrane that faces inward to the lumen. This is particularly evident in epithelial and endothelial cells, but also describes other polarized cells, such as neurons. The basolateral membrane of a polarized cell is the surface of the plasma membrane that forms its basal and lateral surfaces. It faces outwards, towards the interstitium, and away from the lumen. Basolateral membrane is a compound phrase referring to the terms "basal (base) membrane" and "lateral (side) membrane", which, especially in epithelial cells, are identical in composition and activity. Proteins (such as ion channels and pumps) are free to move from the basal to the lateral surface of the cell or vice versa in accordance with the fluid mosaic model. Tight junctions join epithelial cells near their apical surface to prevent the migration of proteins from the basolateral membrane to the apical membrane. The basal and lateral surfaces thus remain roughly equivalent to one another, yet distinct from the apical surface.

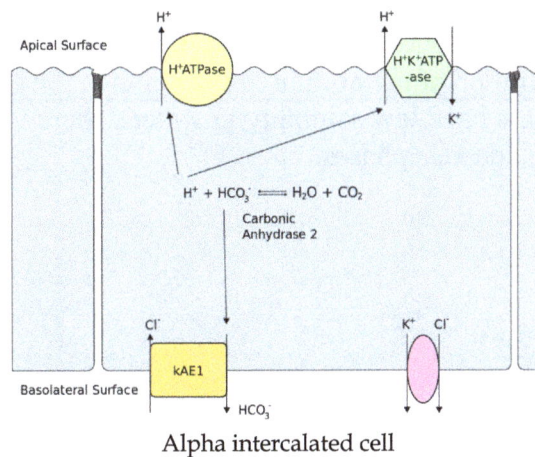

Alpha intercalated cell

Membrane Structures

Cell membrane can form different types of "supramembrane" structures such as caveola, postsynaptic density, podosome, invadopodium, focal adhesion, and different types of cell junctions. These structures are usually responsible for cell adhesion, communication, endocytosis and exocytosis. They can be visualized by electron microscopy or fluorescence microscopy. They are composed of specific proteins, such as integrins and cadherins.

Cytoskeleton

The cytoskeleton is found underlying the cell membrane in the cytoplasm and provides a scaffolding for membrane proteins to anchor to, as well as forming organelles that extend from the cell. Indeed, cytoskeletal elements interact extensively and intimately with the cell membrane. Anchoring proteins restricts them to a particular cell surface — for example, the apical surface of epithelial cells that line the vertebrate gut — and limits how far they may diffuse within the bilayer. The cytoskeleton is able to form appendage-like organelles, such as cilia, which are microtubule-based extensions covered by the cell membrane, and filopodia, which are actin-based extensions. These extensions are ensheathed in membrane and project from the surface of the cell in order to sense the external environment and/or make contact with the substrate or other cells. The apical surfaces of epithelial cells are dense with actin-based finger-like projections known as microvilli, which increase cell surface area and thereby increase the absorption rate of nutrients. Localized decoupling of the cytoskeleton and cell membrane results in formation of a bleb.

Composition

Cell membranes contain a variety of biological molecules, notably lipids and proteins. Material is incorporated into the membrane, or deleted from it, by a variety of mechanisms:

- Fusion of intracellular vesicles with the membrane (exocytosis) not only excretes the contents of the vesicle but also incorporates the vesicle membrane's components into the cell membrane. The membrane may form blebs around extracellular material that pinch off to become vesicles (endocytosis).

- If a membrane is continuous with a tubular structure made of membrane material, then material from the tube can be drawn into the membrane continuously.

- Although the concentration of membrane components in the aqueous phase is low (stable membrane components have low solubility in water), there is an exchange of molecules between the lipid and aqueous phases.

Lipids

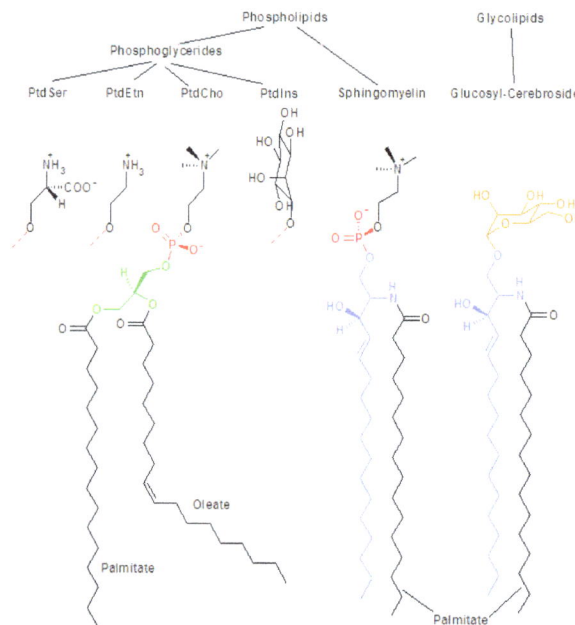

Examples of the major membrane phospholipids and glycolipids: phosphatidylcholine (PtdCho), phosphatidylethanolamine (PtdEtn), phosphatidylinositol (PtdIns), phosphatidylserine (PtdSer).

The cell membrane consists of three classes of amphipathic lipids: phospholipids, glycolipids, and sterols. The amount of each depends upon the type of cell, but in the majority of cases phospholipids are the most abundant. In RBC studies, 30% of the plasma membrane is lipid.

The fatty chains in phospholipids and glycolipids usually contain an even number of carbon atoms, typically between 16 and 20. The 16- and 18-carbon fatty acids are the most common. Fatty acids may be saturated or unsaturated, with the configuration of the double bonds nearly always "cis". The length and the degree of unsaturation of fatty acid chains have a profound effect on membrane fluidity as unsaturated lipids create a kink, preventing the fatty acids from packing together as tightly, thus decreasing the melting temperature (increasing the fluidity) of the membrane. The ability of some organisms to regulate the fluidity of their cell membranes by altering lipid composition is called homeoviscous adaptation.

The entire membrane is held together via non-covalent interaction of hydrophobic tails, however the structure is quite fluid and not fixed rigidly in place. Under physiological conditions phospholipid molecules in the cell membrane are in the liquid crystalline state. It means the lipid molecules are free to diffuse and exhibit rapid lateral diffusion along the layer in which they are present. However, the exchange of phospholipid molecules between intracellular and extracellular leaflets of the bilayer is a very slow process. Lipid rafts and caveolae are examples of cholesterol-enriched microdomains in the cell membrane. Also, a fraction of the lipid in direct contact with integral membrane proteins, which is tightly bound to the protein surface is called annular lipid shell; it behaves as a part of protein complex.

In animal cells cholesterol is normally found dispersed in varying degrees throughout cell membranes, in the irregular spaces between the hydrophobic tails of the membrane lipids, where it confers a stiffening and strengthening effect on the membrane.

Phospholipids Forming Lipid Vesicles

Lipid vesicles or liposomes are circular pockets that are enclosed by a lipid bilayer. These structures are used in laboratories to study the effects of chemicals in cells by delivering these chemicals directly to the cell, as well as getting more insight into cell membrane permeability. Lipid vesicles and liposomes are formed by first suspending a lipid in an aqueous solution then agitating the mixture through sonication, resulting in a vesicle. By measuring the rate of efflux from that of the inside of the vesicle to the ambient solution, allows researcher to better understand membrane permeability. Vesicles can be formed with molecules and ions inside the vesicle by forming the vesicle with the desired molecule or ion present in the solution. Proteins can also be embedded into the membrane through solubilizing the desired proteins in the presence of detergents and attaching them to the phospholipids in which the liposome is formed. These provide researchers with a tool to examine various membrane protein functions.

Carbohydrates

Plasma membranes also contain carbohydrates, predominantly glycoproteins, but with some glycolipids (cerebrosides and gangliosides). For the most part, no glycosylation occurs on membranes within the cell; rather generally glycosylation occurs on the extracellular surface of the plasma membrane. The glycocalyx is an important feature in all cells, especially epithelia with microvilli. Recent data suggest the glycocalyx participates in cell adhesion, lymphocyte homing, and many others. The penultimate sugar is galactose and the terminal sugar is sialic acid, as the sugar backbone is modified in the Golgi apparatus. Sialic acid carries a negative charge, providing an external barrier to charged particles.

Proteins

Type	Description	Examples
Integral proteins or transmembrane proteins	Span the membrane and have a hydrophilic cytosolic domain, which interacts with internal molecules, a hydrophobic membrane-spanning domain that anchors it within the cell membrane, and a hydrophilic extracellular domain that interacts with external molecules. The hydrophobic domain consists of one, multiple, or a combination of α-helices and β sheet protein motifs.	Ion channels, proton pumps, G protein-coupled receptor
Lipid anchored proteins	Covalently bound to single or multiple lipid molecules; hydrophobically insert into the cell membrane and anchor the protein. The protein itself is not in contact with the membrane.	G proteins
Peripheral proteins	Attached to integral membrane proteins, or associated with peripheral regions of the lipid bilayer. These proteins tend to have only temporary interactions with biological membranes, and once reacted, the molecule dissociates to carry on its work in the cytoplasm.	Some enzymes, some hormones

The cell membrane has large content of proteins, typically around 50% of membrane volume These proteins are important for cell because they are responsible for various biological activities. Approximately a third of the genes in yeast code specifically for them, and this number is even higher in multicellular organisms.

The cell membrane, being exposed to the outside environment, is an important site of cell–cell communication. As such, a large variety of protein receptors and identification proteins, such as antigens, are present on the surface of the membrane. Functions of membrane proteins can also include cell–cell contact, surface recognition, cytoskeleton contact, signaling, enzymatic activity, or transporting substances across the membrane.

Most membrane proteins must be inserted in some way into the membrane. For this to occur, an N-terminus "signal sequence" of amino acids directs proteins to the endoplasmic reticulum, which inserts the proteins into a lipid bilayer. Once inserted, the proteins are then transported to their final destination in vesicles, where the vesicle fuses with the target membrane.

Variation

The cell membrane has different lipid and protein compositions in distinct types of cells and may have therefore specific names for certain cell types:

- Sarcolemma in myocytes

- Oolemma in oocytes

- Axolemma in neuronal processes - axons

- Historically, the plasma membrane was also referred to as the plasmalemma

Permeability

The permeability of a membrane is the rate of passive diffusion of molecules through the membrane. These molecules are known as permeant molecules. Permeability depends mainly on the electric charge and polarity of the molecule and to a lesser extent the molar mass of the molecule. Due to the cell membrane's hydrophobic nature, small electrically neutral molecules pass through the membrane more easily than charged, large ones. The inability of charged molecules to pass through the cell membrane results in pH partition of substances throughout the fluid compartments of the body.

Cytoskeleton

A cytoskeleton is present in all cells of all domains of life (archaea, bacteria, eukaryotes). It is a complex network of interlinking filaments and tubules that extend throughout the cytoplasm, from the nucleus to the plasma membrane The cytoskeletal systems of different organisms are composed of similar proteins. In eukaryotes, the cytoskeletal matrix is a dynamic structure composed of three main proteins, which are capable of rapid growth or disassembly dependent on the cell's requirements at a certain period of time.

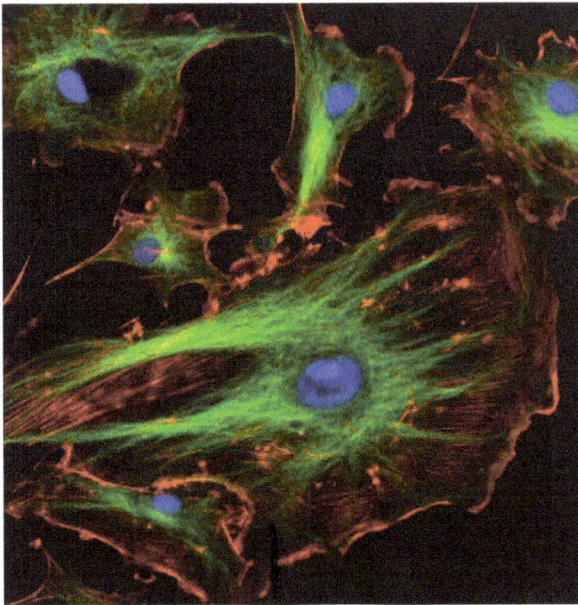

The eukaryotic cytoskeleton. Actin filaments are shown in red, and microtubules composed of beta tubulin are in green.

The structure, function and dynamic behaviour of the cytoskeleton can be very different, depending on organism and cell type. Even within one cell the cytoskeleton can change through association with other proteins and the previous history of the network.

There is a multitude of functions that the cytoskeleton can perform. Primarily, it gives the cell shape and mechanical resistance to deformation, and through association with extracellular connective tissue and other cells it stabilizes entire tissues. The cytoskeleton can also actively contract, thereby deforming the cell and the cell's environment and allowing cells to migrate. Moreover, it is involved in many cell signaling pathways, in the uptake of extracellular material (endocytosis), segregates chromosomes during cellular division, is involved in cytokinesis (the division of a mother cell into two daughter cells), provides a scaffold to organize the contents of the cell in space and for intracellular transport (for example, the movement of vesicles and organelles within the cell); and can be a template for the construction of a cell wall. Furthermore, it forms specialized structures, such as flagella, cilia, lamellipodia and podosomes.

A large-scale example of an action performed by the cytoskeleton is muscle contraction. During contraction of a muscle, within each muscle cell, myosin molecular motors collectively exert forces on parallel actin filaments. This action contracts the muscle cell, and through the synchronous process in many muscle cells, the entire muscle.

History

In 1903, Nikolai K. Koltsov proposed that the shape of cells was determined by a network of tubules that he termed the cytoskeleton. The concept of a protein mosaic that dynamically coordinated cytoplasmic biochemistry was proposed by Rudolph Peters in 1929 while the term (*cytosquelette*, in French) was first introduced by French embryologist Paul Wintrebert in 1931.

Eukaryotic Cytoskeleton

Eukaryotic cells contain three main kinds of cytoskeletal filaments: microfilaments, microtubules, and intermediate filaments. Each cytoskeletal filament has a shape and intracellular distribution. Additionally, the filaments are formed by polymerization of different types of sub-units. The microfilament consist of the polymers of the protein actin which has a diameter of 7 nm. The microtubules are made up of the protein called tubulin which has a diameter of 25 nm. Intermediate filaments are made up of various proteins which varies depending on the cell type. These type of filament normally have diameters ranging from 8-12 nm. The cytoskeleton provides the cell with structure and shape, and by excluding macromolecules from some of the cytosol, it adds to the level of macromolecular crowding in this compartment. Cytoskeletal elements interact extensively and intimately with cellular membranes. A number of small molecule cytoskeletal drugs have been discovered that interact with actin and microtubules. These compounds have proven useful in studying the cytoskeleton and several have clinical applications.

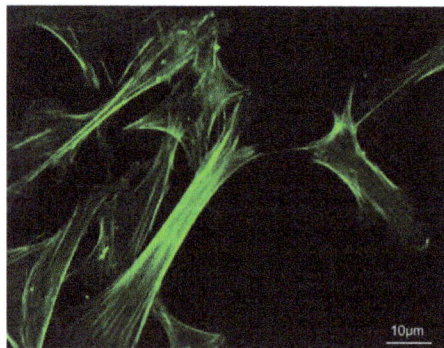

Actin cytoskeleton of mouse embryo fibroblasts, stained with phalloidin

Microfilaments (Actin Filaments)

Microfilaments are composed of linear polymers of G-actin proteins, and generate force when the growing (plus) end of the filament pushes against a barrier, such as the cell membrane.

They also act as tracks for the movement of myosin molecules that attach to the microfilament and "walk" along them. Myosin motoring along F-actin filaments generates contractile forces in so-called actomyosin fibers, both in muscle as well as most non-muscle cell types. Actin structures are controlled by the Rho family of small GTP-binding proteins such as Rho itself for contractile acto-myosin filaments ("stress fibers"), Rac for lamellipodia and Cdc42 for filopodia.

Intermediate Filaments

Intermediate filaments are a part of the cytoskeleton of all animals. These filaments, averaging 10 nanometers in diameter, are more stable (strongly bound) than actin filaments, and heterogeneous constituents of the cytoskeleton. Like actin filaments, they function in the maintenance of cell-shape by bearing tension (microtubules, by contrast, resist compression but can also bear tension during mitosis and during the positioning of the centrosome). Intermediate filaments organize the internal tridimensional structure of the cell, anchoring organelles and serving as structural components of the nuclear lamina. They also participate in some cell-cell and cell-matrix junctions. Nuclear lamina exist in all animals and all tissues. Some animals like the fruit fly do not have any cytoplasmic intermediate filaments. In those animals that express cytoplasmic intermediate filaments, these are tissue specific.

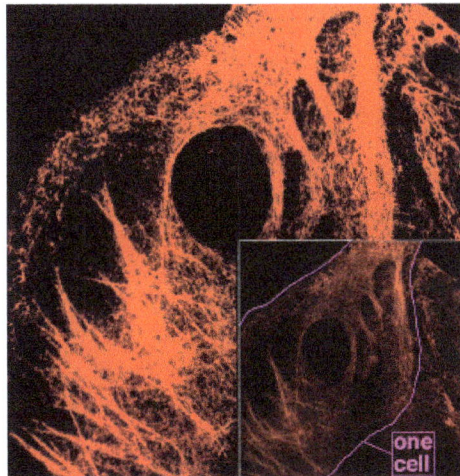

Microscopy of keratin filaments inside cells

Different intermediate filaments are:

- made of vimentins. Vimentin intermediate filaments are in general present in mesenchymal cells.

- made of keratin. Keratin is present in general in epithelial cells.

- neurofilaments of neural cells.

- made of lamin, giving structural support to the nuclear envelope.

- made of desmin, play an important role in structural and mechanical support of muscle cells.

Microtubules

Microtubules are hollow cylinders about 23 nm in diameter (lumen = approximately 15 nm in diameter), most commonly comprising 13 protofilaments that, in turn, are polymers of alpha and beta tubulin. They have a very dynamic behavior, binding GTP for polymerization. They are commonly organized by the centrosome.

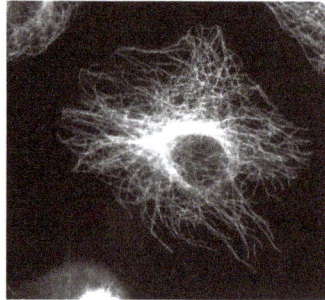

Microtubules in a gel-fixated cell

In nine triplet sets (star-shaped), they form the centrioles, and in nine doublets oriented about two additional microtubules (wheel-shaped), they form cilia and flagella. The latter formation is commonly referred to as a "9+2" arrangement, wherein each doublet is connected to another by the protein dynein. As both flagella and cilia are structural components of the cell, and are maintained by microtubules, they can be considered part of the cytoskeleton.

They play key roles in:

- intracellular transport (associated with dyneins and kinesins, they transport organelles like mitochondria or vesicles).

- the axoneme of cilia and flagella.

- the mitotic spindle.

- synthesis of the cell wall in plants.

Comparison

Cytoskeleton type	Diameter (nm)	Structure	Subunit examples
Microfilaments	6	double helix	actin
Intermediate filaments	10	two anti-parallel helices/ dimers, forming tetramers	• vimentin (mesenchyme) • glial fibrillary acidic protein (glial cells) • neurofilament proteins (neuronal processes) • keratins (epithelial cells) • nuclear lamins
Microtubules	23	protofilaments, in turn consisting of tubulin subunits in complex with stathmin	α- and β-tubulin

Septins

Septins are a group of the highly conserved GTP binding proteins found in eukaryotes. Different septins form protein complexes with each other. These can assemble to filaments and rings. Therefore, septins can be considered part of the cytoskeleton. The function of septins in cells include serving as a localized attachment site for other proteins, and preventing the diffusion of certain molecules from one cell compartment to another. In yeast cells, they build scaffolding to provide structural support during cell division and compartmentalize parts of the cell. Recent research in human cells suggests that septins build cages around bacterial pathogens, immobilizing the harmful microbes and preventing them from invading other cells.

Spectrin

Spectrin is a cytoskeletal protein that lines the intracellular side of the plasma membrane in eukaryotic cells. Spectrin forms pentagonal or hexagonal arrangements, forming a scaffolding and playing an important role in maintenance of plasma membrane integrity and cytoskeletal structure.

Yeast Cytoskeleton

In budding yeast (an important model organism), actin forms cortical patches, actin cables, and a cytokinetic ring and the cap. Cortical patches are discrete actin bodies on the membrane and are important for endocytosis, especially the recycling of glucan synthase which is important for cell wall synthesis. Actin cables are bundles of actin filaments and are involved in the transport of vesicles towards the cap (which contains a number of different proteins to polarize cell growth) and in the positioning of mitochondria. The cytokinetic ring forms and constricts around the site of cell division.

Prokaryotic Cytoskeleton

The cytoskeleton was once thought to be a feature only of eukaryotic cells, but homologues to all the major proteins of the eukaryotic cytoskeleton have been found in prokaryotes. Although the evolutionary relationships are so distant that they are not obvious from protein sequence comparisons alone, the similarity of their three-dimensional structures and similar functions in maintaining cell shape and polarity provides strong evidence that the eukaryotic and prokaryotic cytoskeletons are truly homologous. However, some structures in the bacterial cytoskeleton may not have been identified as of yet.

FtsZ

FtsZ was the first protein of the prokaryotic cytoskeleton to be identified. Like tubulin, FtsZ forms filaments in the presence of guanosine triphosphate (GTP), but these filaments do not group into tubules. During cell division, FtsZ is the first protein to move to the division site, and is essential for recruiting other proteins that synthesize the new cell wall between the dividing cells.

MreB and ParM

Prokaryotic actin-like proteins, such as MreB, are involved in the maintenance of cell shape. All non-spherical bacteria have genes encoding actin-like proteins, and these proteins form a helical network beneath the cell membrane that guides the proteins involved in cell wall biosynthesis.

Some plasmids encode a partitioning system that involves an actin-like protein ParM. Filaments of ParM exhibit dynamic instability, and may partition plasmid DNA into the dividing daughter cells by a mechanism analogous to that used by microtubules during eukaryotic mitosis.

Crescentin

The bacterium *Caulobacter crescentus* contains a third protein, crescentin, that is related to the intermediate filaments of eukaryotic cells. Crescentin is also involved in maintaining cell shape, such as helical and vibrioid forms of bacteria, but the mechanism by which it does this is currently unclear.

Common Features and Differences between Prokaryotes and Eukaryotes

By definition, the cytoskeleton is composed of proteins that can form longitudinal arrays (fibres) in all organisms. These filament forming proteins have been classified into 4 classes. Tubulin-like, actin-like, Walker A cytoskeletal ATPases (WACA-proteins), and intermediate filaments.

Tubulin-like proteins are tubulin in eukaryotes and FtsZ, TubZ, RepX in prokaryotes. Actin-like proteins are actin in eukaryotes and MreB, FtsA in prokaryotes. An example of a WACA-proteins, which are mostly found in prokaryotes, is MinD. Examples for intermediate filaments, which have almost exclusively been found in animals (i.e. eukaryotes) are the lamins, keratins, vimentin, neurofilaments, desmin.

Although tubulin-like proteins share some amino acid sequence similarity, their similarity in protein-fold and the similarity in the GTP binding site is more striking. The same holds true for the actin-like proteins and their structure and ATP binding domain.

Cytoskeletal proteins are usually associated with cell shape, DNA segregation and cell division in prokaryotes and eukaryotes. Which proteins fulfill which task is very different. For example, DNA segregation in all eukaryotes happens through use of tubulin, but in prokaryotes either WACA proteins, actin-like or tubulin-like proteins can be used. Cell division is mediated in eukaryotes by actin, but in prokaryotes usually by tubulin-like (often FtsZ-ring) proteins and sometimes (Crenarchaeota) ESCRT-III, which in eukaryotes still has a role in the last step of division.

Organelle

In cell biology, an organelle is a specialized sub-unit within a cell that has a specific function. Individual organelles are usually separately enclosed within their own lipid bilayers.

The name *organelle* comes from the idea that these structures are to cells what an organ is to the body (hence the name *organelle,* the suffix *-elle* being a diminutive). Organelles are identified by microscopy, and can also be purified by cell fractionation. There are many types of organelles, particularly in eukaryotic cells. While prokaryotes do not possess organelles *per se*, some do contain protein-based microcompartments, which are thought to act as primitive organelles.

History and Terminology

In biology *organs* are defined as confined functional units within an organism. The analogy of bodily organs to microscopic cellular substructures is obvious, as from even early works, authors of respective textbooks rarely elaborate on the distinction between the two.

Credited as the first to use a diminutive of *organ* (i.e., little organ) for cellular structures was German zoologist Karl August Möbius (1884), who used the term *organula* (plural of *organulum*, the diminutive of Latin *organum*). In a footnote, which was published as a correction in the next issue of the journal, he justified his suggestion to call organs of unicellular organisms "organella" since they are only differently formed parts of one cell, in contrast to multicellular organs of multicellular organisms.

Types of Organelles

While most cell biologists consider the term organelle to be synonymous with "cell compartment", other cell biologists choose to limit the term organelle to include only those that are DNA-containing, having originated from formerly autonomous microscopic organisms acquired via endosymbiosis.

Under this definition, there would only be two broad classes of organelles (i.e. those that contain their own DNA, and have originated from endosymbiotic bacteria):

- mitochondria (in almost all eukaryotes)

- plastids (e.g. in plants, algae, and some protists).

Other organelles are also suggested to have endosymbiotic origins, but do not contain their own DNA (notably the flagellum).

Under the more restricted definition of membrane-bound structures, some parts of the cell do not qualify as organelles. Nevertheless, the use of organelle to refer to non-membrane bound structures such as ribosomes is common. This has led some texts to delineate between membrane-bound and non-membrane bound organelles. The non-membrane bound organelles, also called large biomolecular complexes, are large assemblies of macromolecules that carry out particular and specialized functions, but they lack membrane boundaries. Such cell structures include:

- large RNA and protein complexes: ribosome, spliceosome, vault

- large protein complexes: proteasome, DNA polymerase III holoenzyme, RNA polymerase II holoenzyme, symmetric viral capsids, complex of GroEL and GroES; membrane protein complexes: photosystem I, ATP synthase

- large DNA and protein complexes: nucleosome

- centriole and microtubule-organizing center (MTOC)

- cytoskeleton

- flagellum

- Nucleolus

- Stress granule

- Germ cell granule

- Neuronal transport granule

Eukaryotic Organelles

Eukaryotic cells are structurally complex, and by definition are organized, in part, by interior compartments that are themselves enclosed by lipid membranes that resemble the outermost cell membrane. The larger organelles, such as the nucleus and vacuoles, are easily visible with the light microscope. They were among the first biological discoveries made after the invention of the microscope.

Not all eukaryotic cells have each of the organelles listed below. Exceptional organisms have cells that do not include some organelles that might otherwise be considered universal to eukaryotes (such as mitochondria). There are also occasional exceptions to the number of membranes surrounding organelles, listed in the tables below (e.g., some that are listed as double-membrane are sometimes found with single or triple membranes). In addition, the number of individual organelles of each type found in a given cell varies depending upon the function of that cell.

Major eukaryotic organelles				
Organelle	**Main function**	**Structure**	**Organisms**	**Notes**
chloroplast (plastid)	photosynthesis, traps energy from sunlight	double-membrane compartment	plants, protists (rare kleptoplastic organisms)	has own DNA; theorized to be engulfed by the ancestral eukaryotic cell (endosymbiosis)
endoplasmic reticulum	translation and folding of new proteins (rough endoplasmic reticulum), expression of lipids (smooth endoplasmic reticulum)	single-membrane compartment	all eukaryotes	rough endoplasmic reticulum is covered with ribosomes, has folds that are flat sacs; smooth endoplasmic reticulum has folds that are tubular
Flagellum	locomotion, sensory		some eukaryotes	
Golgi apparatus	sorting, packaging, processing and modification of proteins	single-membrane compartment	all eukaryotes	cis-face (convex) nearest to rough endoplasmic reticulum; trans-face (concave) farthest from rough endoplasmic reticulum

mitochondria	energy production from the oxidation of glucose substances and the release of adenosine triphosphate	double-membrane compartment	most eukaryotes	has own DNA; theorized to be engulfed by an ancestral eukaryotic cell (endosymbiosis)
vacuole	storage, transportation, helps maintain homeostasis	single-membrane compartment	eukaryotes	
nucleus	DNA maintenance, controls all activities of the cell, RNA transcription	double-membrane compartment	all eukaryotes	contains bulk of genome

Mitochondria and chloroplasts, which have double-membranes and their own DNA, are believed to have originated from incompletely consumed or invading prokaryotic organisms, which were adopted as a part of the invaded cell. This idea is supported in the Endosymbiotic theory.

Minor eukaryotic organelles and cell components			
Organelle/Macromolecule	Main function	Structure	Organisms
acrosome	helps spermatozoa fuse with ovum	single-membrane compartment	many animals
autophagosome	vesicle that sequesters cytoplasmic material and organelles for degradation	double-membrane compartment	all eukaryotes
centriole	anchor for cytoskeleton, organizes cell division by forming spindle fibers	Microtubule protein	animals
cilium	movement in or of external medium; "critical developmental signaling pathway".	Microtubule protein	animals, protists, few plants
eyespot apparatus	detects light, allowing phototaxis to take place		green algae and other unicellular photosynthetic organisms such as euglenids
glycosome	carries out glycolysis	single-membrane compartment	Some protozoa, such as *Trypanosomes*.
glyoxysome	conversion of fat into sugars	single-membrane compartment	plants
hydrogenosome	energy & hydrogen production	double-membrane compartment	a few unicellular eukaryotes
lysosome	breakdown of large molecules (e.g., proteins + polysaccharides)	single-membrane compartment	animals
melanosome	pigment storage	single-membrane compartment	animals
mitosome	probably plays a role in Iron-sulfur cluster (Fe-S) assembly	double-membrane compartment	a few unicellular eukaryotes that lack mitochondria
myofibril	myocyte contraction	bundled filaments	animals
nematocyst	stinging	coiled hollow tubule	Cnidarians
nucleolus	pre-ribosome production	protein-DNA-RNA	most eukaryotes

parenthesome	not characterized	not characterized	fungi
peroxisome	breakdown of metabolic hydrogen peroxide	single-membrane compartment	all eukaryotes
proteasome	degradation of unneeded or damaged proteins by proteolysis	very large protein complex	All eukaryotes, all archaea, some bacteria
ribosome (80S)	translation of RNA into proteins	RNA-protein	all eukaryotes
vesicle	material transport	single-membrane compartment	all eukaryotes
Stress granule	mRNA storage	membraneless (mRNP complexes)	Most eukaryotes

Other related structures:

- cytosol

- endomembrane system

- nucleosome

- microtubule

- cell membrane

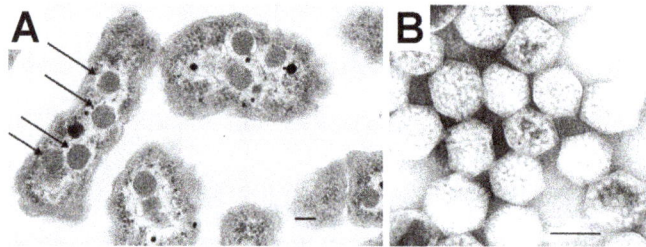

(A) Electron micrograph of *Halothiobacillus neapolitanus* cells, arrows highlight carboxysomes. (B) Image of intact carboxysomes isolated from *H. neapolitanus*. Scale bars are 100 nm.

Prokaryotic Organelles

Prokaryotes are not as structurally complex as eukaryotes, and were once thought not to have any internal structures enclosed by lipid membranes. In the past, they were often viewed as having little internal organization, but slowly, details are emerging about prokaryotic internal structures. An early false turn was the idea developed in the 1970s that bacteria might contain membrane folds termed mesosomes, but these were later shown to be artifacts produced by the chemicals used to prepare the cells for electron microscopy.

However, more recent research has revealed that at least some prokaryotes have microcompartments such as carboxysomes. These subcellular compartments are 100–200 nm in diameter and are enclosed by a shell of proteins. Even more striking is the description of membrane-bound magnetosomes in bacteria, reported in 2006, as well as the nucleus-like structures of the *Plancto-mycetes* that are surrounded by lipid membranes, reported in 2005.

Prokaryotic organelles and cell components			
Organelle/Macromolecule	Main function	Structure	Organisms
carboxysome	carbon fixation	protein-shell compartment	some bacteria
chlorosome	photosynthesis	light harvesting complex	green sulfur bacteria
flagellum	movement in external medium	protein filament	some prokaryotes and eukaryotes
magnetosome	magnetic orientation	inorganic crystal, lipid membrane	magnetotactic bacteria
nucleoid	DNA maintenance, transcription to RNA	DNA-protein	prokaryotes
plasmid	DNA exchange	circular DNA	some bacteria
ribosome (70S)	translation of RNA into proteins	RNA-protein	bacteria and archaea
thylakoid	photosynthesis	photosystem proteins and pigments	mostly cyanobacteria
mesosomes	functions of Golgi bodies, centrioles, etc.	small irregular shaped organelle containing ribosomes	present in most prokaryotic cells
Pilus	Adhesion to other cells for conjugation or to a solid substrate to create motile forces.	a hair-like appendage sticking out (though partially embedded into) the plasma membrane	prokaryotic cells

Proteins and Organelles

The function of a protein is closely correlated with the organelle in which it resides. Some methods were proposed for predicting the organelle in which an uncharacterized protein is located according to its amino acid composition and some methods were based on pseudo amino acid composition.

Cell Growth

The term cell growth is used in the contexts of biological cell development and cell division (reproduction). When used in the context of cell division, it refers to growth of cell populations, where a cell, known as the "mother cell", grows and divides to produce two "daughter cells" (M phase). When used in the context of cell development, the term refers to increase in cytoplasmic and organelle volume (G1 phase), as well as increase in genetic material (G2 phase) following the replication during S phase.

Cell Populations

Cell populations go through a particular type of exponential growth called doubling. Thus, each generation of cells should be twice as numerous as the previous generation. However, the number of generations only gives a maximum figure as not all cells survive in each generation.

Cell Size

Cell size is highly variable among organisms, with some algae such as *Caulerpa taxifolia* being a single cell several meters in length. Plant cells are much larger than animal cells, and protists such as *Paramecium* can be 330 µm long, while a typical human cell might be 10 µm. How these cells "decide" how big they should be before dividing is an open question. Chemical gradients are known to be partly responsible, and it is hypothesized that mechanical stress detection by cyto-skeletal structures is involved. Work on the topic generally requires an organism whose cell cycle is well-characterized.

Yeast Cell Size Regulation

The relationship between cell size and cell division has been extensively studied in yeast. For some cells, there is a mechanism by which cell division is not initiated until a cell has reached a certain size. If the nutrient supply is restricted (after time t = 2 in the diagram, below), and the rate of increase in cell size is slowed, the time period between cell divisions is increased. Yeast cell-size mutants were isolated that begin cell division before reaching a normal/regular size (*wee* mutants).

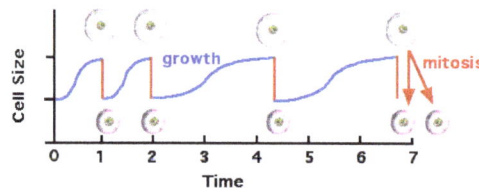

Figure 1:Cell cycle and growth

Wee1 protein is a tyrosine kinase that normally phosphorylates the Cdc2 cell cycle regulatory protein (the homolog of CDK1 in humans), a cyclin-dependent kinase, on a tyrosine residue. Cdc2 drives entry into mitosis by phosphorylating a wide range of targets. This covalent modification of the molecular structure of Cdc2 inhibits the enzymatic activity of Cdc2 and prevents cell division. Wee1 acts to keep Cdc2 inactive during early G2 when cells are still small. When cells have reached sufficient size during G2, the phosphatase Cdc25 removes the inhibitory phosphorylation, and thus activates Cdc2 to allow mitotic entry. A balance of Wee1 and Cdc25 activity with changes in cell size is coordinated by the mitotic entry control system. It has been shown in Wee1 mutants, cells with weakened Wee1 activity, that Cdc2 becomes active when the cell is smaller. Thus, mitosis occurs before the yeast reach their normal size. This suggests that cell division may be regulated in part by dilution of Wee1 protein in cells as they grow larger.

Linking Cdr2 to Wee1

The protein kinase Cdr2 (which negatively regulates Wee1) and the Cdr2-related kinase Cdr1 (which directly phosphorylates and inhibits Wee1 *in vitro*) are localized to a band of cortical nodes in the middle of interphase cells. After entry into mitosis, cytokinesis factors such as myosin II are recruited to similar nodes; these nodes eventually condense to form the cytokinetic ring. A previously uncharacterized protein, Blt1, was found to colocalize with Cdr2 in the medial interphase nodes. Blt1 knockout cells had increased length at division, which is consistent with a delay in mitotic entry. This finding connects a physical location, a band of cortical nodes, with factors that have been shown to directly regulate mitotic entry, namely Cdr1, Cdr2, and Blt1.

Further experimentation with GFP-tagged proteins and mutant proteins indicates that the medial cortical nodes are formed by the ordered, Cdr2-dependent assembly of multiple interacting proteins during interphase. Cdr2 is at the top of this hierarchy and works upstream of Cdr1 and Blt1. Mitosis is promoted by the negative regulation of Wee1 by Cdr2. It has also been shown that Cdr2 recruits Wee1 to the medial cortical node. The mechanism of this recruitment has yet to be discovered. A Cdr2 kinase mutant, which is able to localize properly despite a loss of function in phosphorylation, disrupts the recruitment of Wee1 to the medial cortex and delays entry into mitosis. Thus, Wee1 localizes with its inhibitory network, which demonstrates that mitosis is controlled through Cdr2-dependent negative regulation of Wee1 at the medial cortical nodes.

Cell Polarity Factors

Cell polarity factors positioned at the cell tips provide spatial cues to limit Cdr2 distribution to the cell middle. In fission yeast *Schizosaccharomyces pombe* (*S. Pombe*), cells divide at a defined, reproducible size during mitosis because of the regulated activity of Cdk1. The cell polarity protein kinase Pom1, a member of the dual-specificity tyrosine-phosphorylation regulated kinase (DYRK) family of kinases, localizes to cell ends. In Pom1 knockout cells, Cdr2 was no longer restricted to the cell middle, but was seen diffusely through half of the cell. From this data it becomes apparent that Pom1 provides inhibitory signals that confine Cdr2 to the middle of the cell. It has been further shown that Pom1-dependent signals lead to the phosphorylation of Cdr2. Pom1 knockout cells were also shown to divide at a smaller size than wild-type, which indicates a premature entry into mitosis.

Pom1 forms polar gradients that peak at cell ends, which shows a direct link between size control factors and a specific physical location in the cell. As a cell grows in size, a gradient in Pom1 grows. When cells are small, Pom1 is spread diffusely throughout the cell body. As the cell increases in size, Pom1 concentration decreases in the middle and becomes concentrated at cell ends. Small cells in early G2 which contain sufficient levels of Pom1 in the entirety of the cell have inactive Cdr2 and cannot enter mitosis. It is not until the cells grow into late G2, when Pom1 is confined to the cell ends that Cdr2 in the medial cortical nodes is activated and able to start the inhibition of Wee1. This finding shows how cell size plays a direct role in regulating the start of mitosis. In this model, Pom1 acts as a molecular link between cell growth and mitotic entry through a Cdr2-Cdr1-Wee1-Cdk1 pathway. The Pom1 polar gradient successfully relays information about cell size and geometry to the Cdk1 regulatory system. Through this gradient, the cell ensures it has reached a defined, sufficient size to enter mitosis.

Cell Cycle Regulation in Mammals

Many different types of eukaryotic cells undergo size-dependent transitions during the cell cycle. These transitions are controlled by the cyclin-dependent kinase Cdk1. Though the proteins that control Cdk1 are well understood, their connection to mechanisms monitoring cell size remains elusive. A postulated model for mammalian size control situates mass as the driving force of the cell cycle. A cell is unable to grow to an abnormally large size because at a certain cell size or cell mass, the S phase is initiated. The S phase starts the sequence of events leading to mitosis and cytokinesis. A cell is unable to get too small because the later cell cycle events, such as S, G2, and M, are delayed until mass increases sufficiently to begin S phase.

Many of the signal molecules that convey information to cells during the control of cellular differentiation or growth are called growth factors. The protein mTOR is a serine/threonine kinase that regulates translation and cell division. Nutrient availability influences mTOR so that when cells are not able to grow to normal size they will not undergo cell division. The details of the molecular mechanisms of mammalian cell size control are currently being investigated. The size of post-mitotic neurons depends on the size of the cell body, axon and dendrites. In vertebrates, neuron size is often a reflection of the number of synaptic contacts onto the neuron or from a neuron onto other cells. For example, the size of motoneurons usually reflects the size of the motor unit that is controlled by the motoneuron. Invertebrates often have giant neurons and axons that provide special functions such as rapid action potential propagation. Mammals also use this trick for increasing the speed of signals in the nervous system, but they can also use myelin to accomplish this, so most human neurons are relatively small cells.

Other Experimental Systems for The Study of Cell Size Regulation

One common means to produce very large cells is by cell fusion to form syncytia. For example, very long (several inches) skeletal muscle cells are formed by fusion of thousands of myocytes. Genetic studies of the fruit fly *Drosophila* have revealed several genes that are required for the formation of multinucleated muscle cells by fusion of myoblasts. Some of the key proteins are important for cell adhesion between myocytes and some are involved in adhesion-dependent cell-to-cell signal transduction that allows for a cascade of cell fusion events.

Oocytes can be unusually large cells in species for which embryonic development takes place away from the mother's body. Their large size can be achieved either by pumping in cytosolic components from adjacent cells through cytoplasmic bridges (*Drosophila*) or by internalization of nutrient storage granules (yolk granules) by endocytosis (frogs).

Increases in the size of plant cells are complicated by the fact that almost all plant cells are inside of a solid cell wall. Under the influence of certain plant hormones the cell wall can be remodeled, allowing for increases in cell size that are important for the growth of some plant tissues.

Most unicellular organisms are microscopic in size, but there are some giant bacteria and protozoa that are visible to the naked eye. Dense populations of a giant sulfur bacterium in Namibian shelf sediments— Large protists of the genus *Chaos*, closely related to the genus *Amoeba*

In the rod-shaped bacteria *E. coli*, *Caulobacter crescentus* and *B. subtilis* cell size is controlled by a simple mechanisms in which cell division occurs after a constant volume has been added since the previous division. By always growing by the same amount, cells born smaller or larger than average naturally converge to an average size equivalent to the amount added during each generation.

Cell Division

Cell reproduction is asexual. For most of the constituents of the cell, growth is a steady, continuous process, interrupted only briefly at M phase when the nucleus and then the cell divide in two.

The process of cell division, called cell cycle, has four major parts called phases. The first part, called G_1 phase is marked by synthesis of various enzymes that are required for DNA replication.

The second part of the cell cycle is the S phase, where DNA replication produces two identical sets of chromosomes. The third part is the G_2 phase in which a significant protein synthesis occurs, mainly involving the production of microtubules that are required during the process of division, called mitosis. The fourth phase, M phase, consists of nuclear division (karyokinesis) and cytoplasmic division (cytokinesis), accompanied by the formation of a new cell membrane. This is the physical division of "mother" and "daughter" cells. The M phase has been broken down into several distinct phases, sequentially known as prophase, prometaphase, metaphase, anaphase and telophase leading to cytokinesis.

Cell division is more complex in eukaryotes than in other organisms. Prokaryotic cells such as bacterial cells reproduce by binary fission, a process that includes DNA replication, chromosome segregation, and cytokinesis. Eukaryotic cell division either involves mitosis or a more complex process called meiosis. Mitosis and meiosis are sometimes called the two "nuclear division" processes. Binary fission is similar to eukaryote cell reproduction that involves mitosis. Both lead to the production of two daughter cells with the same number of chromosomes as the parental cell. Meiosis is used for a special cell reproduction process of diploid organisms. It produces four special daughter cells (gametes) which have half the normal cellular amount of DNA. A male and a female gamete can then combine to produce a zygote, a cell which again has the normal amount of chromosomes.

The rest of this article is a comparison of the main features of the three types of cell reproduction that either involve binary fission, mitosis, or meiosis. The diagram below depicts the similarities and differences of these three types of cell reproduction.

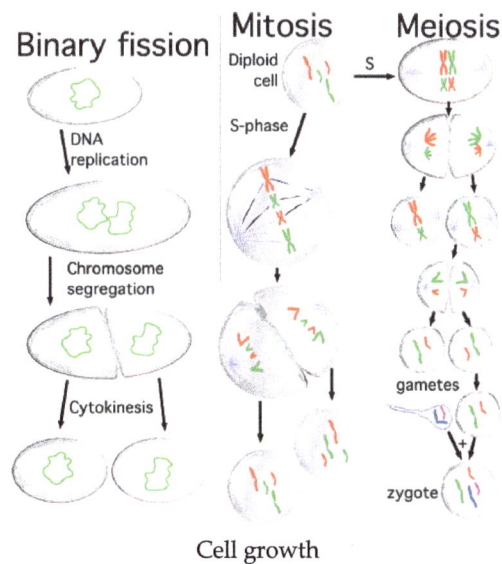

Cell growth

Comparison of The Three Types of Cell Division

The DNA content of a cell is duplicated at the start of the cell reproduction process. Prior to DNA replication, the DNA content of a cell can be represented as the amount Z (the cell has Z chromosomes). After the DNA replication process, the amount of DNA in the cell is 2Z (multiplication: 2 x Z = 2Z). During Binary fission and mitosis the duplicated DNA content of the reproducing parental cell is separated into two equal halves that are destined to end up in the two daughter cells.

The final part of the cell reproduction process is cell division, when daughter cells physically split apart from a parental cell. During meiosis, there are two cell division steps that together produce the four daughter cells.

After the completion of binary fission or cell reproduction involving mitosis, each daughter cell has the same amount of DNA (Z) as what the parental cell had before it replicated its DNA. These two types of cell reproduction produced two daughter cells that have the same number of chromosomes as the parental cell. Chromosomes duplicate prior to cell division when forming new skin cells for reproduction. After meiotic cell reproduction the four daughter cells have half the number of chromosomes that the parental cell originally had. This is the haploid amount of DNA, often symbolized as N. Meiosis is used by diploid organisms to produce haploid gametes. In a diploid organism such as the human organism, most cells of the body have the diploid amount of DNA, 2N. Using this notation for counting chromosomes we say that human somatic cells have 46 chromosomes (2N = 46) while human sperm and eggs have 23 chromosomes (N = 23). Humans have 23 distinct types of chromosomes, the 22 autosomes and the special category of sex chromosomes. There are two distinct sex chromosomes, the X chromosome and the Y chromosome. A diploid human cell has 23 chromosomes from that person's father and 23 from the mother. That is, your body has two copies of human chromosome number 2, one from each of your parents.

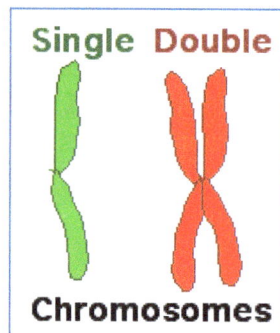

Chromosomes

Immediately after DNA replication a human cell will have 46 "double chromosomes". In each double chromosome there are two copies of that chromosome's DNA molecule. During mitosis the double chromosomes are split to produce 92 "single chromosomes", half of which go into each daughter cell. During meiosis, there are two chromosome separation steps which assure that each of the four daughter cells gets one copy of each of the 23 types of chromosome.

Sexual Reproduction

Though cell reproduction that uses mitosis can reproduce eukaryotic cells, eukaryotes bother with the more complicated process of meiosis because sexual reproduction such as meiosis confers a selective advantage. Notice that when meiosis starts, the two copies of sister chromatids number 2 are adjacent to each other. During this time, there can be genetic recombination events. Parts of the chromosome 2 DNA gained from one parent (red) will swap over to the chromosome 2 DNA molecule that received from the other parent (green). Notice that in mitosis the two copies of chromosome number 2 do not interact. It is these new combinations of parts of chromosomes that provide the major advantage for sexually reproducing organisms by allowing for new combinations of genes and more efficient evolution. However, in organisms with more than one set of chromosomes at the main life cycle stage, sex may also provide an advantage because, under random mating, it produces homozygotes and heterozygotes according to the Hardy-Weinberg ratio.

Cell Growth Disorders

A series of growth disorders can occur at the cellular level and these consequently underpin much of the subsequent course in cancer, in which a group of cells display uncontrolled growth and division beyond the normal limits, *invasion* (intrusion on and destruction of adjacent tissues), and sometimes *metastasis* (spread to other locations in the body via lymph or blood).

Cell Growth Measurement Methods

The cell growth can be detected by a variety of methods. The cell size growth can be visualized by microscopy, using suitable stains. But the increase of cells number is usually more significant. It can be measured by manual counting of cells under microscopy observation, using the dye exclusion method (i.e. trypan blue) to count only viable cells. Less fastidious, scalable, methods include the use of cytometers, while flow cytometry allows combining cell counts ('events') with other specific parameters: fluorescent probes for membranes, cytoplasm or nuclei allow distinguishing dead/viable cells, cell types, cell differentiation, expression of a biomarker such as Ki67.

Beside the increasing number of cells, one can be assessed regarding the metabolic activity growth, that is, the CFDA and calcein-AM measure (fluorimetrically) not only the membrane functionality (dye retention), but also the functionality of cytoplasmic enzymes (esterases). The MTT assays (colorimetric) and the resazurin assay (fluorimetric) dose the mitochondrial redox potential.

All these assays may correlate well, or not, depending on cell growth conditions and desired aspects (activity, proliferation). The task is even more complicated with populations of different cells, furthermore when combining cell growth interferences or toxicity.

Cell Division

Cell division is the process by which a *parent cell* divides into two or more *daughter cells*. Cell division usually occurs as part of a larger cell cycle. In eukaryotes, there are two distinct types of cell division: a vegetative division, whereby each daughter cell is genetically identical to the parent cell (mitosis), and a reproductive cell division, whereby the number of chromosomes in the daughter cells is reduced by half to produce haploid gametes (meiosis). Meiosis results in four haploid daughter cells by undergoing one round of DNA replication followed by two divisions: homologous chromosomes are separated in the first division, and sister chromatids are separated in the second division. Both of these cell division cycles are used in sexually reproducing organisms at some point in their life cycle, and both are believed to be present in the last eukaryotic common ancestor. Prokaryotes also undergo a vegetative cell division known as binary fission, where their genetic material is segregated equally into two daughter cells. All cell divisions, regardless of organism, are preceded by a single round of DNA replication.

For simple unicellular organisms such as the amoeba, one cell division is equivalent to reproduction – an entire new organism is created. On a larger scale, mitotic cell division can create progeny from multicellular organisms, such as plants that grow from cuttings. Mitotic cell division also enables sexually reproducing organisms to develop from the one-celled zygote, which itself was produced by meiotic cell division from gametes. After growth, cell division by mitosis allows for continual construction and repair of the organism. The human body experiences about 10 quadrillion cell divisions in a lifetime.

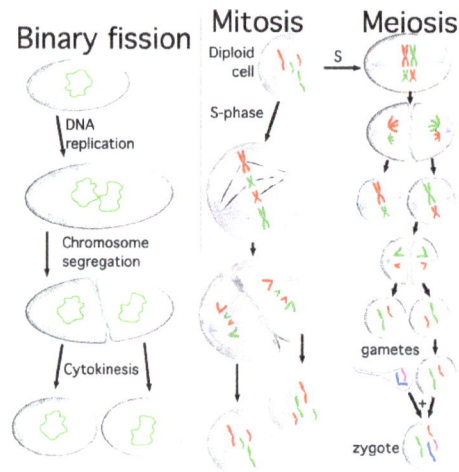

Three types of cell division

The primary concern of cell division is the maintenance of the original cell's genome. Before division can occur, the genomic information that is stored in chromosomes must be replicated, and the duplicated genome must be separated cleanly between cells. A great deal of cellular infrastructure is involved in keeping genomic information consistent between generations.

Variants

Cells are broadly classified into two main categories: simple, non-nucleated prokaryotic cells, and complex, nucleated eukaryotic cells. Owing to their structural differences, eukaryotic and prokaryotic cells do not divide in the same way. Also, the pattern of cell division that transforms eukaryotic stem cells into gametes (sperm cells in males or egg cells in females), termed meiosis, is different from that of the division of somatic cells in the body.

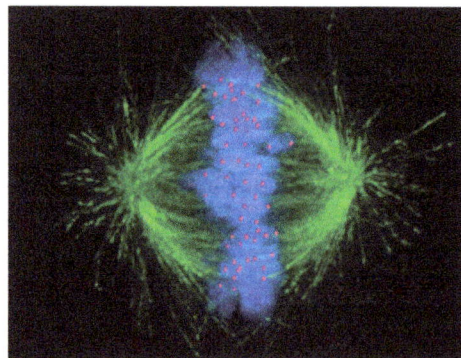

Image of the mitotic spindle in a human cell showing microtubules in green, chromosomes (DNA) in blue, and kinetochores in red.

Cell division over 42 hours. The cells were directly imaged in the cell culture vessel, using non-invasive quantitative phase contrast time-lapse microscopy.

Degradation

Multicellular organisms replace worn-out cells through cell division. In some animals, however, cell division eventually halts. In humans this occurs, on average, after 52 divisions, known as the Hayflick limit. The cell is then referred to as senescent. Cells stop dividing because the telomeres, protective bits of DNA on the end of a chromosome required for replication, shorten with each copy, eventually being consumed. Cancer cells, on the other hand, are not thought to degrade in this way, if at all. An enzyme called telomerase, present in large quantities in cancerous cells, rebuilds the telomeres, allowing division to continue indefinitely.

Cell Cycle

The cell cycle or cell-division cycle is the series of events that take place in a cell leading to its division and duplication of its DNA (DNA replication) to produce two daughter cells. In bacteria, which lack a cell nucleus, the cell cycle is divided into the B, C, and D periods. The B period extends from the end of cell division to the beginning of DNA replication. DNA replication occurs during the C period. The D period refers to the stage between the end of DNA replication and the splitting of the bacterial cell into two daughter cells. In cells with a nucleus, as in eukaryotes, the cell cycle is also divided into three periods: interphase, the mitotic (M) phase, and cytokinesis. During interphase, the cell grows, accumulating nutrients needed for mitosis, preparing it for cell division and duplicating its DNA. During the mitotic phase, the chromosomes separate. During the final stage, cytokinesis, the chromosomes and cytoplasm separate into two new daughter cells. To ensure the proper division of the cell, there are control mechanisms known as cell cycle checkpoints.

The cell-division cycle is a vital process by which a single-celled fertilized egg develops into a mature organism, as well as the process by which hair, skin, blood cells, and some internal organs are renewed. After cell division, each of the daughter cells begin the interphase of a new cycle. Although the various stages of interphase are not usually morphologically distinguishable, each phase of the cell cycle has a distinct set of specialized biochemical processes that prepare the cell for initiation of cell division.

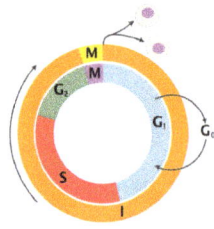

Schematic of the cell cycle. outer ring: I = Interphase, M = Mitosis; inner ring: M = Mitosis, G_1 = Gap 1, G_2 = Gap 2, S = Synthesis; not in ring: G_0 = Gap 0/Resting.

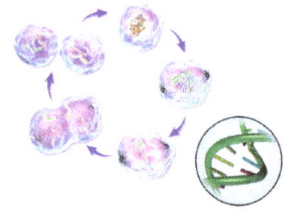

Life Cycle of the cell

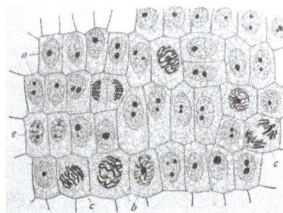

Onion (Allium) cells in different phases of the cell cycle. Growth in an organism is carefully controlled by regulating the cell cycle.

Cell Cycle Phases

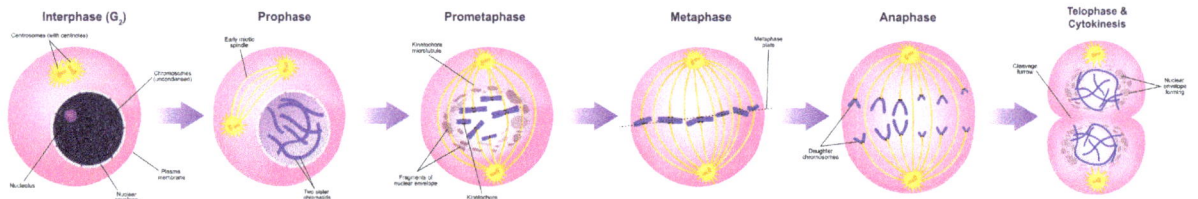

State	Description	Abbreviation	:	
Quiescent senescent/ resting phase	Gap 0	G_0	A resting phase where the cell has left the cycle and has stopped dividing.Cell cycle starts with this phase.	
	Interphase	Gap 1	G_1	Cells increase in size in Gap 1. The G_1 *checkpoint* control mechanism ensures that everything is ready for DNA synthesis.
		Synthesis	S	DNA replication occurs during this phase.
Gap 2			G_2	During the gap between DNA synthesis and mitosis, the cell will continue to grow. The G_2 *checkpoint* control mechanism ensures that everything is ready to enter the M (mitosis) phase and divide.

Cell division	Mitosis	M	Cell growth stops at this stage and cellular energy is focused on the orderly division into two daughter cells. A checkpoint in the middle of mitosis (*Metaphase Checkpoint*) ensures that the cell is ready to complete cell division.	

G_0 Phase (Quiescence)

G_0 is a resting phase where the cell has left the cycle and has stopped dividing. The cell cycle starts with this phase. The word "post-mitotic" is sometimes used to refer to both quiescent and senescent cells. Non proliferative (non-dividing) cells in multicellular eukaryotes generally enter the quiescent G_0 state from G_1 and may remain quiescent for long periods of time, possibly indefinitely (as is often the case for neurons). This is very common for cells that are fully differentiated. Cellular senescence (death) occurs in response to DNA damage or degradation that would make a cell's progeny nonviable. Some cells enter the G_0 phase semi-permanently e.g., some liver, kidney, stomach cells. Many cells do not enter G_0 and continue to divide throughout an organism's life, e.g. epithelial cells.

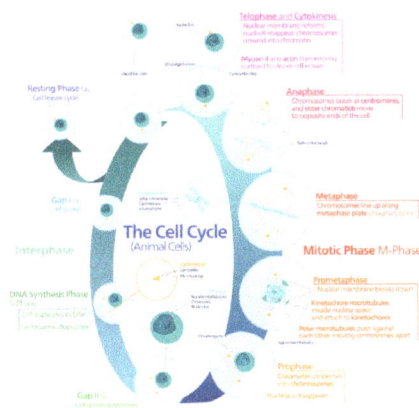

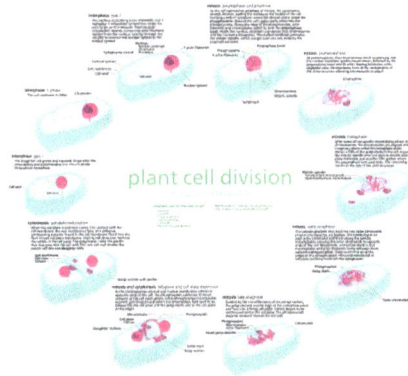

Animal cell cycle   Plant cell cycle

Interphase (Intermitosis)

Before a cell can enter cell division, it needs to take in nutrients. All of the preparations are done during interphase. Interphase is a series of changes that takes place in a newly formed cell and its nucleus, before it becomes capable of division again. It is also called preparatory phase or inter-mitosis. Previously it was called resting stage because there is no apparent activity related to cell division.Typically interphase lasts for at least 90% of the total time required for the cell cycle.

Interphase proceeds in three stages, G_1, S, and G_2, followed by the cycle of mitosis and cytokinesis. The cell's nuclear chromosomes are duplicated during S phase.

G_1 Phase (Growth)

The first phase within interphase, from the end of the previous M phase until the beginning of DNA synthesis, is called G_1 (G indicating *gap*). It is also called the growth phase. During this phase, the biosynthetic activities of the cell, which are considerably slowed down during M phase, resume at a high rate. The duration of G_1 is highly variable, even among different cells of the same species. In this phase, the cell increases its supply of proteins, increases the number of organelles (such as mitochondria, ribosomes), and grows in size.

S Phase (DNA Replication)

The ensuing S phase starts when DNA replication commences; when it is completed, all of the chromosomes have been replicated, i.e., each chromosome has two (sister) chromatids. Thus, during this phase, the amount of DNA in the cell has effectively doubled, though the ploidy of the cell remains the same. During this phase, synthesis is completed as quickly as possible due to the exposed base pairs being sensitive to harmful external factors such as mutagens.

G_2 Phase (Growth)

G2 phase is a period of protein synthesis and rapid cell growth to prepare the cell for mitosis.

Mitotic Phase (Chromosome Separation)

The relatively brief *M phase* consists of nuclear division (karyokinesis). It is a relatively short period of the cell cycle. M phase is complex and highly regulated. The sequence of events is divided into phases, corresponding to the completion of one set of activities and the start of the next. These phases are sequentially known as:

- prophase,
- metaphase,
- anaphase,
- telophase

Mitosis is the process by which a eukaryotic cell separates the chromosomes in its cell nucleus into two identical sets in two nuclei. During the process of mitosis the pairs of chromosomes condense and attach to fibers that pull the sister chromatids to opposite sides of the cell.

Mitosis occurs exclusively in eukaryotic cells, but occurs in different ways in different species. For example, animals undergo an "open" mitosis, where the nuclear envelope breaks down before the chromosomes separate, while fungi such as *Aspergillus nidulans* and *Saccharomyces cerevisiae* (yeast) undergo a "closed" mitosis, where chromosomes divide within an intact cell nucleus. Prokaryotic cells, which lack a nucleus, divide by a process called binary fission.

Cytokinesis Phase (Separation of All Cell Components)

Mitosis is immediately followed by cytokinesis, which divides the nuclei, cytoplasm, organelles and cell membrane into two cells containing roughly equal shares of these cellular components. Mitosis and cytokinesis together define the division of the mother cell into two daughter cells, genetically identical to each other and to their parent cell. This accounts for approximately 10% of the cell cycle.

10changeably with "M phase". However, there are many cells where mitosis and cytokinesis occur separately, forming single cells with multiple nuclei in a process called endoreplication. This occurs most notably among the fungi and slime moulds, but is found in various groups. Even in animals, cytokinesis and mitosis may occur independently, for instance during certain stages of fruit fly embryonic development. Errors in mitosis can either kill a cell through apoptosis or cause mutations that may lead to cancer.

Regulation of Eukaryotic Cell Cycle

Regulation of the cell cycle involves processes crucial to the survival of a cell, including the detection and repair of genetic damage as well as the prevention of uncontrolled cell division. The molecular events that control the cell cycle are ordered and directional; that is, each process occurs in a sequential fashion and it is impossible to "reverse" the cycle.

Role of Cyclins and CDKs

Two key classes of regulatory molecules, cyclins and cyclin-dependent kinases (CDKs), determine a cell's progress through the cell cycle. Leland H. Hartwell, R. Timothy Hunt, and Paul M. Nurse won the 2001 Nobel Prize in Physiology or Medicine for their discovery of these central molecules. Many of the genes encoding cyclins and CDKs are conserved among all eukaryotes, but in general more complex organisms have more elaborate cell cycle control systems that incorporate more individual components. Many of the relevant genes were first identified by studying yeast, especially *Saccharomyces cerevisiae*; genetic nomenclature in yeast dubs many of these genes *cdc* (for "cell division cycle") followed by an identifying number, e.g. *cdc25* or *cdc20*.

Nobel Laureate Paul Nurse

Cyclins form the regulatory subunits and CDKs the catalytic subunits of an activated heterodimer; cyclins have no catalytic activity and CDKs are inactive in the absence of a partner cyclin. When activated by a bound cyclin, CDKs perform a common biochemical reaction called phosphorylation that activates or inactivates target proteins to orchestrate coordinated entry into the next phase of the cell cycle. Different cyclin-CDK combinations determine the downstream proteins targeted. CDKs are constitutively expressed in cells whereas cyclins are synthesised at specific stages of the cell cycle, in response to various molecular signals.

General Mechanism of Cyclin-CDK Interaction

Upon receiving a pro-mitotic extracellular signal, G_1 cyclin-CDK complexes become active to prepare the cell for S phase, promoting the expression of transcription factors that in turn promote the expression of S cyclins and of enzymes required for DNA replication. The G_1 cyclin-CDK complexes also promote the degradation of molecules that function as S phase inhibitors by targeting them for ubiquitination. Once a protein has been ubiquitinated, it is targeted for proteolytic degradation by the proteasome. However, results from a recent study of E2F transcriptional dynamics at the single-cell level argue that the role of G1 cyclin-CDK activities, in particular cyclin D-CDK4/6, is to tune the timing rather than the commitment of cell cycle entry.

Active S cyclin-CDK complexes phosphorylate proteins that make up the pre-replication complexes assembled during G_1 phase on DNA replication origins. The phosphorylation serves two purposes: to activate each already-assembled pre-replication complex, and to prevent new complexes from forming. This ensures that every portion of the cell's genome will be replicated once and only once. The reason for prevention of gaps in replication is fairly clear, because daughter cells that are missing all or part of crucial genes will die. However, for reasons related to gene copy number effects, possession of extra copies of certain genes is also deleterious to the daughter cells.

Mitotic cyclin-CDK complexes, which are synthesized but inactivated during S and G_2 phases, promote the initiation of mitosis by stimulating downstream proteins involved in chromosome condensation and mitotic spindle assembly. A critical complex activated during this process is a ubiquitin ligase known as the anaphase-promoting complex (APC), which promotes degradation of structural proteins associated with the chromosomal kinetochore. APC also targets the mitotic cyclins for degradation, ensuring that telophase and cytokinesis can proceed.

Specific Action of Cyclin-CDK Complexes

Cyclin D is the first cyclin produced in the cell cycle, in response to extracellular signals (e.g. growth factors). Cyclin D binds to existing CDK4, forming the active cyclin D-CDK4 complex. Cyclin D-CDK4 complex in turn phosphorylates the retinoblastoma susceptibility protein (Rb). The hyperphosphorylated Rb dissociates from the E2F/DP1/Rb complex (which was bound to the E2F responsive genes, effectively "blocking" them from transcription), activating E2F. Activation of E2F results in transcription of various genes like cyclin E, cyclin A, DNA polymerase, thymidine kinase, etc. Cyclin E thus produced binds to CDK2, forming the cyclin E-CDK2 complex, which pushes the cell from G_1 to S phase (G_1/S, which initiates the G_2/M transition). Cyclin B-cdk1 complex activation causes breakdown of nuclear envelope and initiation of prophase, and subsequently, its deactivation causes the cell to exit mitosis.

A quantitative study of E2F transcriptional dynamics at the single-cell level by using engineered fluorescent reporter cells provided a quantitative framework for understanding the control logic of cell cycle entry, challenging the canonical textbook model. Genes that regulate the amplitude of E2F accumulation, such as Myc, determine the commitment into cell cycle and S phase entry. G1 cyclin-CDK activities are not the driver of cell cycle entry. Instead, they primarily tune the timing of E2F increase, thereby modulating the pace of cell cycle progression.

Inhibitors

Two families of genes, the *cip/kip* (*CDK interacting protein/Kinase inhibitory protein*) family and the INK4a/ARF (*Inhibitor of Kinase 4/Alternative Reading Frame*) family, prevent the progression of the cell cycle. Because these genes are instrumental in prevention of tumor formation, they are known as tumor suppressors.

Overview of signal transduction pathways involved in apoptosis, also known as "programmed cell death".

The *cip/kip* family includes the genes p21, p27 and p57. They halt cell cycle in G_1 phase, by binding to, and inactivating, cyclin-CDK complexes. p21 is activated by p53 (which, in turn, is triggered by DNA damage e.g. due to radiation). p27 is activated by Transforming Growth Factor of β (TGF β), a growth inhibitor.

The INK4a/ARF family includes p16[INK4a], which binds to CDK4 and arrests the cell cycle in G_1 phase, and p14[ARF] which prevents p53 degradation.

Synthetic inhibitors of Cdc25 could also be useful for the arrest of cell cycle and therefore be useful as antineoplastic and anticancer agents.

Transcriptional Regulatory Network

Current evidence suggests that a semi-autonomous transcriptional network acts in concert with the CDK-cyclin machinery to regulate the cell cycle. Several gene expression studies in *Saccharomyces cerevisiae* have identified 800-1200 genes that change expression over the course of the cell cycle. They are transcribed at high levels at specific points in the cell cycle, and remain at lower levels throughout the rest of the cycle. While the set of identified genes differs between studies due to the computational methods and criteria used to identify them, each study indicates that a large portion of yeast genes are temporally regulated.

Many periodically expressed genes are driven by transcription factors that are also periodically expressed. One screen of single-gene knockouts identified 48 transcription factors (about 20% of all non-essential transcription factors) that show cell cycle progression defects. Genome-wide studies using high throughput technologies have identified the transcription factors that bind to the promoters of yeast genes, and correlating these findings with temporal expression patterns have allowed the identification of transcription factors that drive phase-specific gene expression. The expression profiles of these transcription factors are driven by the transcription factors that peak in the prior phase, and computational models have shown that a CDK-autonomous network of these transcription factors is sufficient to produce steady-state oscillations in gene expression).

Experimental evidence also suggests that gene expression can oscillate with the period seen in dividing wild-type cells independently of the CDK machinery. Orlando *et al.* used microarrays to measure the expression of a set of 1,271 genes that they identified as periodic in both wild type cells and cells lacking all S-phase and mitotic cyclins (*clb1,2,3,4,5,6*). Of the 1,271 genes assayed, 882 continued to be expressed in the cyclin-deficient cells at the same time as in the wild type cells, despite the fact that the cyclin-deficient cells arrest at the border between G_1 and S phase. However, 833 of the genes assayed changed behavior between the wild type and mutant cells, indicating that these genes are likely directly or indirectly regulated by the CDK-cyclin machinery. Some genes that continued to be expressed on time in the mutant cells were also expressed at different levels in the mutant and wild type cells. These findings suggest that while the transcriptional network may oscillate independently of the CDK-cyclin oscillator, they are coupled in a manner that requires both to ensure the proper timing of cell cycle events. Other work indicates that phosphorylation, a post-translational modification, of cell cycle transcription factors by Cdk1 may alter the localization or activity of the transcription factors in order to tightly control timing of target genes.

While oscillatory transcription plays a key role in the progression of the yeast cell cycle, the CDK-cyclin machinery operates independently in the early embryonic cell cycle. Before the midblastula transition, zygotic transcription does not occur and all needed proteins, such as the B-type cyclins, are translated from maternally loaded mRNA.

DNA Replication and DNA Replication Origin Activity

Analyses of synchronized cultures of *Saccharomyces cerevisiae* under conditions that prevent DNA replication initiation without delaying cell cycle progression showed that origin licensing decreases the expression of genes with origins near their 3' ends, revealing that downstream origins can regulate the expression of upstream genes. This confirms previous predictions from mathematical modeling of a global causal coordination between DNA replication origin activity and mRNA expression, and shows that mathematical modeling of DNA microarray data can be used to correctly predict previously unknown biological modes of regulation.

Checkpoints

Cell cycle checkpoints are used by the cell to monitor and regulate the progress of the cell cycle. Checkpoints prevent cell cycle progression at specific points, allowing verification of necessary phase processes and repair of DNA damage. The cell cannot proceed to the next phase until checkpoint requirements have been met. Checkpoints typically consist of a network of regulatory proteins that monitor and dictate the progression of the cell through the different stages of the cell cycle.

There are several checkpoints to ensure that damaged or incomplete DNA is not passed on to daughter cells. Three main checkpoints exist: the G_1/S checkpoint, the G_2/M checkpoint and the metaphase (mitotic) checkpoint.

G_1/S transition is a rate-limiting step in the cell cycle and is also known as restriction point. This is where the cell checks whether it has enough raw materials to fully replicate its DNA (nucleotide bases, DNA synthase, chromatin, etc.). An unhealthy or malnourished cell will get stuck at this checkpoint.

The G_2/M checkpoint is where the cell ensures that it has enough cytoplasm and phospholipids for two daughter cells. But sometimes more importantly, it checks to see if it is the right time to replicate. There are some situations where many cells need to all replicate simultaneously (for example, a growing embryo should have a symmetric cell distribution until it reaches the mid-blastula transition). This is done by controlling the G_2/M checkpoint.

The metaphase checkpoint is a fairly minor checkpoint, in that once a cell is in metaphase, it has committed to undergoing mitosis. However that's not to say it isn't important. In this checkpoint, the cell checks to ensure that the spindle has formed and that all of the chromosomes are aligned at the spindle equator before anaphase begins.

While these are the three "main" checkpoints, not all cells have to pass through each of these checkpoints in this order to replicate. Many types of cancer are caused by mutations that allow the cells to speed through the various checkpoints or even skip them altogether. Going from S to M to S phase almost consecutively. Because these cells have lost their checkpoints, any DNA mutations that may have occurred are disregarded and passed on to the daughter cells. This is one reason why cancer cells have a tendency to exponentially accrue mutations. Aside from cancer cells, many fully differentiated cell types no longer replicate so they leave the cell cycle and stay in G_0 until their death. Thus removing the need for cellular checkpoints. An alternative model of the cell cycle response to DNA damage has also been proposed, known as the postreplication checkpoint.

Checkpoint regulation plays an important role in an organism's development. In sexual reproduction, when egg fertilization occurs, when the sperm binds to the egg, it releases signalling factors that notify the egg that it has been fertilized. Among other things, this induces the now fertilized oocyte to return from its previously dormant, G_0, state back into the cell cycle and on to mitotic replication and division.

p53 plays an important role in triggering the control mechanisms at both G_1/S and G_2/M checkpoints. In addition to p53, checkpoint regulators are being heavily researched for their roles in cancer growth and proliferation.

Fluorescence Imaging of The Cell Cycle

Fluorescent proteins visualize the cell cycle progression. IFP2.0-hGem(1/110) fluorescence is shown in green and highlights the S/G_2/M phases. smURFP-hCdtI(30/120) fluorescence is shown in red and highlights the G_0/G_1 phases.

Pioneering work by Atsushi Miyawaki and coworkers developed the fluorescent ubiquitination-based cell cycle indicator (FUCCI), which enables fluorescence imaging of the cell cycle. Originally, a green fluorescent protein, mAG, was fused to hGem(1/110) and an orange fluorescent protein (mKO$_2$) was fused to hCdt1(30/120). Note, these fusions are fragments that contain a nuclear localization signal and ubiquitination sites for degradation, but are not functional proteins. The green fluorescent protein is made during the S, G$_2$, or M phase and degraded during the G$_0$ or G$_1$ phase, while the orange fluorescent protein is made during the G$_0$ or G$_1$ phase and destroyed during the S, G$_2$, or M phase. A far-red and near-infrared FUCCI was developed using a cyanobacteria-derived fluorescent protein (smURFP) and a bacteriophytochrome-derived fluorescent protein (movie found at this link).

Role in Tumor Formation

A disregulation of the cell cycle components may lead to tumor formation. As mentioned above, when some genes like the cell cycle inhibitors, RB, p53 etc. mutate, they may cause the cell to multiply uncontrollably, forming a tumor. Although the duration of cell cycle in tumor cells is equal to or longer than that of normal cell cycle, the proportion of cells that are in active cell division (versus quiescent cells in G$_0$ phase) in tumors is much higher than that in normal tissue. Thus there is a net increase in cell number as the number of cells that die by apoptosis or senescence remains the same.

The cells which are actively undergoing cell cycle are targeted in cancer therapy as the DNA is relatively exposed during cell division and hence susceptible to damage by drugs or radiation. This fact is made use of in cancer treatment; by a process known as debulking, a significant mass of the tumor is removed which pushes a significant number of the remaining tumor cells from G$_0$ to G$_1$ phase (due to increased availability of nutrients, oxygen, growth factors etc.). Radiation or chemotherapy following the debulking procedure kills these cells which have newly entered the cell cycle.

The fastest cycling mammalian cells in culture, crypt cells in the intestinal epithelium, have a cycle time as short as 9 to 10 hours. Stem cells in resting mouse skin may have a cycle time of more than 200 hours. Most of this difference is due to the varying length of G$_1$, the most variable phase of the cycle. M and S do not vary much.

In general, cells are most radiosensitive in late M and G$_2$ phases and most resistant in late S phase.

For cells with a longer cell cycle time and a significantly long G$_1$ phase, there is a second peak of resistance late in G$_1$.

The pattern of resistance and sensitivity correlates with the level of sulfhydryl compounds in the cell. Sulfhydryls are natural substances that protect cells from radiation damage and tend to be at their highest levels in S and at their lowest near mitosis.

Mitosis

In cell biology, mitosis is a part of the cell cycle when replicated chromosomes are separated into two new nuclei. In general, mitosis (division of the nucleus) is preceded by the S stage of interphase (during which the DNA is replicated) and is often accompanied or followed by cytokinesis, which divides the cytoplasm, organelles and cell membrane into two new cells containing roughly equal shares of these cellular components. Mitosis and cytokinesis together define the mitotic (M) phase of an animal cell cycle—the division of the mother cell into two daughter cells genetically identical to each other.

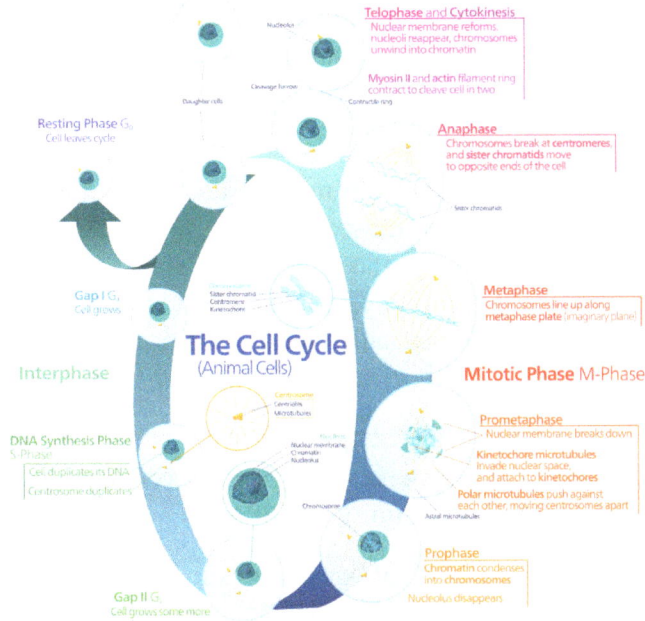

Mitosis in an animal cell (phases ordered counter-clockwise).

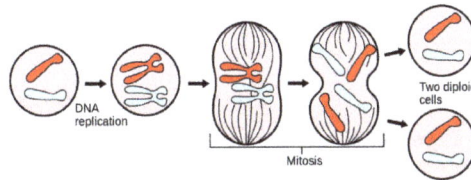

Mitosis divides the chromosomes in a cell nucleus.

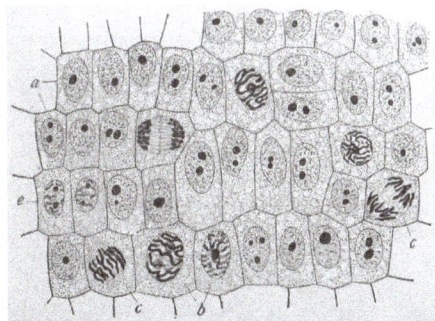

Onion (Allium) cells in different phases of the cell cycle enlarged 800 diameters.
a. non-dividing cells
b. nuclei preparing for division (spireme-stage)
c. dividing cells showing mitotic figures
e. pair of daughter-cells shortly after division

The process of mitosis is divided into stages corresponding to the completion of one set of activities and the start of the next. These stages are prophase, prometaphase, metaphase, anaphase, and telophase. During mitosis, the chromosomes, which have already duplicated, condense and attach to spindle fibers that pull one copy of each chromosome to opposite sides of the cell. The result is two genetically identical daughter nuclei. The rest of the cell may then continue to divide by cytokinesis to produce two daughter cells. Producing three or more daughter cells instead of normal two is a mitotic error called tripolar mitosis or multipolar mitosis (direct cell triplication / multiplication). Other errors during mitosis can induce apoptosis (programmed cell death) or cause mutations. Certain types of cancer can arise from such mutations.

Mitosis occurs only in eukaryotic cells and the process varies in different organisms. For example, animals undergo an "open" mitosis, where the nuclear envelope breaks down before the chromosomes separate, while fungi undergo a "closed" mitosis, where chromosomes divide within an intact cell nucleus. Furthermore, most animal cells undergo a shape change, known as mitotic cell rounding, to adopt a near spherical morphology at the start of mitosis. Prokaryotic cells, which lack a nucleus, divide by a different process called binary fission.

Discovery

German zoologist Otto Bütschli might have claimed the discovery of the process presently known as "mitosis", a term coined by Walther Flemming in 1882.

Mitosis was discovered in frog, rabbit, and cat cornea cells in 1873 and described for the first time by the Polish histologist Wacław Mayzel in 1875. The term is derived from the Greek word μίτος *mitos* "warp thread".

Overview of Mitosis

Types of mitosis

The primary result of mitosis and cytokinesis is the transfer of a parent cell's genome into two daughter cells. The genome is composed of a number of chromosomes—complexes of tightly coiled DNA that contain genetic information vital for proper cell function. Because each resultant daughter cell should be genetically identical to the parent cell, the parent cell must make a copy of each chromosome before mitosis. This occurs during the S phase of interphase. Chromosome duplication results in two identical *sister chromatids* bound together by cohesin proteins at the *centromere*.

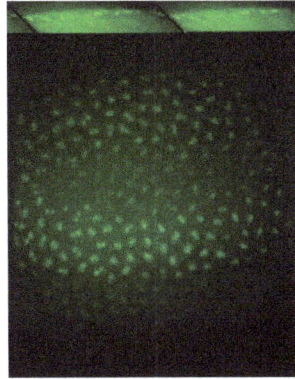

Time-lapse video of mitosis in a *Drosophila melanogaster* embryo.

When mitosis begins, the chromosomes condense and become visible. In some eukaryotes, for example animals, the nuclear envelope, which segregates the DNA from the cytoplasm, disintegrates into small vesicles. The nucleolus, which makes ribosomes in the cell, also disappears. Microtubules project from opposite ends of the cell, attach to the centromeres, and align the chromosomes centrally within the cell. The microtubules then contract to pull the sister chromatids of each chromosome apart. Sister chromatids at this point are called *daughter chromosomes*. As the cell elongates, corresponding daughter chromosomes are pulled toward opposite ends of the cell and condense maximally in late anaphase. A new nuclear envelope forms around the separated daughter chromosomes, which decondense to form interphase nuclei.

During mitotic progression, typically after anaphase onset, the cell may undergo cytokinesis. In animal cells, a cell membrane pinches inward between the two developing nuclei to produce two new cells. In plant cells, a cell plate forms between the two nuclei. Cytokinesis does not always occur; coenocytic (a type of multinucleate condition) cells undergo mitosis without cytokinesis.

Phases of Cell Cycle and Mitosis

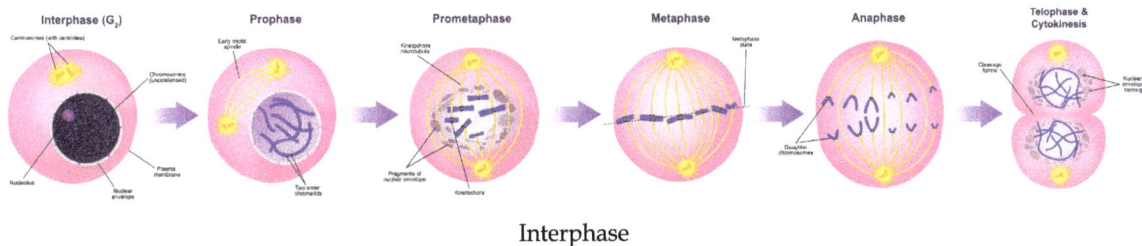

Interphase

The mitotic phase is a relatively short period of the cell cycle. It alternates with the much longer *interphase*, where the cell prepares itself for the process of cell division. Interphase is divided into three phases: G_1 (first gap), S (synthesis), and G_2 (second gap). During all three phases, the cell grows by producing proteins and cytoplasmic organelles. However, chromosomes are replicated only during the S phase. Thus, a cell grows (G_1), continues to grow as it duplicates its chromosomes (S), grows more and prepares for mitosis (G_2), and finally divides (M) before restarting the cycle. All these phases in the cell cycle are highly regulated by cyclins, cyclin-dependent kinases, and other cell cycle proteins. The phases follow one another in strict order and there are "checkpoints"

that give the cell cues to proceed from one phase to another. Cells may also temporarily or perma-
nently leave the cell cycle and enter G_0 phase to stop dividing. This can occur when cells become
overcrowded (density-dependent inhibition) or when they differentiate to carry out specific functions for the organism, as is the case for human heart muscle cells and neurons. Some G_0 cells have
the ability to re-enter the cell cycle.

Preprophase (Plant Cells)

In plant cells only, prophase is preceded by a pre-prophase stage. In highly vacuolated plant cells,
the nucleus has to migrate into the center of the cell before mitosis can begin. This is achieved
through the formation of a phragmosome, a transverse sheet of cytoplasm that bisects the cell
along the future plane of cell division. In addition to phragmosome formation, preprophase is
characterized by the formation of a ring of microtubules and actin filaments (called preprophase
band) underneath the plasma membrane around the equatorial plane of the future mitotic spindle.
This band marks the position where the cell will eventually divide. The cells of higher plants (such
as the flowering plants) lack centrioles; instead, microtubules form a spindle on the surface of the
nucleus and are then organized into a spindle by the chromosomes themselves, after the nuclear
envelope breaks down. The preprophase band disappears during nuclear envelope breakdown and
spindle formation in prometaphase.

Prophase

During prophase, which occurs after G_2 interphase, the cell prepares to divide by tightly condensing its chromosomes and initiating mitotic spindle formation, this process is called chromosome
condensation. During interphase, the genetic material in the nucleus consists of loosely packed
chromatin. At the onset of prophase, chromatin fibers condense into discrete chromosomes that
are typically visible at high magnification through a light microscope. In this stage, chromosomes
are long, thin and thread-like. Each chromosome has two chromatids. The two chromatids are
joined at a place called centromere.

Gene transcription ceases during prophase and does not resume until late anaphase to early G1
phase. The nucleolus also disappears during early prophase.

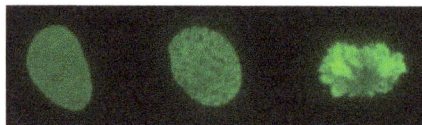

Condensing chromosomes. Interphase nucleus (left), condensing chromosomes (middle) and condensed chromosomes
(right).

Close to the nucleus of animal cells are structures called centrosomes, consisting of a pair of centrioles surrounded by a loose collection of proteins. The centrosome is the coordinating center for
the cell's microtubules. A cell inherits a single centrosome at cell division, which is duplicated by
the cell before a new round of mitosis begins, giving a pair of centrosomes. The two centrosomes
polymerize tubulin to help form a microtubule spindle apparatus. Motor proteins then push the
centrosomes along these microtubules to opposite sides of the cell. Although centrosomes help
organize microtubule assembly, they are not essential for the formation of the spindle apparatus,
since they are absent from plants, and are not absolutely required for animal cell mitosis.

Prometaphase

At the beginning of prometaphase in animal cells, phosphorylation of nuclear lamins causes the nuclear envelope to disintegrate into small membrane vesicles. As this happens, microtubules invade the nuclear space. This is called *open mitosis*, and it occurs in some multicellular organisms. Fungi and some protists, such as algae or trichomonads, undergo a variation called *closed mitosis* where the spindle forms inside the nucleus, or the microtubules penetrate the intact nuclear envelope.

In late prometaphase, *kinetochore microtubules* begin to search for and attach to chromosomal kinetochores. A *kinetochore* is a proteinaceous microtubule-binding structure that forms on the chromosomal centromere during late prophase. A number of *polar microtubules* find and interact with corresponding polar microtubules from the opposite centrosome to form the mitotic spindle. Although the kinetochore structure and function are not fully understood, it is known that it contains some form of molecular motor. When a microtubule connects with the kinetochore, the motor activates, using energy from ATP to "crawl" up the tube toward the originating centrosome. This motor activity, coupled with polymerisation and depolymerisation of microtubules, provides the pulling force necessary to later separate the chromosome's two chromatids.

Metaphase

After the microtubules have located and attached to the kinetochores in prometaphase, the two centrosomes begin pulling the chromosomes towards opposite ends of the cell. The resulting tension causes the chromosomes to align along the *metaphase plate* or *equatorial plane*, an imaginary line that is centrally located between the two centrosomes (at approximately the midline of the cell). To ensure equitable distribution of chromosomes at the end of mitosis, the *metaphase checkpoint* guarantees that kinetochores are properly attached to the mitotic spindle and that the chromosomes are aligned along the metaphase plate. If the cell successfully passes through the metaphase checkpoint, it proceeds to anaphase.

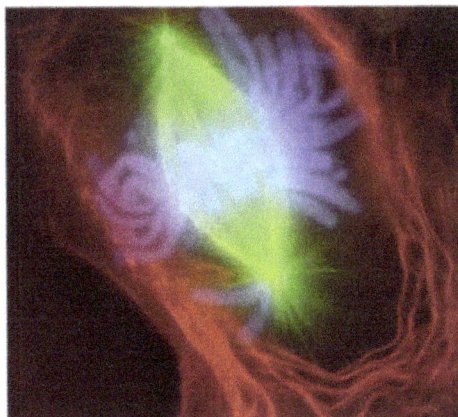

A cell in late metaphase. All chromosomes (blue) but one have arrived at the metaphase plate.

Anaphase

During *anaphase A*, the cohesins that bind sister chromatids together are cleaved, forming two identical daughter chromosomes. Shortening of the kinetochore microtubules pulls the newly formed daughter chromosomes to opposite ends of the cell. During *anaphase B*, polar microtubules push against each other, causing the cell to elongate. In late anaphase, chromosomes also reach their overall maximal condensation level, to help chromosome segregation and the re-formation of the nucleus. In most animal cells, anaphase A precedes anaphase B, but some vertebrate egg cells demonstrate the opposite order of events.

Telophase

Telophase is a reversal of prophase and prometaphase events. At telophase, the polar microtubules continue to lengthen, elongating the cell even more. If the nuclear envelope has broken down, a new nuclear envelope forms using the membrane vesicles of the parent cell's old nuclear envelope. The new envelope forms around each set of separated daughter chromosomes (though the membrane does not enclose the centrosomes) and the nucleolus reappears. Both sets of chromosomes, now surrounded by new nuclear membrane, begin to "relax" or decondense. Mitosis is complete. Each daughter nucleus has an identical set of chromosomes. Cell division may or may not occur at this time depending on the organism.

Cytokinesis

Cytokinesis is not a phase of mitosis but rather a separate process, necessary for completing cell division. In animal cells, a cleavage furrow (pinch) containing a contractile ring develops where the metaphase plate used to be, pinching off the separated nuclei. In both animal and plant cells, cell division is also driven by vesicles derived from the Golgi apparatus, which move along microtubules to the middle of the cell. In plants, this structure coalesces into a cell plate at the center of the phragmoplast and develops into a cell wall, separating the two nuclei. The phragmoplast is a microtubule structure typical for higher plants, whereas some green algae use a phycoplast microtubule array during cytokinesis. Each daughter cell has a complete copy of the genome of its parent cell. The end of cytokinesis marks the end of the M-phase.

Cilliate undergoing cytokinesis, with the cleavage furrow being clearly visible

There are many cells where mitosis and cytokinesis occur separately, forming single cells with multiple nuclei. The most notable occurrence of this is among the fungi, slime molds, and coenocytic algae, but the phenomenon is found in various other organisms. Even in animals, cytokinesis and mitosis may occur independently, for instance during certain stages of fruit fly embryonic development.

Significance

Mitosis is important for the maintenance of the chromosomal set; each cell formed receives chromosomes that are alike in composition and equal in number to the chromosomes of the parent cell.

Mitosis occurs in the following circumstances:

Development and growth

The number of cells within an organism increases by mitosis. This is the basis of the development of a multicellular body from a single cell, i.e., zygote and also the basis of the growth of a multicellular body.

Cell replacement

In some parts of body, e.g. skin and digestive tract, cells are constantly sloughed off and replaced by new ones. New cells are formed by mitosis and so are exact copies of the cells being replaced. In like manner, red blood cells have short lifespan (only about 4 months) and new RBCs are formed by mitosis.

Regeneration

Some organisms can regenerate body parts. The production of new cells in such instances is achieved by mitosis. For example, starfish regenerate lost arms through mitosis.

Asexual Reproduction

Some organisms produce genetically similar offspring through asexual reproduction. For example, the hydra reproduces asexually by budding. The cells at the surface of hydra undergo mitosis and form a mass called a bud. Mitosis continues in the cells of the bud and this grows into a new individual. The same division happens during asexual reproduction or vegetative propagation in plants.

Cell Shape Changes During Mitosis

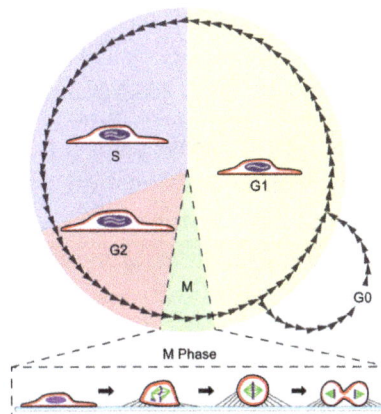

Cell shape changes through mitosis for a typical animal cell cultured on a flat surface. The cell undergoes mitotic cell rounding during spindle assembly and then divides via cytokinesis. The actomyosin cortex is depicted in red, DNA/ chromosomes purple, microtubules green, and membrane and retraction fibers in black. Rounding also occurs in live tissue, as described in the text.

In animal tissue, most cells round up to a near-spherical shape during mitosis. In epithelia and epidermis, an efficient rounding process is correlated with proper mitotic spindle alignment and subsequent correct positioning of daughter cells. Moreover, researchers have found that if rounding is heavily suppressed it may result in spindle defects, primarily pole splitting and failure to efficiently capture chromosomes. Therefore, mitotic cell rounding is thought to play a protective role in ensuring accurate mitosis.

Rounding forces are driven by reorganization of F-actin and myosin (actomyosin) into a contractile homogeneous cell cortex that 1) rigidifies the cell periphery and 2) facilitates generation of intracellular hydrostatic pressure (up to 10 fold higher than interphase). The generation of intracellular pressure is particularly critical under confinement, such as would be important in a tissue scenario, where outward forces must be produced to round up against surrounding cells and/or the extracellular matrix. Generation of pressure is dependent on formin-mediated F-actin nucleation and Rho kinase (ROCK)-mediated myosin II contraction, both of which are governed upstream by signaling pathways RhoA and ECT2 through the activity of Cdk1. Due to its importance in mitosis, the molecular components and dynamics of the mitotic actomyosin cortex is an area of active research.

Errors and Variations of Mitosis

Errors can occur during mitosis, especially during early embryonic development in humans. Mitotic errors can create aneuploid cells that have too few or too many of one or more chromosomes, a condition associated with cancer. Early human embryos, cancer cells, infected or intoxicated cells can also suffer from pathological division into three or more daughter cells (tripolar or multipolar mitosis), resulting in severe errors in their chromosomal complements.

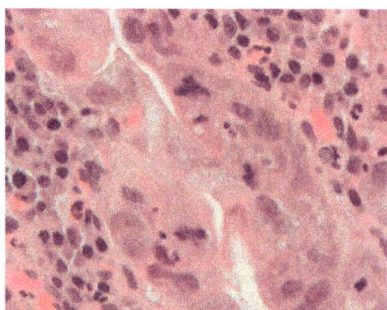

An abnormal (tripolar) mitosis (12 o'clock position) in a precancerous lesion of the stomach (H&E stain)

In *nondisjunction*, sister chromatids fail to separate during anaphase. One daughter cell receives both sister chromatids from the nondisjoining chromosome and the other cell receives none. As a result, the former cell gets three copies of the chromosome, a condition known as *trisomy*, and the latter will have only one copy, a condition known as *monosomy*. On occasion, when cells experience nondisjunction, they fail to complete cytokinesis and retain both nuclei in one cell, resulting in binucleated cells.

Anaphase lag occurs when the movement of one chromatid is impeded during anaphase. This may be caused by a failure of the mitotic spindle to properly attach to the chromosome. The lagging chromatid is excluded from both nuclei and is lost. Therefore, one of the daughter cells will be monosomic for that chromosome.

Endoreduplication (or endoreplication) occurs when chromosomes duplicate but the cell does not subsequently divide. This results in polyploid cells or, if the chromosomes duplicates repeatedly, polytene chromosomes. Endoreduplication is found in many species and appears to be a normal part of development. Endomitosis is a variant of endoreduplication in which cells replicate their chromosomes during S phase and enter, but prematurely terminate, mitosis. Instead of being divided into two new daughter nuclei, the replicated chromosomes are retained within the original nucleus. The cells then re-enter G_1 and S phase and replicate their chromosomes again. This may occur multiple times, increasing the chromosome number with each round of replication and endomitosis. Platelet-producing megakaryocytes go through endomitosis during cell differentiation.

Meiosis

Meiosis [i]/ma☐☐o☐s☐s/ is a specialized type of cell division that reduces the chromosome number by half, creating four haploid cells, each genetically distinct from the parent cell that gave rise to them. This process occurs in all sexually reproducing single-celled and multicellular eukaryotes, including animals, plants, and fungi. Errors in meiosis resulting in aneuploidy are the leading known cause of miscarriage and the most frequent genetic cause of developmental disabilities.

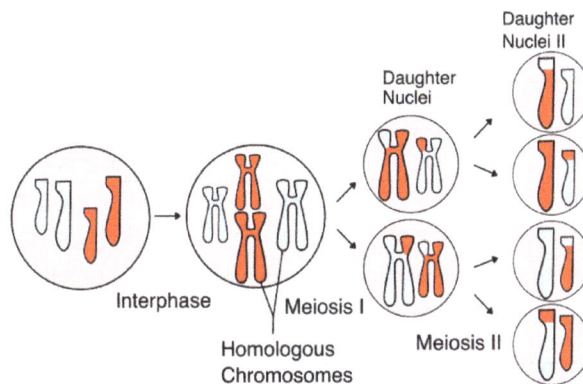

In meiosis, the chromosomes duplicate (during interphase) and homologous chromosomes exchange genetic information (chromosomal crossover) during the first division, called meiosis I. The daughter cells divide again in meiosis II, splitting up sister chromatids to form haploid gametes. Male and female gametes fuse during fertilization, creating a diploid cell with a complete set of paired chromosomes.

In meiosis, DNA replication is followed by two rounds of cell division to produce four potential daughter cells, each with half the number of chromosomes as the original parent cell. The two meiotic divisions are known as *Meiosis I* and *Meiosis II*. Before meiosis begins, during S phase of the cell cycle, the DNA of each chromosome is replicated so that it consists of two identical sister chromatids, which remain held together through sister chromatid cohesion. This S-phase can be referred to as "premeiotic S-phase" or "meiotic S-phase." Immediately following DNA replication, meiotic cells enter a prolonged G2-like stage known as meiotic prophase. During this time, homologous chromosomes pair with each other and undergo genetic recombination, a programmed process in which DNA is cut and then repaired, which allows them to exchange some of their genetic information. A subset of recombination events results in crossovers, which create physical links known as chiasmata (singular: chiasma, for the Greek letter Chi (X)) between the homologous chromosomes. In most organisms, these links are essential to direct each pair of homologous chromosomes to segregate away from each other during Meiosis I, resulting in two haploid cells that have half the number of chromosomes as the parent cell. During Meiosis II, the cohesion between sister chromatids is released and they segregate from one another, as during mitosis. In some cases all four of the meiotic products form gametes such as sperm, spores, or pollen. In female animals, three of the four meiotic products are typically eliminated by extrusion into polar bodies, and only one cell develops to produce an ovum.

Because the number of chromosomes is halved during meiosis, gametes can fuse (i.e. fertilization) to form a diploid zygote that contains two copies of each chromosome, one from each parent. Thus, alternating cycles of meiosis and fertilization enable sexual reproduction, with successive generations maintaining the same number of chromosomes. For example, diploid human cells contain 23 pairs of chromosomes (46 total), half of maternal origin and half of paternal origin. Meiosis produces haploid gametes (ova or sperm) that contain one set of 23 chromosomes. When two gametes (an egg and a sperm) fuse, the resulting zygote is once again diploid, with the mother and father each contributing 23 chromosomes. This same pattern, but not the same number of chromosomes, occurs in all organisms that utilize meiosis.

Overview

While the process of meiosis is related to the more general cell division process of mitosis, it differs in two important respects:

recombination	meiosis	shuffles the genes between the two chromosomes in each pair (one received from each parent), producing recombinant chromosomes with unique genetic combinations in every gamete
	mitosis	occurs only if needed to repair DNA damage; usually occurs between identical sister chromatids and does not result in genetic changes
chromosome number (ploidy)	meiosis	produces four genetically unique cells, each with half the number of chromosomes as in the parent
	mitosis	produces two genetically identical cells, each with the same number of chromosomes as in the parent

Meiosis begins with a diploid cell, which contains two copies of each chromosome, termed homologs. First, the cell undergoes DNA replication, so each homolog now consists of two identical sister chromatids. Then each set of homologs pair with each other and exchange DNA by homologous recombination leading to physical connections (crossovers) between the homologs. In the first meiotic division, the homologs are segregated to separate daughter cells by the spindle apparatus. The cells then proceed to a second division without an intervening round of DNA replication. The sister chromatids are segregated to separate daughter cells to produce a total of four haploid cells. Female animals employ a slight variation on this pattern and produce one large ovum and two small polar bodies. Because of recombination, an individual chromatid can consist of a new combination of maternal and paternal DNA, resulting in offspring that are genetically distinct from either parent. Furthermore, an individual gamete can include an assortment of maternal, paternal, and recombinant chromatids. This genetic diversity resulting from sexual reproduction contributes to the variation in traits upon which natural selection can act.

Meiosis uses many of the same mechanisms as mitosis, the type of cell division used by eukaryotes to divide one cell into two identical daughter cells. In some plants, fungi, and protists meiosis results in the formation of spores: haploid cells that can divide vegetatively without undergoing fertilization. Some eukaryotes, like bdelloid rotifers, do not have the ability to carry out meiosis and have acquired the ability to reproduce by parthenogenesis.

Meiosis does not occur in archaea or bacteria, which generally reproduce via asexual processes such as binary fission. However, a "sexual" process known as horizontal gene transfer involves the transfer of DNA from one bacterium or archaeon to another and recombination of these DNA molecules of different parental origin.

History

Meiosis was discovered and described for the first time in sea urchin eggs in 1876 by the German biologist Oscar Hertwig. It was described again in 1883, at the level of chromosomes, by the Belgian zoologist Edouard Van Beneden, in *Ascaris* roundworm eggs. The significance of meiosis for reproduction and inheritance, however, was described only in 1890 by German biologist August Weismann, who noted that two cell divisions were necessary to transform one diploid cell into four haploid cells if the number of chromosomes had to be maintained. In 1911 the American geneticist Thomas Hunt Morgan detected crossovers in meiosis in the fruit fly *Drosophila melanogaster*, which helped to establish that genetic traits are transmitted on chromosomes.

The term meiosis (originally spelled "maiosis") was introduced to biology by J.B. Farmer and J.E.S. Moore in 1905:

We propose to apply the terms Maiosis or Maiotic phase to cover the whole series of nuclear changes included in the two divisions that were designated as Heterotype and Homotype by Flemming.

Occurrence in Eukaryotic Life Cycles

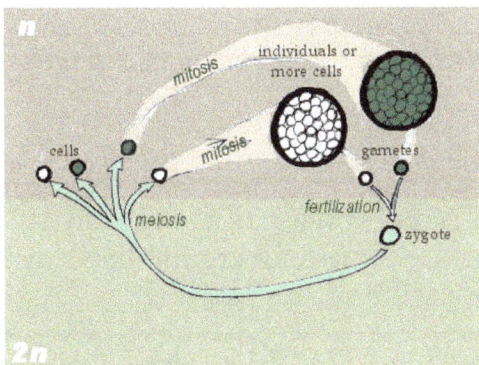

Zygotic life cycle.

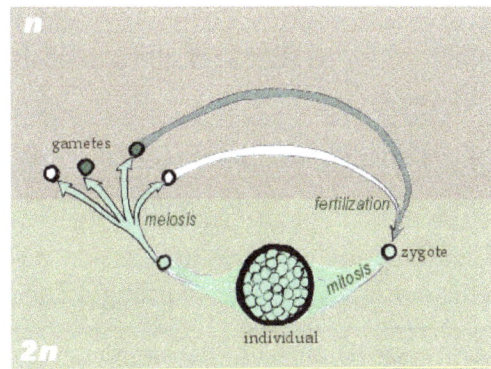

Gametic life cycle.

Meiosis occurs in eukaryotic life cycles involving sexual reproduction, consisting of the constant cyclical process of meiosis and fertilization. This takes place alongside normal mitotic cell division. In multicellular organisms, there is an intermediary step between the diploid and haploid transition where the organism grows. At certain stages of the life cycle, germ cells produce gametes. Somatic cells make up the body of the organism and are not involved in gamete production.

Cycling meiosis and fertilization events produces a series of transitions back and forth between alternating haploid and diploid states. The organism phase of the life cycle can occur either during the diploid state (*gametic* or *diploid* life cycle), during the haploid state (*zygotic* or *haploid* life cycle), or both (*sporic* or *haplodiploid* life cycle, in which there are two distinct organism phases, one during the haploid state and the other during the diploid state). In this sense there are three types of life cycles that utilize sexual reproduction, differentiated by the location of the organism phase(s).

In the *gametic life cycle* or " diplontic life cycle", of which humans are a part, the organism is diploid, grown from a diploid cell called the zygote. The organism's diploid germ-line stem cells undergo meiosis to create haploid gametes (the spermatozoa for males and ova for females), which fertilize to form the zygote. The diploid zygote undergoes repeated cellular division by mitosis to grow into the organism.

In the *zygotic life cycle* the organism is haploid instead, spawned by the proliferation and differentiation of a single haploid cell called the gamete. Two organisms of opposing sex contribute their haploid gametes to form a diploid zygote. The zygote undergoes meiosis immediately, creating four haploid cells. These cells undergo mitosis to create the organism. Many fungi and many protozoa utilize the zygotic life cycle.

Finally, in the *sporic life cycle*, the living organism alternates between haploid and diploid states. Consequently, this cycle is also known as the alternation of generations. The diploid organism's germ-line cells undergo meiosis to produce spores. The spores proliferate by mitosis, growing into a haploid organism. The haploid organism's gamete then combines with another haploid organism's gamete, creating the zygote. The zygote undergoes repeated mitosis and differentiation to become a diploid organism again. The sporic life cycle can be considered a fusion of the gametic and zygotic life cycles.

Process

The preparatory steps that lead up to meiosis are identical in pattern and name to interphase of the mitotic cell cycle.

Interphase is divided into three phases:

- Growth 1 (G_1) phase: In this very active phase, the cell synthesizes its vast array of proteins, including the enzymes and structural proteins it will need for growth. In G_1, each of the chromosomes consists of a single linear molecule of DNA.

- Synthesis (S) phase: The genetic material is replicated; each of the cell's chromosomes duplicates to become two identical sister chromatids attached at a centromere. This replication does not change the ploidy of the cell since the centromere number remains the same.

The identical sister chromatids have not yet condensed into the densely packaged chromosomes visible with the light microscope. This will take place during prophase I in meiosis.

- Growth 2 (G_2) phase: G_2 phase as seen before mitosis is not present in meiosis. Meiotic prophase corresponds most closely to the G_2 phase of the mitotic cell cycle.

Interphase is followed by meiosis I and then meiosis II. Meiosis I separates homologous chromosomes, each still made up of two sister chromatids, into two daughter cells, thus reducing the chromosome number by half. During meiosis II, sister chromatids decouple and the resultant daughter chromosomes are segregated into four daughter cells. For diploid organisms, the daughter cells resulting from meiosis are haploid and contain only one copy of each chromosome. In some species, cells enter a resting phase known as interkinesis between meiosis I and meiosis II.

Meiosis I and II are each divided into prophase, metaphase, anaphase, and telophase stages, similar in purpose to their analogous subphases in the mitotic cell cycle. Therefore, meiosis includes the stages of meiosis I (prophase I, metaphase I, anaphase I, telophase I) and meiosis II (prophase II, metaphase II, anaphase II, telophase II).

Meiosis generates gamete genetic diversity in two ways: (1) the independent orientation of homologous chromosome pairs along the metaphase plate during metaphase I and the subsequent separation of homologs during anaphase I allows a random and independent distribution of chromosomes to each daughter cell (and ultimately to gametes); and (2) physical exchange of homologous chromosomal regions by homologous recombination during prophase I results in new combinations of DNA within chromosomes.

During meiosis, specific genes are more highly transcribed. In addition to strong meiotic stage-specific expression of mRNA, there are also pervasive translational controls (e.g. selective usage of preformed mRNA), regulating the ultimate meiotic stage-specific protein expression of genes during meiosis. Thus, both transcriptional and translational controls determine the broad restructuring of meiotic cells needed to carry out meiosis.

Phases

Meiosis is divided into meiosis I and meiosis II which are further divided into Karyokinesis I and Cytokinesis I & Karyokinesis II and Cytokinesis II respectively.

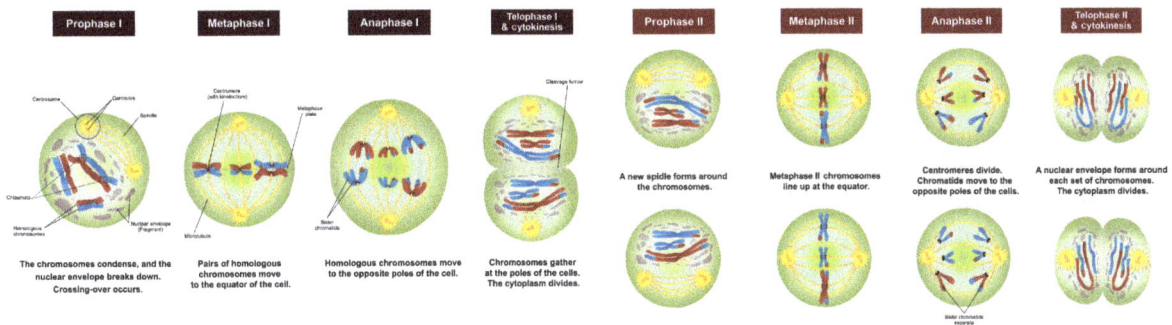

Diagram of the meiotic phases

Meiosis I

Meiosis I segregates homologous chromosomes, which are joined as tetrads (2n, 4c), producing two haploid cells (n chromosomes, 23 in humans) which each contain chromatid pairs (1n, 2c). Because the ploidy is reduced from diploid to haploid, meiosis I is referred to as a *reductional division*. Meiosis II is an *equational division* analogous to mitosis, in which the sister chromatids are segregated, creating four haploid daughter cells (1n, 1c).

Prophase I

Prophase I is typically the longest phase of meiosis. During prophase I, homologous chromosomes pair and exchange DNA in a process called homologous recombination. This often results in chromosomal crossover. This process is critical for pairing between homologous chromosomes and hence for accurate segregation of the chromosomes at the first meiosis division. The new combinations of DNA created during crossover are a significant source of genetic variation, and result in new combinations of alleles, which may be beneficial. The paired and replicated chromosomes are called bivalents or tetrads, which have two chromosomes and four chromatids, with one chromosome coming from each parent. The process of pairing the homologous chromosomes is called synapsis. At this stage, non-sister chromatids may cross-over at points called chiasmata (plural; singular chiasma). Prophase I has historically been divided into a series of substages which are named according to the appearance of chromosomes.

Leptotene

The first stage of prophase I is the *leptotene* stage, also known as *leptonema*, from Greek words meaning "thin threads". In this stage of prophase I, individual chromosomes—each consisting of two sister chromatids—become "individualized" to form visible strands within the nucleus. The two sister chromatids closely associate and are visually indistinguishable from one another. During leptotene, lateral elements of the synaptonemal complex assemble. Leptotene is of very short duration and progressive condensation and coiling of chromosome fibers takes place.

Zygotene

The *zygotene* stage, also known as *zygonema*, from Greek words meaning "paired threads", occurs as the chromosomes approximately line up with each other into homologous chromosome pairs. In some organisms, this is called the bouquet stage because of the way the telomeres cluster at one end of the nucleus. At this stage, the synapsis (pairing/coming together) of homologous chromosomes takes place, facilitated by assembly of central element of the synaptonemal complex. Pairing is brought about in a zipper-like fashion and may start at the centromere (procentric), at the chromosome ends (proterminal), or at any other portion (intermediate). Individuals of a pair are equal in length and in position of the centromere. Thus pairing is highly specific and exact. The paired chromosomes are called bivalent or tetrad chromosomes.

Pachytene

The *pachytene* stage, also known as *pachynema*, from Greek words meaning "thick threads",. At this point a tetrad of the chromosomes has formed known as a bivalent. This is the stage when chromosomal crossover (crossing over) occurs. Nonsister chromatids of homologous chromosomes may exchange segments over regions of homology. Sex chromosomes, however, are not wholly identical, and only exchange information over a small region of homology. At the sites where exchange happens, chiasmata form. The exchange of information between the non-sister chromatids results in a recombination of information; each chromosome has the complete set of information it had before, and there are no gaps formed as a result of the process. Because the chromosomes cannot be distinguished in the synaptonemal complex, the actual act of crossing over is not perceivable through the microscope, and chiasmata are not visible until the next stage.

Diplotene

During the *diplotene* stage, also known as *diplonema*, from Greek words meaning "two threads", the synaptonemal complex degrades and homologous chromosomes separate from one another a little. The chromosomes themselves uncoil a bit, allowing some transcription of DNA. However, the homologous chromosomes of each bivalent remain tightly bound at chiasmata, the regions where crossing-over occurred. The chiasmata remain on the chromosomes until they are severed at the transition to anaphase I.

In mammalian and human fetal oogenesis all developing oocytes develop to this stage and are arrested before birth. This suspended state is referred to as the *dictyotene stage* or dictyate. It lasts until meiosis is resumed to prepare the oocyte for ovulation, which happens at puberty or even later.

Diakinesis

Chromosomes condense further during the *diakinesis* stage, from Greek words meaning "moving through". This is the first point in meiosis where the four parts of the tetrads are actually visible. Sites of crossing over entangle together, effectively overlapping, making chiasmata clearly visible. Other than this observation, the rest of the stage closely resembles prometaphase of mitosis; the nucleoli disappear, the nuclear membrane disintegrates into vesicles, and the meiotic spindle begins to form.

Synchronous Processes

During these stages, two centrosomes, containing a pair of centrioles in animal cells, migrate to the two poles of the cell. These centrosomes, which were duplicated during S-phase, function as microtubule organizing centers nucleating microtubules, which are essentially cellular ropes and poles. The microtubules invade the nuclear region after the nuclear envelope disintegrates, attaching to the chromosomes at the kinetochore. The kinetochore functions as a motor, pulling the chromosome along the attached microtubule toward the originating centrosome, like a train on a track. There are four kinetochores on each tetrad, but the pair of kinetochores on each sister chromatid fuses and functions as a unit during meiosis I.

Microtubules that attach to the kinetochores are known as *kinetochore microtubules*. Other microtubules will interact with microtubules from the opposite centrosome: these are called *nonkinetochore microtubules* or *polar microtubules*. A third type of microtubules, the aster microtubules, radiates from the centrosome into the cytoplasm or contacts components of the membrane skeleton.

Metaphase I

Homologous pairs move together along the metaphase plate: As *kinetochore microtubules* from both centrosomes attach to their respective kinetochores, the paired homologous chromosomes align along an equatorial plane that bisects the spindle, due to continuous counterbalancing forces exerted on the bivalents by the microtubules emanating from the two kinetochores of homologous chromosomes. The physical basis of the independent assortment of chromosomes is the random orientation of each bivalent along the metaphase plate, with respect to the orientation of the other bivalents along the same equatorial line. The protein complex cohesin holds sister chromatids together from the time of their replication until anaphase. In mitosis, the force of kinetochore microtubules pulling in opposite directions creates tension. The cell senses this tension and does not progress with anaphase until all the chromosomes are properly bi-oriented. In meiosis, establishing tension requires at least one crossover per chromosome pair in addition to cohesin between sister chromatids.

Anaphase I

Kinetochore microtubules shorten, pulling homologous chromosomes (which consist of a pair of sister chromatids) to opposite poles. Nonkinetochore microtubules lengthen, pushing the centrosomes farther apart. The cell elongates in preparation for division down the center. Unlike in mitosis, only the cohesin from the chromosome arms is degraded while the cohesin surrounding the centromere remains protected. This allows the sister chromatids to remain together while homologs are segregated.

Telophase I

The first meiotic division effectively ends when the chromosomes arrive at the poles. Each daughter cell now has half the number of chromosomes but each chromosome consists of a pair of chromatids. The microtubules that make up the spindle network disappear, and a new nuclear membrane surrounds each haploid set. The chromosomes uncoil back into chromatin. Cytokinesis, the pinching of the cell membrane in animal cells or the formation of the cell wall in plant cells, occurs, completing the creation of two daughter cells. Sister chromatids remain attached during telophase I.

Cells may enter a period of rest known as interkinesis or interphase II. No DNA replication occurs during this stage.

Meiosis II

Meiosis II is the second meiotic division, and usually involves equational segregation, or separation of sister chromatids. Mechanically, the process is similar to mitosis, though its genetic results are fundamentally different. The end result is production of four haploid cells (n chromosomes, 23 in humans) from the two haploid cells (with n chromosomes, each consisting of two sister chromatids) produced in meiosis I. The four main steps of Meiosis II are: Prophase II, Metaphase II, Anaphase II, and Telophase II.

In prophase II we see the disappearance of the nucleoli and the nuclear envelope again as well as the shortening and thickening of the chromatids. Centrosomes move to the polar regions and arrange spindle fibers for the second meiotic division.

In metaphase II, the centromeres contain two kinetochores that attach to spindle fibers from the centrosomes at opposite poles. The new equatorial metaphase plate is rotated by 90 degrees when compared to meiosis I, perpendicular to the previous plate.

This is followed by anaphase II, in which the remaining centromeric cohesin is cleaved allowing the sister chromatids to segregate. The sister chromatids by convention are now called sister chromosomes as they move toward opposing poles.

The process ends with telophase II, which is similar to telophase I, and is marked by decondensation and lengthening of the chromosomes and the disassembly of the spindle. Nuclear envelopes reform and cleavage or cell plate formation eventually produces a total of four daughter cells, each with a haploid set of chromosomes.

Meiosis is now complete and ends up with four new daughter cells.

Origin and Function

The origin and function of meiosis are fundamental to understanding the evolution of sexual reproduction in Eukaryotes. There is no current consensus among biologists on the questions of how sex in Eukaryotes arose in evolution, what basic function sexual reproduction serves, and why it is maintained, given the basic two-fold cost of sex. It is clear that it evolved over 1.2 billion years ago, and that almost all species which are descendents of the original sexually reproducing species are still sexual reproducers, including plants, fungi, and animals.

Meiosis is ubiquitous among eukaryotes. It occurs in single-celled organisms such as yeast, as well as in multicellular organisms, such as humans. Eukaryotes arose from prokaryotes more than 1.5 billion years ago, and the earliest eukaryotes were likely single-celled organisms. To understand sex in eukaryotes, it is necessary to understand (1) how meiosis arose in single celled eukaryotes, and (2) the function of meiosis.

Nondisjunction

The normal separation of chromosomes in meiosis I or sister chromatids in meiosis II is termed *disjunction*. When the segregation is not normal, it is called *nondisjunction*. This results in the production of gametes which have either too many or too few of a particular chromosome, and is a common mechanism for trisomy or monosomy. Nondisjunction can occur in the meiosis I or meiosis II, phases of cellular reproduction, or during mitosis.

Most monosomic and trisomic human embryos are not viable, but some aneuploidies can be tolerated, such as trisomy for the smallest chromosome, chromosome 21. Phenotypes of these aneuploidies range from severe developmental disorders to asymptomatic. Medical conditions include but are not limited to:

- Down Syndrome - trisomy of chromosome 21

- Patau Syndrome - trisomy of chromosome 13

- Edward Syndrome - trisomy of chromosome 18

- Klinefelter Syndrome - extra X chromosomes in males - i.e. XXY, XXXY, XXXXY, etc.

- Turner Syndrome - lacking of one X chromosome in females - i.e. Xo

- Triple X syndrome - an extra X chromosome in females

- XYY Syndrome - an extra Y chromosome in males.

The probability of nondisjunction in human oocytes increases with increasing maternal age, presumably due to loss of cohesin over time.

Meiosis in Plants and Animals

Meiosis occurs in all animals and plants. The end result, the production of gametes with half the number of chromosomes as the parent cell, is the same, but the detailed process is different. In animals, meiosis produces gametes directly. In land plants and some algae, there is an alternation of generations such that meiosis in the diploid sporophyte generation produces haploid spores. These spores multiply by mitosis, developing into the haploid gametophyte generation, which then gives rise to gametes directly (i.e. without further meiosis). In both animals and plants, the final stage is for the gametes to fuse, restoring the original number of chromosomes.

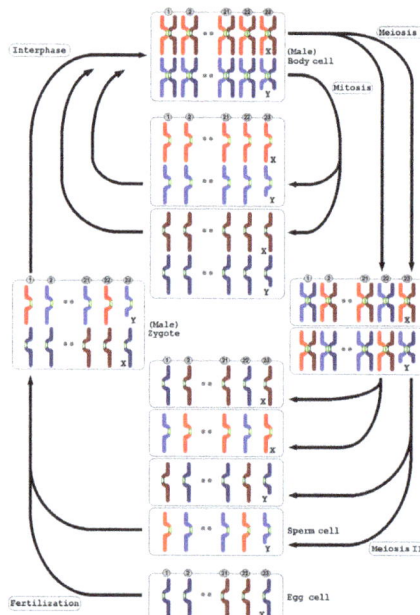

Overview of chromatides' and chromosomes' distribution within the mitotic and meiotic cycle of a male human cell

Meiosis In Mammals

In females, meiosis occurs in cells known as oocytes (singular: oocyte). Each oocyte that initiates meiosis divides twice, unequally in each case. The first division produces a daughter cell that will undergo a second division, and a much smaller "polar body" that is extruded from the surface of the cell and does not divide further. Following Meiosis II, a "second polar body" is extruded, and the single remaining haploid cell enlarges to become an ovum. Since the first polar body normally disintegrates rather than dividing again, meiosis in female mammals results in three products, the oocyte and two polar bodies. However, before these divisions occur, these cells stop at the diplotene stage of meiosis I and lie dormant within a protective shell of somatic cells called the follicle. Follicles begin growth at a steady pace in a process known as folliculogenesis, and a small number enter the menstrual cycle. Menstruated oocytes continue meiosis I and arrest at meiosis II until fertilization. The process of meiosis in females occurs during oogenesis, and differs from the typical meiosis in that it features a long period of meiotic arrest known as the dictyate stage and lacks the assistance of centrosomes.

In males, meiosis occurs during spermatogenesis in the seminiferous tubules of the testicles. Meiosis during spermatogenesis is specific to a type of cell called spermatocytes that will later mature to become spermatozoa.

In female mammals, meiosis begins immediately after primordial germ cells migrate to the ovary in the embryo, but in the males, meiosis begins later, at the time of puberty. It is retinoic acid, derived from the primitive kidney (mesonephros) that stimulates meiosis in ovarian oogonia. Tissues of the male testis suppress meiosis by degrading retinoic acid, a stimulator of meiosis. This is overcome at puberty when cells within seminiferous tubules called Sertoli cells start making their own retinoic acid. Sensitivity to retinoic acid is also adjusted by proteins called nanos and DAZL.

Meiosis Vs. Mitosis

In order to understand meiosis, a comparison to mitosis is helpful. The table below shows the differences between meiosis and mitosis.

	Meiosis	Mitosis
End result	Normally four cells, each with half the number of chromosomes as the parent	Two cells, having the same number of chromosomes as the parent
Function	Sexual reproduction, production of gametes (sex cells)	Cellular reproduction, growth, repair, asexual reproduction
Where does it happen?	Reproductive cells of almost all eukaryotes (animals, plants, fungi, and protists)	All proliferating cells in all eukaryotes
Steps	Prophase I, Metaphase I, Anaphase I, Telophase I, Prophase II, Metaphase II, Anaphase II, Telophase II	Prophase, Prometaphase, Metaphase, Anaphase, Telophase
Genetically same as parent?	No	Yes
Crossing over happens?	Yes, normally occurs between each pair of homologous chromosomes	Very rarely
Pairing of homologous chromosomes?	Yes	No
Cytokinesis	Occurs in Telophase I and Telophase II	Occurs in Telophase
Centromeres split	Does not occur in Anaphase I, but occurs in Anaphase II	Occurs in Anaphase

References

- Cristianini, N. and Hahn, M. Introduction to Computational Genomics, Cambridge University Press, 2006. (ISBN 9780521671910 | ISBN 0-521-67191-4)

- Campbell, Neil A.; Brad Williamson; Robin J. Heyden (2006). Biology: Exploring Life. Boston, Massachusetts: Pearson Prentice Hall. ISBN 0-13-250882-6.

- Peter Hamilton Raven; George Brooks Johnson (2002). Biology. McGraw-Hill Education. p. 68. ISBN 978-0-07-112261-0. Retrieved 7 July 2013.

- Alberts B, Johnson A, Lewis J, et al. (2002). Molecular Biology of the Cell (4th ed.). New York: Garland Science. ISBN 0-8153-3218-1.

- Lodish H, Berk A, Zipursky LS, et al. (2004). Molecular Cell Biology (4th ed.). New York: Scientific American Books. ISBN 0-7167-3136-3.

- Hardin, Jeff; Bertoni, Gregory; Kleinsmith, Lewis J. (2015). Becker's World of the Cell (8th ed.). New York: Pearson. pp. 422–446. ISBN 978013399939-6.

- Unless else specified in boxes, then ref is:Walter F., PhD. Boron (2003). Medical Physiology: A Cellular And Molecular Approaoch. Elsevier/Saunders. p. 1300. ISBN 1-4160-2328-3. Page 25

- Robert.S Hine, ed. (2008). Oxford Dictionary Biology (6th ed.). New York: Oxford University Press. p. 113. ISBN 978-0-19-920462-5.

- Griffiths, Anthony J.F.; Wessler, Susan R.; Carroll, Sean B.; Doebley, John (2012). Introduction to Genetic Analysis (10 ed.). New York: W.H. Freeman and Company. p. 35. ISBN 978-1-4292-2943-2.

- Cooper GM (2000). "Chapter 14: The Eukaryotic Cell Cycle". The cell: a molecular approach (2nd ed.). Washington, D.C: ASM Press. ISBN 0-87893-106-6.

- Robbins, Stanley L; Cotran, Ramzi S (2004). Vinay Kumar; Abul K Abbas; Nelson Fausto, eds. Pathological Basis of Disease. Elsevier. ISBN 81-8147-528-3.

- Norbury C (1995). "Cdk2 protein kinase (vertebrates)". In Hardie, D. Grahame; Hanks, Steven. Protein kinase factsBook. Boston: Academic Press. p. 184. ISBN 0-12-324719-5.

- Morgan DO (2007). "2–3". The Cell Cycle: Principles of Control. London: New Science Press. p. 18. ISBN 0-9539181-2-2.

- Maton, A.; Hopkins, J. J.; LaHart, S; Quon Warner, D.; Wright, M.; Jill, D. (1997). Cells: Building Blocks of Life. New Jersey: Prentice Hall. pp. 70–4. ISBN 0-13-423476-6.

- Olson, Mark O. J. (2011). The Nucleolus. Volume 15 of Protein Reviews. Berlin: Springer Science & Business Media. p. 15. ISBN 9781461405146.

Botany: Plant Biology

Botany is the study of plants. Plant anatomy is the study of the structure of plants. The themes covered in this chapter are plant breeding, plant propagation, plant morphology, plant ecology, plant pathology etc. The chapter serves as a source to understand the major categories related to plant biology.

Botany

Botany, also called plant science(s), plant biology or phytology, is the science of plant life and a branch of biology. A botanist or plant scientist is a scientist who specialises in this field. Traditionally, botany has also included the study of fungi and algae by mycologists and phycologists respectively, with the study of these three groups of organisms remaining within the sphere of interest of the International Botanical Congress. Nowadays, botanists study approximately 400,000 species of living organisms of which some 260,000 species are vascular plants and about 248,000 are flowering plants.

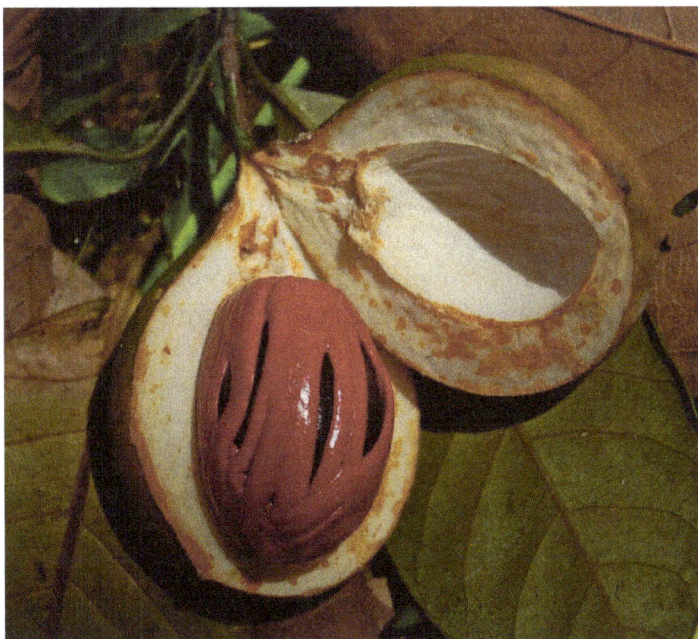

The fruit of *Myristica fragrans*, a species native to Indonesia, is the source of two valuable spices, the red aril (mace) enclosing the dark brown nutmeg.

Botany originated in prehistory as herbalism with the efforts of early humans to identify – and later cultivate – edible, medicinal and poisonous plants, making it one of the oldest branches of science. Medieval physic gardens, often attached to monasteries, contained plants of medical importance. They were forerunners of the first botanical gardens attached to universities, founded from the 1540s onwards. One of the earliest was the Padua botanical garden. These gardens facilitated the academic study of plants. Efforts to catalogue and describe their collections were the beginnings of plant taxonomy, and led in 1753 to the binomial system of Carl Linnaeus that remains in use to this day.

In the 19th and 20th centuries, new techniques were developed for the study of plants, including methods of optical microscopy and live cell imaging, electron microscopy, analysis of chromosome number, plant chemistry and the structure and function of enzymes and other proteins. In the last two decades of the 20th century, botanists exploited the techniques of molecular genetic analysis, including genomics and proteomics and DNA sequences to classify plants more accurately.

Modern botany is a broad, multidisciplinary subject with inputs from most other areas of science and technology. Research topics include the study of plant structure, growth and differentiation, reproduction, biochemistry and primary metabolism, chemical products, development, diseases, evolutionary relationships, systematics, and plant taxonomy. Dominant themes in 21st century plant science are molecular genetics and epigenetics, which are the mechanisms and control of gene expression during differentiation of plant cells and tissues. Botanical research has diverse applications in providing staple foods, materials such as timber, oil, rubber, fibre and drugs, in modern horticulture, agriculture and forestry, plant propagation, breeding and genetic modification, in the synthesis of chemicals and raw materials for construction and energy production, in environmental management, and the maintenance of biodiversity.

History

Early Botany

Botany originated as herbalism, the study and use of plants for their medicinal properties. Many records of the Holocene period date early botanical knowledge as far back as 10,000 years ago. This early unrecorded knowledge of plants was discovered in ancient sites of human occupation within Tennessee, which make up much of the Cherokee land today. The early recorded history of botany includes many ancient writings and plant classifications. Examples of early botanical works have been found in ancient texts from India dating back to before 1100 BC, in archaic Avestan writings, and in works from China before it was unified in 221 BC.

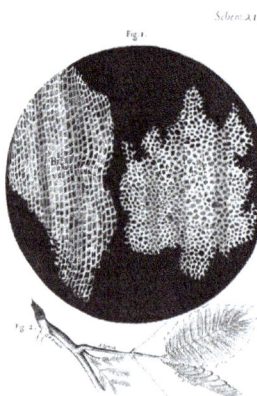

An engraving of the cells of cork, from Robert Hooke's *Micrographia*, 1665

Modern botany traces its roots back to Ancient Greece specifically to Theophrastus (c. 371–287 BC), a student of Aristotle who invented and described many of its principles and is widely regarded in the scientific community as the "Father of Botany". His major works, *Enquiry into Plants* and *On the Causes of Plants*, constitute the most important contributions to botanical science until the Middle Ages, almost seventeen centuries later.

Another work from Ancient Greece that made an early impact on botany is *De Materia Medica*, a five-volume encyclopedia about herbal medicine written in the middle of the first century by Greek physician and pharmacologist Pedanius Dioscorides. *De Materia Medica* was widely read for more than 1,500 years. Important contributions from the medieval Muslim world include Ibn Wahshiyya's *Nabatean Agriculture*, Abū Ḥanīfa Dīnawarī's (828–896) the *Book of Plants*, and Ibn Bassal's *The Classification of Soils*. In the early 13th century, Abu al-Abbas al-Nabati, and Ibn al-Baitar (d. 1248) wrote on botany in a systematic and scientific manner.

In the mid-16th century, "botanical gardens" were founded in a number of Italian universities – the Padua botanical garden in 1545 is usually considered to be the first which is still in its original location. These gardens continued the practical value of earlier "physic gardens", often associated with monasteries, in which plants were cultivated for medical use. They supported the growth of botany as an academic subject. Lectures were given about the plants grown in the gardens and their medical uses demonstrated. Botanical gardens came much later to northern Europe; the first in England was the University of Oxford Botanic Garden in 1621. Throughout this period, botany remained firmly subordinate to medicine.

German physician Leonhart Fuchs (1501–1566) was one of "the three German fathers of botany", along with theologian Otto Brunfels (1489–1534) and physician Hieronymus Bock (1498–1554) (also called Hieronymus Tragus). Fuchs and Brunfels broke away from the tradition of copying earlier works to make original observations of their own. Bock created his own system of plant classification.

Physician Valerius Cordus (1515–1544) authored a botanically and pharmacologically important herbal *Historia Plantarum* in 1544 and a pharmacopoeia of lasting importance, the *Dispensatorium* in 1546. Naturalist Conrad von Gesner (1516–1565) and herbalist John Gerard (1545–c. 1611) published herbals covering the medicinal uses of plants. Naturalist Ulisse Aldrovandi (1522–1605) was considered the *father of natural history*, which included the study of plants. In 1665, using an early microscope, Polymath Robert Hooke discovered cells, a term he coined, in cork, and a short time later in living plant tissue.

Early Modern Botany

During the 18th century, systems of plant identification were developed comparable to dichotomous keys, where unidentified plants are placed into taxonomic groups (e.g. family, genus and species) by making a series of choices between pairs of characters. The choice and sequence of the characters may be artificial in keys designed purely for identification (diagnostic keys) or more closely related to the natural or phyletic order of the taxa in synoptic keys. By the 18th century, new plants for study were arriving in Europe in increasing numbers from newly discovered countries and the European colonies worldwide. In 1753 Carl von Linné (Carl Linnaeus) published his Species Plantarum, a hierarchical classification of plant species that remains the reference point for modern botanical nomenclature. This established a standardised binomial or two-part naming scheme where the first name represented the genus and the second identified the species within the genus. For the purposes of identification, Linnaeus's *Systema Sexuale* classified plants into 24 groups according to the number of their male sexual organs. The 24th group, *Cryptogamia*, included all plants with concealed reproductive parts, mosses, liverworts, ferns, algae and fungi.

The Linnaean Garden of Linnaeus' residence in Uppsala, Sweden, was planted according to his *Systema sexuale*.

Increasing knowledge of plant anatomy, morphology and life cycles led to the realisation that there were more natural affinities between plants than the artificial sexual system of Linnaeus. Adanson (1763), de Jussieu (1789), and Candolle (1819) all proposed various alternative natural systems of classification that grouped plants using a wider range of shared characters and were widely followed. The Candollean system reflected his ideas of the progression of morphological complexity and the later classification by Bentham and Hooker, which was influential until the mid-19th century, was influenced by Candolle's approach. Darwin's publication of the *Origin of Species* in 1859 and his concept of common descent required modifications to the Candollean system to reflect evolutionary relationships as distinct from mere morphological similarity.

Botany was greatly stimulated by the appearance of the first "modern" textbook, Matthias Schleiden's *Grundzüge der Wissenschaftlichen Botanik*, published in English in 1849 as *Principles of Scientific Botany*. Schleiden was a microscopist and an early plant anatomist who co-founded the cell theory with Theodor Schwann and Rudolf Virchow and was among the first to grasp the significance of the cell nucleus that had been described by Robert Brown in 1831. In 1855, Adolf Fick formulated Fick's laws that enabled the calculation of the rates of molecular diffusion in biological systems.

Echeveria glauca in a Connecticut greenhouse. Botany uses Latin names for identification, here, the specific name *glauca* means blue.

Modern Botany

Building upon the gene-chromosome theory of heredity that originated with Gregor Mendel (1822–1884), August Weismann (1834–1914) proved that inheritance only takes place through gametes. No other cells can pass on inherited characters. The work of Katherine Esau (1898–1997) on plant anatomy is still a major foundation of modern botany. Her books *Plant Anatomy* and *Anatomy of Seed Plants* have been key plant structural biology texts for more than half a century.

Biologist and statistician Ronald Fisher

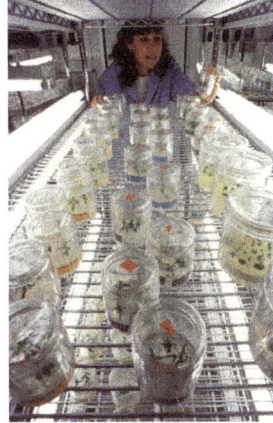

Micropropagation of transgenic plants

The discipline of plant ecology was pioneered in the late 19th century by botanists such as Eugenius Warming, who produced the hypothesis that plants form communities, and his mentor and successor Christen C. Raunkiær whose system for describing plant life forms is still in use today. The concept that the composition of plant communities such as temperate broadleaf forest changes by a process of ecological succession was developed by Henry Chandler Cowles, Arthur Tansley and Frederic Clements. Clements is credited with the idea of climax vegetation as the most complex vegetation that an environment can support and Tansley introduced the concept of ecosystems to biology. Building on the extensive earlier work of Alphonse de Candolle, Nikolai Vavilov (1887–1943) produced accounts of the biogeography, centres of origin, and evolutionary history of economic plants.

Particularly since the mid-1960s there have been advances in understanding of the physics of plant physiological processes such as transpiration (the transport of water within plant tissues), the temperature dependence of rates of water evaporation from the leaf surface and the molecular diffusion of water vapour and carbon dioxide through stomatal apertures. These developments, coupled with new methods for measuring the size of stomatal apertures, and the rate of photosynthesis have enabled precise description of the rates of gas exchange between plants and the atmosphere. Innovations in statistical analysis by Ronald Fisher, Frank Yates and others at Rothamsted Experimental Station facilitated rational experimental design and data analysis in botanical research. The discovery and identification of the auxin plant hormones by Kenneth V. Thimann in 1948 enabled regulation of plant growth by externally applied chemicals. Frederick Campion Steward pioneered techniques of micropropagation and plant tissue culture controlled by plant hormones. The synthetic auxin 2,4-Dichlorophenoxyacetic acid or 2,4-D was one of the first commercial synthetic herbicides.

20th century developments in plant biochemistry have been driven by modern techniques of organic chemical analysis, such as spectroscopy, chromatography and electrophoresis. With the rise of the related molecular-scale biological approaches of molecular biology, genomics, proteomics and metabolomics, the relationship between the plant genome and most aspects of the biochemistry, physiology, morphology and behaviour of plants can be subjected to detailed experimental analysis. The concept originally stated by Gottlieb Haberlandt in 1902 that all plant cells are totipotent and can be grown *in vitro* ultimately enabled the use of genetic engineering experimentally to knock out a gene or genes responsible for a specific trait, or to add genes such as GFP that report when a gene of interest is being expressed. These technologies enable the biotechnological use of whole plants or plant cell cultures grown in bioreactors to synthesise pesticides, antibiotics or other pharmaceuticals, as well as the practical application of genetically modified crops designed for traits such as improved yield.

Modern morphology recognises a continuum between the major morphological categories of root, stem (caulome), leaf (phyllome) and trichome. Furthermore, it emphasises structural dynamics. Modern systematics aims to reflect and discover phylogenetic relationships between plants. Modern Molecular phylogenetics largely ignores morphological characters, relying on DNA sequences as data. Molecular analysis of DNA sequences from most families of flowering plants enabled the Angiosperm Phylogeny Group to publish in 1998 a phylogeny of flowering plants, answering many of the questions about relationships among angiosperm families and species. The theoretical possibility of a practical method for identification of plant species and commercial varieties by DNA barcoding is the subject of active current research.

Scope and Importance

Botany involves the recording and description of plants, such as this herbarium specimen of the lady fern *Athyrium filix-femina*.

The study of plants is vital because they underpin almost all animal life on Earth by generating a large proportion of the oxygen and food that provide humans and other organisms with aerobic respiration with the chemical energy they need to exist. Plants, algae and cyanobacteria are the major groups of organisms that carry out photosynthesis, a process that uses the energy of sunlight to convert water and carbon dioxide into sugars that can be used both as a source of chemical energy and of organic molecules that are used in the structural components of cells. As a by-product of photosynthesis, plants release oxygen into the atmosphere, a gas that is required by nearly all living things to carry out cellular respiration. In addition, they are influential in the global carbon and water cycles and plant roots bind and stabilise soils, preventing soil erosion. Plants are crucial to the future of human society as they provide food, oxygen, medicine, and products for people, as well as creating and preserving soil.

Historically, all living things were classified as either animals or plants and botany covered the study of all organisms not considered animals. Botanists examine both the internal functions and processes within plant organelles, cells, tissues, whole plants, plant populations and plant communities. At each of these levels, a botanist may be concerned with the classification (taxonomy), phylogeny and evolution, structure (anatomy and morphology), or function (physiology) of plant life.

The strictest definition of "plant" includes only the "land plants" or embryophytes, which include seed plants (gymnosperms, including the pines, and flowering plants) and the free-sporing cryptogams including ferns, clubmosses, liverworts, hornworts and mosses. Embryophytes are multicellular eukaryotes descended from an ancestor that obtained its energy from sunlight by photosynthesis. They have life cycles with alternating haploid and diploid phases. The sexual haploid phase of embryophytes, known as the gametophyte, nurtures the developing diploid embryo sporophyte within its tissues for at least part of its life, even in the seed plants, where the gametophyte itself is nurtured by its parent sporophyte. Other groups of organisms that were previously studied by botanists include bacteria (now studied in bacteriology), fungi (mycology) – including lichen-forming fungi (lichenology), non-chlorophyte algae (phycology), and viruses (virology). However, attention is still given to these groups by botanists, and fungi (including lichens) and photosynthetic protists are usually covered in introductory botany courses.

Palaeobotanists study ancient plants in the fossil record to provide information about the evolutionary history of plants. Cyanobacteria, the first oxygen-releasing photosynthetic organisms on Earth, are thought to have given rise to the ancestor of plants by entering into an endosymbiotic relationship with an early eukaryote, ultimately becoming the chloroplasts in plant cells. The new photosynthetic plants (along with their algal relatives) accelerated the rise in atmospheric oxygen started by the cyanobacteria, changing the ancient oxygen-free, reducing, atmosphere to one in which free oxygen has been abundant for more than 2 billion years.

Among the important botanical questions of the 21st century are the role of plants as primary producers in the global cycling of life's basic ingredients: energy, carbon, oxygen, nitrogen and water, and ways that our plant stewardship can help address the global environmental issues of resource management, conservation, human food security, biologically invasive organisms, carbon sequestration, climate change, and sustainability.

Human Nutrition

Virtually all staple foods come either directly from primary production by plants, or indirectly from animals that eat them. Plants and other photosynthetic organisms are at the base of most food chains because they use the energy from the sun and nutrients from the soil and atmosphere, converting them into a form that can be used by animals. This is what ecologists call the first trophic level. The modern forms of the major staple foods, such as maize, rice, wheat and other cereal grasses, pulses, bananas and plantains, as well as flax and cotton grown for their fibres, are the outcome of prehistoric selection over thousands of years from among wild ancestral plants with the most desirable characteristics.

The food we eat comes directly or indirectly from plants such as rice.

Botanists study how plants produce food and how to increase yields, for example through plant breeding, making their work important to mankind's ability to feed the world and provide food security for future generations. Botanists also study weeds, which are a considerable problem in agriculture, and the biology and control of plant pathogens in agriculture and natural ecosystems. Ethnobotany is the study of the relationships between plants and people. When applied to the investigation of historical plant–people relationships ethnobotany may be referred to as archaeobotany or palaeoethnobotany. Some of the earliest plant-people relationships arose between the indigenous people of Canada in identifying edible plants from inedible plants. This relationship the indigenous people had with plants was recorded by ethnobotanists.

Plant Biochemistry

Plant biochemistry is the study of the chemical processes used by plants. Some of these processes are used in their primary metabolism like the photosynthetic Calvin cycle and crassulacean acid metabolism. Others make specialised materials like the cellulose and lignin used to build their bodies, and secondary products like resins and aroma compounds.

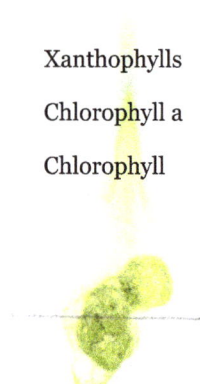

Xanthophylls

Chlorophyll a

Chlorophyll

Plants make various photosynthetic pigments, some of which can be seen here through paper chromatography.

Plants and various other groups of photosynthetic eukaryotes collectively known as "algae" have unique organelles known as chloroplasts. Chloroplasts are thought to be descended from cyanobacteria that formed endosymbiotic relationships with ancient plant and algal ancestors. Chloroplasts and cyanobacteria contain the blue-green pigment chlorophyll a. Chlorophyll a (as well as its plant and green algal-specific cousin chlorophyll b) absorbs light in the blue-violet and orange/red parts of the spectrum while reflecting and transmitting the green light that we see as the characteristic colour of these organisms. The energy in the red and blue light that these pigments absorb is used by chloroplasts to make energy-rich carbon compounds from carbon dioxide and water by oxygenic photosynthesis, a process that generates molecular oxygen (O_2) as a by-product.

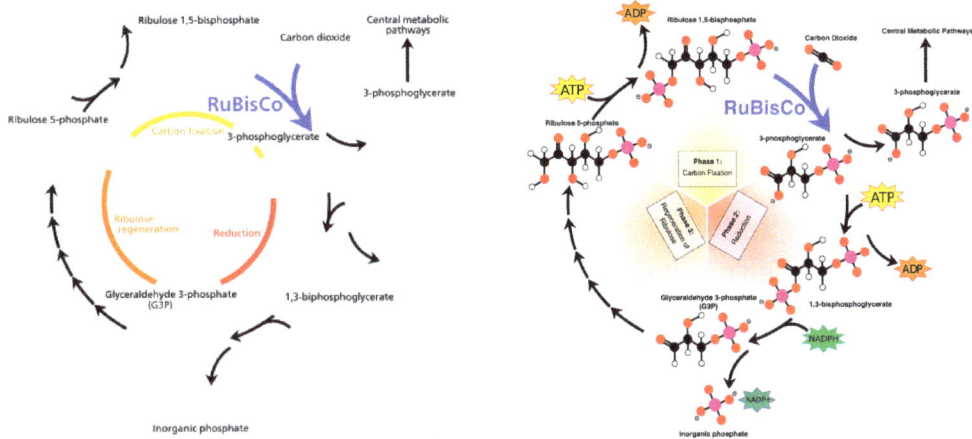

The Calvin cycle *(Interactive diagram)* The Calvin cycle incorporates carbon dioxide into sugar molecules.

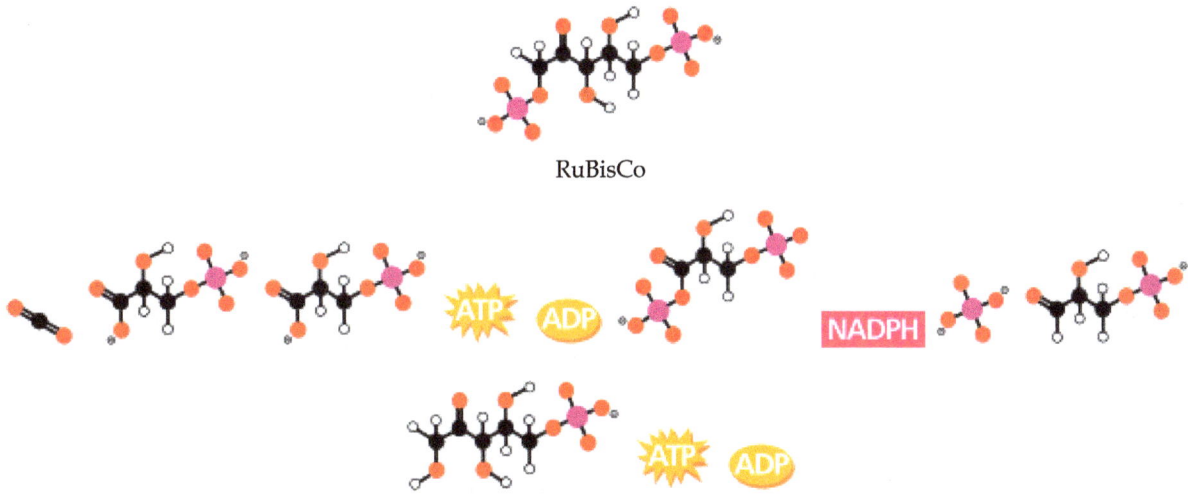

RuBisCo

Carbon fixation
Reduction
3-phosphoglycerate
3-phosphoglycerate
Carbon dioxide
1,3-biphosphoglycerate
Glyceraldehyde-3-phosphate
(G3P)
Inorganic phosphate
Ribulose 5-phosphate
Ribulose-1,5-bisphosphate
Edit · Source image

The light energy captured by chlorophyll *a* is initially in the form of electrons (and later a proton gradient) that's used to make molecules of ATP and NADPH which temporarily store and transport energy. Their energy is used in the light-independent reactions of the Calvin cycle by the enzyme rubisco to produce molecules of the 3-carbon sugar glyceraldehyde 3-phosphate (G3P). Glyceraldehyde 3-phosphate is the first product of photosynthesis and the raw material from which glucose and almost all other organic molecules of biological origin are synthesised. Some of the glucose is converted to starch which is stored in the chloroplast. Starch is the characteristic energy store of most land plants and algae, while inulin, a polymer of fructose is used for the same purpose in the sunflower family Asteraceae. Some of the glucose is converted to sucrose (common table sugar) for export to the rest of the plant.

Unlike in animals (which lack chloroplasts), plants and their eukaryote relatives have delegated many biochemical roles to their chloroplasts, including synthesising all their fatty acids, and most amino acids. The fatty acids that chloroplasts make are used for many things, such as providing material to build cell membranes out of and making the polymer cutin which is found in the plant cuticle that protects land plants from drying out.

Plants synthesise a number of unique polymers like the polysaccharide molecules cellulose, pectin and xyloglucan from which the land plant cell wall is constructed. Vascular land plants make lignin, a polymer used to strengthen the secondary cell walls of xylem tracheids and vessels to keep them from collapsing when a plant sucks water through them under water stress. Lignin is also used in other cell types like sclerenchyma fibres that provide structural support for a plant and is a major constituent of wood. Sporopollenin is a chemically resistant polymer found in the outer cell walls of spores and pollen of land plants responsible for the survival of early land plant spores and the pollen of seed plants in the fossil record. It is widely regarded as a marker for the start of land plant evolution during the Ordovician period. The concentration of carbon dioxide in the atmosphere today is much lower than it was when plants emerged onto land during the Ordovician and Silurian periods. Many monocots like maize and the pineapple and some dicots like the Asteraceae have since independently evolved pathways like Crassulacean acid metabolism and the C_4 carbon fixation pathway for photosynthesis which avoid the losses resulting from photorespiration in the more common C_3 carbon fixation pathway. These biochemical strategies are unique to land plants.

Medicine and Materials

Phytochemistry is a branch of plant biochemistry primarily concerned with the chemical substances produced by plants during secondary metabolism. Some of these compounds are toxins such as the alkaloid coniine from hemlock. Others, such as the essential oils peppermint oil and lemon oil are useful for their aroma, as flavourings and spices (e.g., capsaicin), and in medicine as pharmaceuticals as in opium from opium poppies. Many medicinal and recreational drugs, such as tetrahydrocannabinol (active ingredient in cannabis), caffeine, morphine and nicotine come directly from plants. Others are simple derivatives of botanical natural products. For example, the pain killer aspirin is the acetyl ester of salicylic acid, originally isolated from the bark of willow trees, and a wide range of opiate painkillers like heroin are obtained by chemical modification of morphine obtained from the opium poppy. Popular stimulants come from plants, such as caffeine from coffee, tea and chocolate, and nicotine from tobacco. Most alcoholic beverages come from fermentation of carbohydrate-rich plant products such as barley (beer), rice (sake) and grapes (wine).

Native Americans have used various plants as ways of treating illness or disease for thousands of years. This knowledge Native Americans have on plants has been recorded by enthnobotanists and then in turn has been used by pharmaceutical companies as a way of drug discovery.

Tapping a rubber tree in Thailand

Plants can synthesise useful coloured dyes and pigments such as the anthocyanins responsible for the red colour of red wine, yellow weld and blue woad used together to produce Lincoln green, indoxyl, source of the blue dye indigo traditionally used to dye denim and the artist's pigments gamboge and rose madder.

Sugar, starch, cotton, linen, hemp, some types of rope, wood and particle boards, papyrus and paper, vegetable oils, wax, and natural rubber are examples of commercially important materials made from plant tissues or their secondary products. Charcoal, a pure form of carbon made by pyrolysis of wood, has a long history as a metal-smelting fuel, as a filter material and adsorbent and as an artist's material and is one of the three ingredients of gunpowder. Cellulose, the world's most abundant organic polymer, can be converted into energy, fuels, materials and chemical feedstock. Products made from cellulose include rayon and cellophane, wallpaper paste, biobutanol and gun cotton. Sugarcane, rapeseed and soy are some of the plants with a highly fermentable sugar or oil content that are used as sources of biofuels, important alternatives to fossil fuels, such as biodiesel. Sweetgrass was used by NativeAmericanse to ward of bugs like mosquitoes. These bug repelling properties of sweetgrass were later found by the American Chemical Society in the molecules phytol and coumarin.

Plant Ecology

Holdridge life zones model relationships between vegetation type, moisture availability and temperature.

Plant ecology is the science of the functional relationships between plants and their habitats—the environments where they complete their life cycles. Plant ecologists study the composition of local and regional floras, their biodiversity, genetic diversity and fitness, the adaptation of plants to their environment, and their competitive or mutualistic interactions with other species. Some ecologists even rely on empirical data from indigenous people that is gathered by ethnobotanists. This information can relay a great deal of information on how the land once was thousands of years ago and how it has changed over that time. The goals of plant ecology are to understand the causes of their distribution patterns, productivity, environmental impact, evolution, and responses to environmental change.

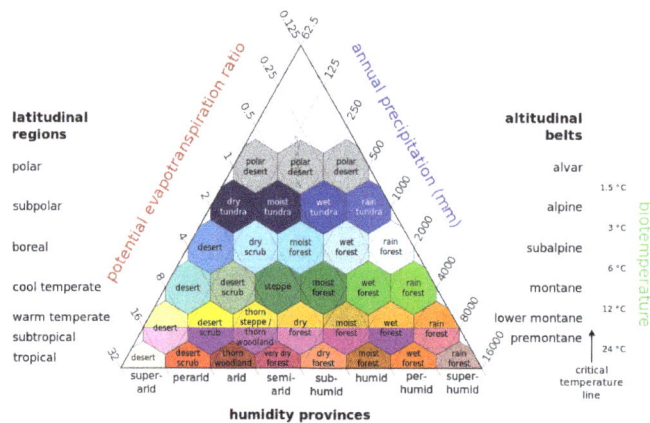

Plants depend on certain edaphic (soil) and climatic factors in their environment but can modify these factors too. For example, they can change their environment's albedo, increase runoff interception, stabilise mineral soils and develop their organic content, and affect local temperature. Plants compete with other organisms in their ecosystem for resources. They interact with their neighbours at a variety of spatial scales in groups, populations and communities that collectively constitute vegetation. Regions with characteristic vegetation types and dominant plants as well as similar abiotic and biotic factors, climate, and geography make up biomes like tundra or tropical rainforest.

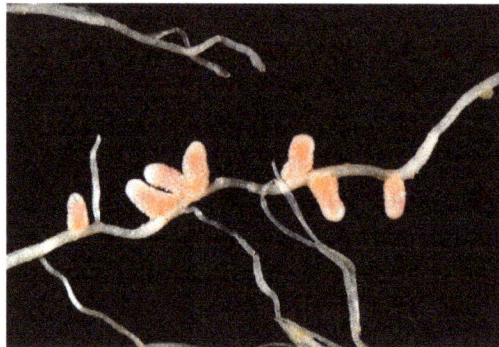

The nodules of *Medicago italica* contain the nitrogen fixing bacterium *Sinorhizobium meliloti*. The plant provides the bacteria with nutrients and an anaerobic environment, and the bacteria fix nitrogen for the plant.

Herbivores eat plants, but plants can defend themselves and some species are parasitic or even carnivorous. Other organisms form mutually beneficial relationships with plants. For example, mycorrhizal fungi and rhizobia provide plants with nutrients in exchange for food, ants are recruited by ant plants to provide protection, honey bees, bats and other animals pollinate flowers and humans and other animals act as dispersal vectors to spread spores and seeds.

Plants, Climate and Environmental Change

Plant responses to climate and other environmental changes can inform our understanding of how these changes affect ecosystem function and productivity. For example, plant phenology can be a useful proxy for temperature in historical climatology, and the biological impact of climate change and global warming. Palynology, the analysis of fossil pollen deposits in sediments from thousands or millions of years ago allows the reconstruction of past climates. Estimates of atmospheric CO_2 concentrations since the Palaeozoic have been obtained from stomatal densities and the leaf shapes and sizes of ancient land plants. Ozone depletion can expose plants to higher levels of ultraviolet radiation-B (UV-B), resulting in lower growth rates. Moreover, information from studies of community ecology, plant systematics, and taxonomy is essential to understanding vegetation change, habitat destruction and species extinction.

Genetics

Inheritance in plants follows the same fundamental principles of genetics as in other multicellular organisms. Gregor Mendel discovered the genetic laws of inheritance by studying inherited traits such as shape in *Pisum sativum* (peas). What Mendel learned from studying plants has had far reaching benefits outside of botany. Similarly, "jumping genes" were discovered by Barbara McClintock while she was studying maize. Nevertheless, there are some distinctive genetic differences between plants and other organisms.

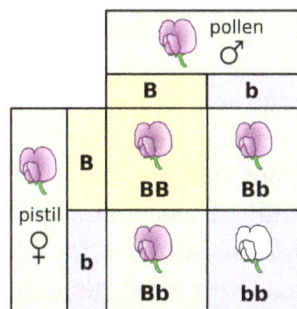

A Punnett square depicting a cross between two pea plants heterozygous for purple (B) and white (b) blossoms

Species boundaries in plants may be weaker than in animals, and cross species hybrids are often possible. A familiar example is peppermint, *Mentha × piperita*, a sterile hybrid between *Mentha aquatica* and spearmint, *Mentha spicata*. The many cultivated varieties of wheat are the result of multiple inter- and intra-specific crosses between wild species and their hybrids. Angiosperms with monoecious flowers often have self-incompatibility mechanisms that operate between the pollen and stigma so that the pollen either fails to reach the stigma or fails to germinate and produce male gametes. This is one of several methods used by plants to promote outcrossing. In many land plants the male and female gametes are produced by separate individuals. These species are said to be dioecious when referring to vascular plant sporophytes and dioicous when referring to bryophyte gametophytes.

Unlike in higher animals, where parthenogenesis is rare, asexual reproduction may occur in plants by several different mechanisms. The formation of stem tubers in potato is one example. Particularly in arctic or alpine habitats, where opportunities for fertilisation of flowers by animals are rare, plantlets or bulbs, may develop instead of flowers, replacing sexual reproduction with asexual reproduction and giving rise to clonal populations genetically identical to the parent. This is one of several types of apomixis that occur in plants. Apomixis can also happen in a seed, producing a seed that contains an embryo genetically identical to the parent.

Most sexually reproducing organisms are diploid, with paired chromosomes, but doubling of their chromosome number may occur due to errors in cytokinesis. This can occur early in development to produce an autopolyploid or partly autopolyploid organism, or during normal processes of cellular differentiation to produce some cell types that are polyploid (endopolyploidy), or during gamete formation. An allopolyploid plant may result from a hybridisation event between two different species. Both autopolyploid and allopolyploid plants can often reproduce normally, but may be unable to cross-breed successfully with the parent population because there is a mismatch in chromosome numbers. These plants that are reproductively isolated from the parent species but live within the same geographical area, may be sufficiently successful to form a new species. Some otherwise sterile plant polyploids can still reproduce vegetatively or by seed apomixis, forming clonal populations of identical individuals. Durum wheat is a fertile tetraploid allopolyploid, while bread wheat is a fertile hexaploid. The commercial banana is an example of a sterile, seedless triploid hybrid. Common dandelion is a triploid that produces viable seeds by apomictic seed.

As in other eukaryotes, the inheritance of endosymbiotic organelles like mitochondria and chloroplasts in plants is non-Mendelian. Chloroplasts are inherited through the male parent in gymnosperms but often through the female parent in flowering plants.

Molecular Genetics

A considerable amount of new knowledge about plant function comes from studies of the molecular genetics of model plants such as the Thale cress, *Arabidopsis thaliana*, a weedy species in the mustard family (Brassicaceae). The genome or hereditary information contained in the genes of this species is encoded by about 135 million base pairs of DNA, forming one of the smallest genomes among flowering plants. *Arabidopsis* was the first plant to have its genome sequenced, in 2000. The sequencing of some other relatively small genomes, of rice (*Oryza sativa*) and *Brachypodium distachyon*, has made them important model species for understanding the genetics, cellular and molecular biology of cereals, grasses and monocots generally.

Thale cress, *Arabidopsis thaliana*, the first plant to have its genome sequenced, remains the most important model organism.

Model plants such as *Arabidopsis thaliana* are used for studying the molecular biology of plant cells and the chloroplast. Ideally, these organisms have small genomes that are well known or completely sequenced, small stature and short generation times. Corn has been used to study mechanisms of photosynthesis and phloem loading of sugar in C_4 plants. The single celled green alga *Chlamydomonas reinhardtii*, while not an embryophyte itself, contains a green-pigmented chloroplast related to that of land plants, making it useful for study. A red alga *Cyanidioschyzon merolae* has also been used to study some basic chloroplast functions. Spinach, peas, soybeans and a moss *Physcomitrella patens* are commonly used to study plant cell biology.

Agrobacterium tumefaciens, a soil rhizosphere bacterium, can attach to plant cells and infect them with a callus-inducing Ti plasmid by horizontal gene transfer, causing a callus infection called crown gall disease. Schell and Van Montagu (1977) hypothesised that the Ti plasmid could be a natural vector for introducing the Nif gene responsible for nitrogen fixation in the root nodules of legumes and other plant species. Today, genetic modification of the Ti plasmid is one of the main techniques for introduction of transgenes to plants and the creation of genetically modified crops.

Epigenetics

Epigenetics is the study of heritable changes in gene function that cannot be explained by changes in the underlying DNA sequence but cause the organism's genes to behave (or "express themselves") differently. One example of epigenetic change is the marking of the genes by DNA methylation which determines whether they will be expressed or not. Gene expression can also be controlled by repressor proteins that attach to silencer regions of the DNA and prevent that region of the DNA code from being expressed. Epigenetic marks may be added or removed from the DNA during programmed stages of development of the plant, and are responsible, for example, for the differences between anthers, petals and normal leaves, despite the fact that they all have the same underlying genetic code. Epigenetic changes may be temporary or may remain through successive cell divisions for the remainder of the cell's life. Some epigenetic changes have been shown to be heritable, while others are reset in the germ cells.

Epigenetic changes in eukaryotic biology serve to regulate the process of cellular differentiation. During morphogenesis, totipotent stem cells become the various pluripotent cell lines of the embryo, which in turn become fully differentiated cells. A single fertilised egg cell, the zygote, gives rise to the many different plant cell types including parenchyma, xylem vessel elements, phloem sieve tubes, guard cells of the epidermis, etc. as it continues to divide. The process results from the epigenetic activation of some genes and inhibition of others.

Unlike animals, many plant cells, particularly those of the parenchyma, do not terminally differentiate, remaining totipotent with the ability to give rise to a new individual plant. Exceptions include highly lignified cells, the sclerenchyma and xylem which are dead at maturity, and the phloem sieve tubes which lack nuclei. While plants use many of the same epigenetic mechanisms as animals, such as chromatin remodelling, an alternative hypothesis is that plants set their gene expression patterns using positional information from the environment and surrounding cells to determine their developmental fate.

Plant Evolution

The chloroplasts of plants have a number of biochemical, structural and genetic similarities to cyanobacteria, (commonly but incorrectly known as "blue-green algae") and are thought to be derived from an ancient endosymbiotic relationship between an ancestral eukaryotic cell and a cyanobacterial resident.

The algae are a polyphyletic group and are placed in various divisions, some more closely related to plants than others. There are many differences between them in features such as cell wall composition, biochemistry, pigmentation, chloroplast structure and nutrient reserves. The algal division Charophyta, sister to the green algal division Chlorophyta, is considered to contain the ancestor of true plants. The Charophyte class Charophyceae and the land plant sub-kingdom Embryophyta together form the monophyletic group or clade Streptophytina.

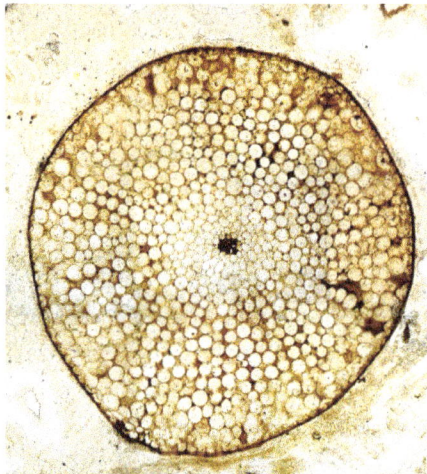

Transverse section of a fossil stem of the Devonian vascular plant *Rhynia gwynne-vaughani*

Nonvascular land plants are embryophytes that lack the vascular tissues xylem and phloem. They include mosses, liverworts and hornworts. Pteridophytic vascular plants with true xylem and phloem that reproduced by spores germinating into free-living gametophytes evolved during the Silurian period and diversified into several lineages during the late Silurian and early Devonian. Representatives of the lycopods have survived to the present day. By the end of the Devonian period, several groups, including the lycopods, sphenophylls and progymnosperms, had independently evolved "megaspory" – their spores were of two distinct sizes, larger megaspores and smaller microspores. Their reduced gametophytes developed from megaspores retained within the spore-producing organs (megasporangia) of the sporophyte, a condition known as endospory. Seeds consist of an endosporic megasporangium surrounded by one or two sheathing layers (integuments). The young sporophyte develops within the seed, which on germination splits to release it. The earliest known seed plants date from the latest Devonian Famennian stage. Following the evolution of the seed habit, seed plants diversified, giving rise to a number of now-extinct groups, including seed ferns, as well as the modern gymnosperms and angiosperms. Gymnosperms produce "naked seeds" not fully enclosed in an ovary; modern representatives include conifers, cycads, *Ginkgo*, and Gnetales. Angiosperms produce seeds enclosed in a structure such as a carpel or an ovary. Ongoing research on the molecular phylogenetics of living plants appears to show that the angiosperms are a sister clade to the gymnosperms.

Plant Physiology

Plant physiology encompasses all the internal chemical and physical activities of plants associated with life. Chemicals obtained from the air, soil and water form the basis of all plant metabolism. The energy of sunlight, captured by oxygenic photosynthesis and released by cellular respiration, is the basis of almost all life. Photoautotrophs, including all green plants, algae and cyanobacteria gather energy directly from sunlight by photosynthesis. Heterotrophs including all animals, all fungi, all completely parasitic plants, and non-photosynthetic bacteria take in organic molecules produced by photoautotrophs and respire them or use them in the construction of cells and tissues. Respiration is the oxidation of carbon compounds by breaking them down into simpler structures to release the energy they contain, essentially the opposite of photosynthesis.

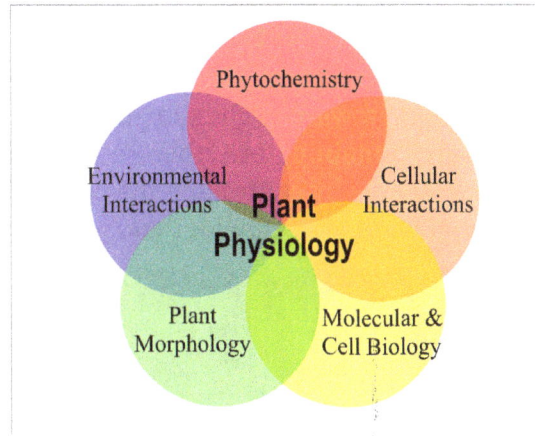

Five of the key areas of study within plant physiology

Molecules are moved within plants by transport processes that operate at a variety of spatial scales. Subcellular transport of ions, electrons and molecules such as water and enzymes occurs across cell membranes. Minerals and water are transported from roots to other parts of the plant in the transpiration stream. Diffusion, osmosis, and active transport and mass flow are all different ways transport can occur. Examples of elements that plants need to transport are nitrogen, phosphorus, potassium, calcium, magnesium, and sulphur. In vascular plants, these elements are extracted from the soil as soluble ions by the roots and transported throughout the plant in the xylem. Most of the elements required for plant nutrition come from the chemical breakdown of soil minerals. Sucrose produced by photosynthesis is transported from the leaves to other parts of the plant in the phloem and plant hormones are transported by a variety of processes.

Plant Hormones

Plants are not passive, but respond to external signals such as light, touch, and injury by moving or growing towards or away from the stimulus, as appropriate. Tangible evidence of touch sensitivity is the almost instantaneous collapse of leaflets of *Mimosa pudica*, the insect traps of Venus flytrap and bladderworts, and the pollinia of orchids.

1 An oat coleoptile with the sun overhead. Auxin (pink) is evenly distributed in its tip.
2 With the sun at an angle and only shining on one side of the shoot, auxin moves to the opposite side and stimulates cell elongation there.
3 and 4 Extra growth on that side causes the shoot to bend towards the sun.

The hypothesis that plant growth and development is coordinated by plant hormones or plant growth regulators first emerged in the late 19th century. Darwin experimented on the movements of plant shoots and roots towards light and gravity, and concluded "It is hardly an exaggeration to say that the tip of the radicle . . acts like the brain of one of the lower animals . . directing the several movements". About the same time, the role of auxins in control of plant growth was first outlined by the Dutch scientist Frits Went. The first known auxin, indole-3-acetic acid (IAA), which promotes cell growth, was only isolated from plants about 50 years later. This compound mediates the tropic responses of shoots and roots towards light and gravity. The finding in 1939 that plant callus could be maintained in culture containing IAA, followed by the observation in 1947 that it could be induced to form roots and shoots by controlling the concentration of growth hormones were key steps in the development of plant biotechnology and genetic modification.

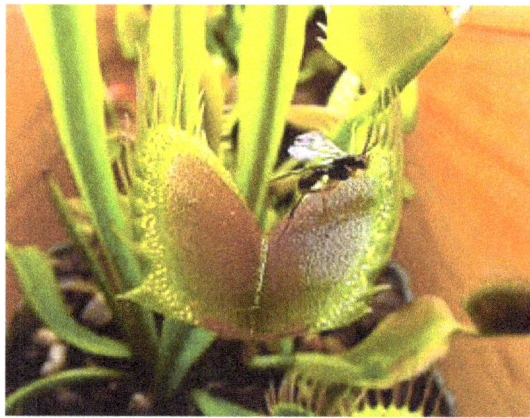

Venus's fly trap, *Dionaea muscipula*, showing the touch-sensitive insect trap in action

Cytokinins are a class of plant hormones named for their control of cell division or cytokinesis. The natural cytokinin zeatin was discovered in corn, *Zea mays*, and is a derivative of the purine adenine. Zeatin is produced in roots and transported to shoots in the xylem where it promotes cell division, bud development, and the greening of chloroplasts. The gibberelins, such as Gibberelic acid are diterpenes synthesised from acetyl CoA via the mevalonate pathway. They are involved in the promotion of germination and dormancy-breaking in seeds, in regulation of plant height by controlling stem elongation and the control of flowering. Abscisic acid (ABA) occurs in all land plants except liverworts, and is synthesised from carotenoids in the chloroplasts and other plastids. It inhibits cell division, promotes seed maturation, and dormancy, and promotes stomatal closure. It was so named because it was originally thought to control abscission. Ethylene is a gaseous hormone that is produced in all higher plant tissues from methionine. It is now known to be the hormone that stimulates or regulates fruit ripening and abscission, and it, or the synthetic growth regulator ethephon which is rapidly metabolised to produce ethylene, are used on industrial scale to promote ripening of cotton, pineapples and other climacteric crops.

Another class of phytohormones is the jasmonates, first isolated from the oil of *Jasminum grandiflorum* which regulates wound responses in plants by unblocking the expression of genes required in the systemic acquired resistance response to pathogen attack.

In addition to being the primary energy source for plants, light functions as a signalling device, providing information to the plant, such as how much sunlight the plant receives each day. This can result in adaptive changes in a process known as photomorphogenesis. Phytochromes are the photoreceptors in a plant that are sensitive to light.

Plant Anatomy and Morphology

Plant anatomy is the study of the structure of plant cells and tissues, whereas plant morphology is the study of their external form. All plants are multicellular eukaryotes, their DNA stored in nuclei. The characteristic features of plant cells that distinguish them from those of animals and fungi include a primary cell wall composed of the polysaccharides cellulose, hemicellulose and pectin, larger vacuoles than in animal cells and the presence of plastids with unique photosynthetic and biosynthetic functions as in the chloroplasts. Other plastids contain storage products such as starch (amyloplasts) or lipids (elaioplasts). Uniquely, streptophyte cells and those of the green algal order Trentepohliales divide by construction of a phragmoplast as a template for building a cell plate late in cell division.

A nineteenth-century illustration showing the morphology of the roots, stems, leaves and flowers of the rice plant *Oryza sativa*

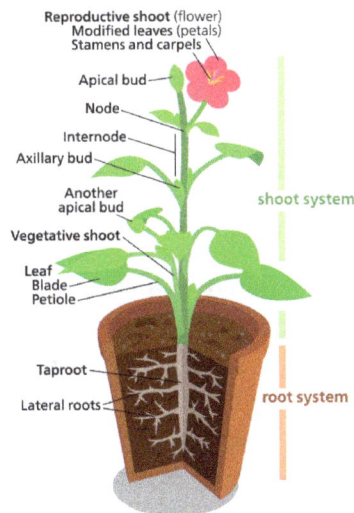

A diagram of a "typical" eudicot, the most common type of plant (three-fifths of all plant species). No plant actually looks exactly like this though.

The bodies of vascular plants including clubmosses, ferns and seed plants (gymnosperms and an-giosperms) generally have aerial and subterranean subsystems. The shoots consist of stems bearing green photosynthesising leaves and reproductive structures. The underground vascularised roots bear root hairs at their tips and generally lack chlorophyll. Non-vascular plants, the liverworts, hornworts and mosses do not produce ground-penetrating vascular roots and most of the plant participates in photosynthesis. The sporophyte generation is nonphotosynthetic in liverworts but may be able to contribute part of its energy needs by photosynthesis in mosses and hornworts.

The root system and the shoot system are interdependent – the usually nonphotosynthetic root system depends on the shoot system for food, and the usually photosynthetic shoot system de-pends on water and minerals from the root system. Cells in each system are capable of creating cells of the other and producing adventitious shoots or roots. Stolons and tubers are examples of shoots that can grow roots. Roots that spread out close to the surface, such as those of willows, can produce shoots and ultimately new plants. In the event that one of the systems is lost, the other can often regrow it. In fact it is possible to grow an entire plant from a single leaf, as is the case with *Saintpaulia*, or even a single cell – which can dedifferentiate into a callus (a mass of unspecialised cells) that can grow into a new plant. In vascular plants, the xylem and phloem are the conductive tissues that transport resources between shoots and roots. Roots are often adapted to store food such as sugars or starch, as in sugar beets and carrots.

Stems mainly provide support to the leaves and reproductive structures, but can store water in succulent plants such as cacti, food as in potato tubers, or reproduce vegetatively as in the stolons of strawberry plants or in the process of layering. Leaves gather sunlight and carry out photo-synthesis. Large, flat, flexible, green leaves are called foliage leaves. Gymnosperms, such as coni-fers, cycads, *Ginkgo*, and gnetophytes are seed-producing plants with open seeds. Angiosperms are seed-producing plants that produce flowers and have enclosed seeds. Woody plants, such as azaleas and oaks, undergo a secondary growth phase resulting in two additional types of tissues: wood (secondary xylem) and bark (secondary phloem and cork). All gymnosperms and many an-giosperms are woody plants. Some plants reproduce sexually, some asexually, and some via both means.

Although reference to major morphological categories such as root, stem, leaf, and trichome are useful, one has to keep in mind that these categories are linked through intermediate forms so that a continuum between the categories results. Furthermore, structures can be seen as processes, that is, process combinations.

Systematic Botany

Systematic botany is part of systematic biology, which is concerned with the range and diversity of organisms and their relationships, particularly as determined by their evolutionary history. It involves, or is related to, biological classification, scientific taxonomy and phylogenetics. Biological classification is the method by which botanists group organisms into categories such as genera or species. Biological classification is a form of scientific taxonomy. Modern taxonomy is rooted in the work of Carl Linnaeus, who grouped species according to shared physical characteristics. These groupings have since been revised to align better with the Darwinian principle of common descent – grouping organisms by ancestry rather than superficial characteristics. While scientists do not always agree on how to classify organisms, molecular phylogenetics, which uses DNA sequences as data, has driven many recent revisions along evolutionary lines and is likely to continue to do so. The dominant classification system is called Linnaean taxonomy. It includes ranks and binomial nomenclature. The nomenclature of botanical organisms is codified in the International Code of Nomenclature for algae, fungi, and plants (ICN) and administered by the International Botanical Congress.

A botanist preparing a plant specimen for mounting in the herbarium

Kingdom Plantae belongs to Domain Eukarya and is broken down recursively until each species is separately classified. The order is: Kingdom; Phylum (or Division); Class; Order; Family; Genus (plural *genera*); Species. The scientific name of a plant represents its genus and its species within the genus, resulting in a single worldwide name for each organism. For example, the tiger lily is *Lilium columbianum*. *Lilium* is the genus, and *columbianum* the specific epithet. The combination is the name of the species. When writing the scientific name of an organism, it is proper to capitalise the first letter in the genus and put all of the specific epithet in lowercase. Additionally, the entire term is ordinarily italicised (or underlined when italics are not available).

The evolutionary relationships and heredity of a group of organisms is called its phylogeny. Phylogenetic studies attempt to discover phylogenies. The basic approach is to use similarities based on shared inheritance to determine relationships. As an example, species of *Pereskia* are trees or bushes with prominent leaves. They do not obviously resemble a typical leafless cactus such as an *Echinocactus*. However, both *Pereskia* and *Echinocactus* have spines produced from areoles (highly specialised pad-like structures) suggesting that the two genera are indeed related.

Two Cacti of Very Different Appearance

Judging relationships based on shared characters requires care, since plants may resemble one another through convergent evolution in which characters have arisen independently. Some euphorbias have leafless, rounded bodies adapted to water conservation similar to those of globular cacti, but characters such as the structure of their flowers make it clear that the two groups are not closely related. The cladistic method takes a systematic approach to characters, distinguishing between those that carry no information about shared evolutionary history – such as those evolved separately in different groups (homoplasies) or those left over from ancestors (plesiomorphies) – and derived characters, which have been passed down from innovations in a shared ancestor (apomorphies). Only derived characters, such as the spine-producing areoles of cacti, provide evidence for descent from a common ancestor. The results of cladistic analyses are expressed as cladograms: tree-like diagrams showing the pattern of evolutionary branching and descent.

Echinocactus grusonii

Pereskia aculeata

Although *Pereskia* is a tree with leaves, it has spines and areoles like a more typical cactus, such as *Echinocactus*.

From the 1990s onwards, the predominant approach to constructing phylogenies for living plants has been molecular phylogenetics, which uses molecular characters, particularly DNA sequences, rather than morphological characters like the presence or absence of spines and areoles. The difference is that the genetic code itself is used to decide evolutionary relationships, instead of being used indirectly via the characters it gives rise to. Clive Stace describes this as having "direct access to the genetic basis of evolution." As a simple example, prior to the use of genetic evidence, fungi were thought either to be plants or to be more closely related to plants than animals. Genetic evidence suggests that the true evolutionary relationship of multicelled organisms is as shown in the cladogram below – fungi are more closely related to animals than to plants.

In 1998 the Angiosperm Phylogeny Group published a phylogeny for flowering plants based on an analysis of DNA sequences from most families of flowering plants. As a result of this work, many questions, such as which families represent the earliest branches of angiosperms, have now been answered. Investigating how plant species are related to each other allows botanists to better understand the process of evolution in plants. Despite the study of model plants and increasing use of DNA evidence, there is ongoing work and discussion among taxonomists about how best to classify plants into various taxa. Technological developments such as computers and electron microscopes have greatly increased the level of detail studied and speed at which data can be analysed.

Plant Anatomy

Plant anatomy or phytotomy is the general term for the study of the internal structure of plants. While originally it included plant morphology, which is the description of the physical form and external structure of plants, since the mid-20th century the investigations of plant anatomy are considered a separate, distinct field, and plant anatomy refers to *just* the internal plant structures. Plant anatomy is now frequently investigated at the cellular level, and often involves the sectioning of tissues and microscopy.

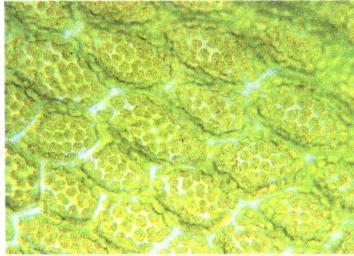

Chloroplasts in leaf cells of the moss *Mnium stellare*

Structural Divisions

Plant anatomy is often divided into the following categories:

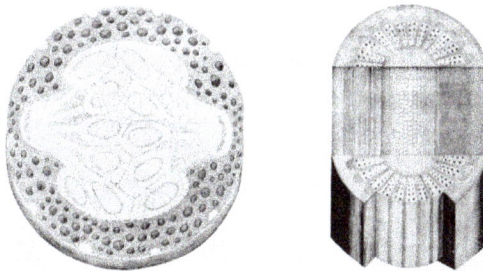

Vascular tissue of a gooseberry vine from Grew's *Anatomy of Plants*

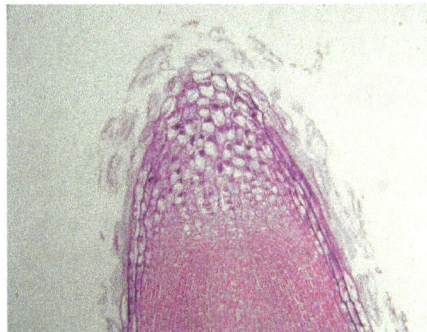

Root tip
Flower anatomy
Calyx
Corolla
Androecium
Gynoecium
Leaf anatomy
Stem anatomy
Stem structure
Fruit/Seed anatomy
Ovule
Seed structure
Pericarp
Accessory fruit
Wood anatomy
Bark
Cork

Phloem
Vascular cambium
Heartwood and sapwood
branch collar
Root anatomy
Root structure

History

About 300 BC Theophrastus wrote a number of plant treatises, only two of which survive, *Enquiry into Plants* (Περ☐ φυτ☐ν ☐στορία), and *On the Causes of Plants* (Περ☐ φυτ☐ν α☐τι☐ν). He developed concepts of plant morphology and classification, which did not withstand the scientific scrutiny of the Renaissance.

A Swiss physician and botanist, Gaspard Bauhin, introduced binomial nomenclature into plant taxonomy. He published *Pinax theatri botanici* in 1596, which was the first to use this convention for naming of species. His criteria for classification included natural relationships, or 'affinities', which in many cases were structural.

It was in the late 1600s that plant anatomy became refined into a modern science. Italian doctor and microscopist, Marcello Malpighi, was one of the two founders of plant anatomy. In 1671 he published his *Anatomia Plantarum*, the first major advance in plant physiogamy since Aristotle. The other founder was the British doctor Nehemiah Grew. He published *An Idea of a Philosophical History of Plants* in 1672 and *The Anatomy of Plants* in 1682. Grew is credited with the recognition of plant cells, although he called them 'vesicles' and 'bladders'. He correctly identified and described the sexual organs of plants (flowers) and their parts.

In the Eighteenth Century, Carolus Linnaeus established taxonomy based on structure, and his early work was with plant anatomy. While the exact structural level which is to be considered to be scientifically valid for comparison and differentiation has changed with the growth of knowledge, the basic principles were established by Linnaeus. He published his master work, *Species Plantarum* in 1753.

In 1802, French botanist Charles-François Brisseau de Mirbel, published *Traité d'anatomie et de physiologie végétale* (*Treatise on Plant Anatomy and Physiology*) establishing the beginnings of the science of plant cytology.

In 1812, Johann Jacob Paul Moldenhawer published *Beyträge zur Anatomie der Pflanzen*, describing microscopic studies of plant tissues.

In 1813 a Swiss botanist, Augustin Pyrame de Candolle, published *Théorie élémentaire de la botanique*, in which he argued that plant anatomy, not physiology, ought to be the sole basis for plant classification. Using a scientific basis, he established structural criteria for defining and separating plant genera.

In 1830, Franz Meyen published *Phytotomie*, the first comprehensive review of plant anatomy.

In 1838 German botanist Matthias Jakob Schleiden, published *Contributions to Phytogenesis*, stating, "the lower plants all consist of one cell, while the higher plants are composed of (many) individual cells" thus confirming and continuing Mirbel's work.

A German-Polish botanist, Eduard Strasburger, described the mitotic process in plant cells and further demonstrated that new cell nuclei can only arise from the division of other pre-existing nuclei. His *Studien über Protoplasma* was published in 1876.

Gottlieb Haberlandt, a German botanist, studied plant physiology and classified plant tissue based upon function. On this basis, in 1884 he published *Physiologische Pflanzenanatomie* (*Physiological Plant Anatomy*) in which he described twelve types of tissue systems (absorptive, mechanical, photosynthetic, etc.).

British paleobotanists Dunkinfield Henry Scott and William Crawford Williamson described the structures of fossilized plants at the end of the Nineteenth Century. Scott's *Studies in Fossil Botany* was published in 1900.

Following Charles Darwin's *Origin of Species* a Canadian botanist, Edward Charles Jeffrey, who was studying the comparative anatomy and phylogeny of different vascular plant groups, applied the theory to plants using the form and structure of plants to establish a number of evolutionary lines. He published his *The Anatomy of Woody Plants* in 1917.

The growth of comparative plant anatomy was spearheaded by British botanist Agnes Arber. She published *Water Plants: A Study of Aquatic Angiosperms* in 1920, *Monocotyledons: A Morphological Study* in 1925, and *The Gramineae: A Study of Cereal, Bamboo and Grass* in 1934.

Following World War II, Katherine Esau published, *Plant Anatomy* (1953), which became the definitive textbook on plant structure in North American universities and elsewhere, it was still in print as of 2006. She followed up with her *Anatomy of seed plants* in 1960.

Plant Breeding

Plant breeding is the art and science of changing the traits of plants in order to produce desired characteristics. Plant breeding can be accomplished through many different techniques ranging from simply selecting plants with desirable characteristics for propagation, to more complex mo-lecular techniques.

The Yecoro wheat (right) cultivar is sensitive to salinity, plants resulting from a hybrid cross with cultivar W4910 (left) show greater tolerance to high salinity

Plant breeding has been practiced for thousands of years, since near the beginning of human civilization. It is practiced worldwide by individuals such as gardeners and farmers, or by professional plant breeders employed by organizations such as government institutions, universities, crop-specific industry associations or research centers.

International development nation agencies believe that breeding new crops is important for ensuring food security by developing new varieties that are higher-yielding, disease resistant, drought-resistant or regionally adapted to different environments and growing conditions.

History

Plant breeding started with sedentary agriculture and particularly the domestication of the first agricultural plants, a practice which is estimated to date back 9,000 to 11,000 years. Initially early farmers simply selected food plants with particular desirable characteristics, and employed these as progenitors for subsequent generations, resulting in an accumulation of valuable traits over time.

Gregor Mendel's experiments with plant hybridization led to his establishing laws of inheritance. Once this work became well known, it formed the basis of the new science of genetics, which stimulated research by many plant scientists dedicated to improving crop production through plant breeding.

Modern plant breeding is applied genetics, but its scientific basis is broader, covering molecular biology, cytology, systematics, physiology, pathology, entomology, chemistry, and statistics (biometrics). It has also developed its own technology.

Classical Plant Breeding

One major technique of plant breeding is selection, the process of selectively propagating plants with desirable characteristics and eliminating or "culling" those with less desirable characteristics.

Another technique is the deliberate interbreeding (crossing) of closely or distantly related individuals to produce new crop varieties or lines with desirable properties. Plants are crossbred to introduce traits/genes from one variety or line into a new genetic background. For example, a mildew-resistant pea may be crossed with a high-yielding but susceptible pea, the goal of the cross being to introduce mildew resistance without losing the high-yield characteristics. Progeny from the cross would then be crossed with the high-yielding parent to ensure that the progeny were most like the high-yielding parent, (backcrossing). The progeny from that cross would then be tested for yield (selection, as described above) and mildew resistance and high-yielding resistant plants would be further developed. Plants may also be crossed with themselves to produce inbred varieties for breeding. Pollinators may be excluded through the use of pollination bags.

Classical breeding relies largely on homologous recombination between chromosomes to generate genetic diversity. The classical plant breeder may also make use of a number of *in vitro* techniques such as protoplast fusion, embryo rescue or mutagenesis (see below) to generate diversity and produce hybrid plants that would not exist in nature.

Traits that breeders have tried to incorporate into crop plants include:

1.   Improved quality, such as increased nutrition, improved flavor, or greater beauty

2. Increased yield of the crop

3. Increased tolerance of environmental pressures (salinity, extreme temperature, drought)

4. Resistance to viruses, fungi and bacteria

5. Increased tolerance to insect pests

6. Increased tolerance of herbicides

7. Longer storage period for the harvested crop

Before World War II

Successful commercial plant breeding concerns were founded from the late 19th century. Gartons Agricultural Plant Breeders in England was established in the 1890s by John Garton, who was one of the first to commercialize new varieties of agricultural crops created through cross-pollination. The firm's first introduction was Abundance Oat, one of the first agricultural grain varieties bred from a *controlled* cross, introduced to commerce in 1892.

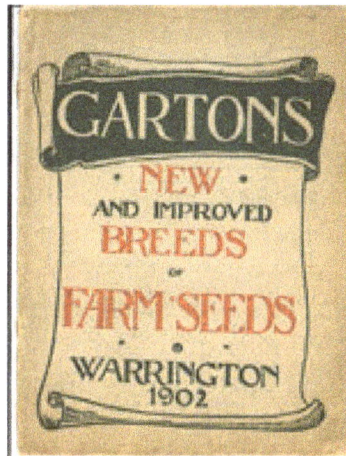

Garton's catalogue from 1902

In the early 20th century, plant breeders realized that Mendel's findings on the non-random nature of inheritance could be applied to seedling populations produced through deliberate pollinations to predict the frequencies of different types. Wheat hybrids were bred to increase the crop production of Italy during the so-called "Battle for Grain" (1925–1940). Heterosis was explained by George Harrison Shull. It describes the tendency of the progeny of a specific cross to outperform both parents. The detection of the usefulness of heterosis for plant breeding has led to the development of inbred lines that reveal a heterotic yield advantage when they are crossed. Maize was the first species where heterosis was widely used to produce hybrids.

Statistical methods were also developed to analyze gene action and distinguish heritable variation from variation caused by environment. In 1933 another important breeding technique, cytoplasmic male sterility (CMS), developed in maize, was described by Marcus Morton Rhoades. CMS is a maternally inherited trait that makes the plant produce sterile pollen. This enables the production of hybrids without the need for labor-intensive detasseling.

These early breeding techniques resulted in large yield increase in the United States in the early 20th century. Similar yield increases were not produced elsewhere until after World War II, the Green Revolution increased crop production in the developing world in the 1960s.

After World War II

Following World War II a number of techniques were developed that allowed plant breeders to hybridize distantly related species, and artificially induce genetic diversity.

In vitro-culture of Vitis (grapevine), Geisenheim Grape Breeding Institute

When distantly related species are crossed, plant breeders make use of a number of plant tissue culture techniques to produce progeny from otherwise fruitless mating. Interspecific and intergeneric hybrids are produced from a cross of related species or genera that do not normally sexually reproduce with each other. These crosses are referred to as *Wide crosses*. For example, the cereal triticale is a wheat and rye hybrid. The cells in the plants derived from the first generation created from the cross contained an uneven number of chromosomes and as result was sterile. The cell division inhibitor colchicine was used to double the number of chromosomes in the cell and thus allow the production of a fertile line.

Failure to produce a hybrid may be due to pre- or post-fertilization incompatibility. If fertilization is possible between two species or genera, the hybrid embryo may abort before maturation. If this does occur the embryo resulting from an interspecific or intergeneric cross can sometimes be rescued and cultured to produce a whole plant. Such a method is referred to as Embryo Rescue. This technique has been used to produce new rice for Africa, an interspecific cross of Asian rice *(Oryza sativa)* and African rice *(Oryza glaberrima)*.

Hybrids may also be produced by a technique called protoplast fusion. In this case protoplasts are fused, usually in an electric field. Viable recombinants can be regenerated in culture.

Chemical mutagens like EMS and DMS, radiation and transposons are used to generate mutants with desirable traits to be bred with other cultivars - a process known as *Mutation Breeding*. Classical plant breeders also generate genetic diversity within a species by exploiting a process called somaclonal variation, which occurs in plants produced from tissue culture, particularly plants derived from callus. Induced polyploidy, and the addition or removal of chromosomes using a technique called chromosome engineering may also be used.

When a desirable trait has been bred into a species, a number of crosses to the favored parent are made to make the new plant as similar to the favored parent as possible. Returning to the example of the mildew resistant pea being crossed with a high-yielding but susceptible pea, to make the mildew resistant progeny of the cross most like the high-yielding parent, the progeny will be crossed back to that parent for several generations. This process removes most of the genetic contribution of the mildew resistant parent. Classical breeding is therefore a cyclical process.

With classical breeding techniques, the breeder does not know exactly what genes have been introduced to the new cultivars. Some scientists therefore argue that plants produced by classical breeding methods should undergo the same safety testing regime as genetically modified plants. There have been instances where plants bred using classical techniques have been unsuitable for human consumption, for example the poison solanine was unintentionally increased to unacceptable levels in certain varieties of potato through plant breeding. New potato varieties are often screened for solanine levels before reaching the marketplace.

Modern Plant Breeding

Modern plant breeding may use techniques of molecular biology to select, or in the case of genetic modification, to insert, desirable traits into plants. Application of biotechnology or molecular biol-ogy is also known as molecular breeding.

Modern facilities in molecular biology have converted classical plant breeding to molecular plant breeding

Steps of Plant Breeding

The following are the major activities of plant breeding:

1. Collection of variation

2. Selection

3. Evaluation

4. Release

5. Multiplication

6. Distribution of the new variety

7. Selling to people

Marker Assisted Selection

Sometimes many different genes can influence a desirable trait in plant breeding. The use of tools such as molecular markers or DNA fingerprinting can map thousands of genes. This allows plant breeders to screen large populations of plants for those that possess the trait of interest. The screening is based on the presence or absence of a certain gene as determined by laboratory procedures, rather than on the visual identification of the expressed trait in the plant.

Reverse Breeding and Doubled Haploids (DH)

A method for efficiently producing homozygous plants from a heterozygous starting plant, which has all desirable traits. This starting plant is induced to produce doubled haploid from haploid cells, and later on creating homozygous/doubled haploid plants from those cells. While in natural offspring genetic recombination occurs and traits can be unlinked from each other, in doubled haploid cells and in the resulting DH plants recombination is no longer an issue. There, a recombination between two corresponding chromosomes does not lead to un-linkage of alleles or traits, since it just leads to recombination with its identical copy. Thus, traits on one chromosome stay linked. Selecting those offspring having the desired set of chromosomes and crossing them will result in a final F1 hybrid plant, having exactly the same set of chromosomes, genes and traits as the starting hybrid plant. The homozygous parental lines can reconstitute the original heterozygous plant by crossing, if desired even in a large quantity. An individual heterozygous plant can be converted into a heterozygous variety (F1 hybrid) without the necessity of vegetative propagation but as the result of the cross of two homozygous/doubled haploid lines derived from the originally selected plant. patent

Genetic Modification

Genetic modification of plants is achieved by adding a specific gene or genes to a plant, or by knocking down a gene with RNAi, to produce a desirable phenotype. The plants resulting from adding a gene are often referred to as transgenic plants. If for genetic modification genes of the species or of a crossable plant are used under control of their native promoter, then they are called cisgenic plants. Sometimes genetic modification can produce a plant with the desired trait or traits faster than classical breeding because the majority of the plant's genome is not altered.

To genetically modify a plant, a genetic construct must be designed so that the gene to be added or removed will be expressed by the plant. To do this, a promoter to drive transcription and a termination sequence to stop transcription of the new gene, and the gene or genes of interest must be introduced to the plant. A marker for the selection of transformed plants is also included. In the laboratory, antibiotic resistance is a commonly used marker: Plants that have been successfully transformed will grow on media containing antibiotics; plants that have not been transformed will die. In some instances markers for selection are removed by backcrossing with the parent plant prior to commercial release.

The construct can be inserted in the plant genome by genetic recombination using the bacteria *Agrobacterium tumefaciens* or *A. rhizogenes*, or by direct methods like the gene gun or micro-injection. Using plant viruses to insert genetic constructs into plants is also a possibility, but the technique is limited by the host range of the virus. For example, Cauliflower mosaic virus (CaMV) only infects cauliflower and related species. Another limitation of viral vectors is that the virus is not usually passed on the progeny, so every plant has to be inoculated.

The majority of commercially released transgenic plants are currently limited to plants that have introduced resistance to insect pests and herbicides. Insect resistance is achieved through incorporation of a gene from *Bacillus thuringiensis* (Bt) that encodes a protein that is toxic to some insects. For example, the cotton bollworm, a common cotton pest, feeds on Bt cotton it will ingest the toxin and die. Herbicides usually work by binding to certain plant enzymes and inhibiting their action. The enzymes that the herbicide inhibits are known as the herbicides *target site*. Herbicide resistance can be engineered into crops by expressing a version of *target site* protein that is not inhibited by the herbicide. This is the method used to produce glyphosate resistant crop plants.

Genetic modification of plants that can produce pharmaceuticals (and industrial chemicals), sometimes called *pharming*, is a rather radical new area of plant breeding.

Issues and Concerns

Modern plant breeding, whether classical or through genetic engineering, comes with issues of concern, particularly with regard to food crops. The question of whether breeding can have a negative effect on nutritional value is central in this respect. Although relatively little direct research in this area has been done, there are scientific indications that, by favoring certain aspects of a plant's development, other aspects may be retarded. A study published in the *Journal of the American College of Nutrition* in 2004, entitled *Changes in USDA Food Composition Data for 43 Garden Crops, 1950 to 1999*, compared nutritional analysis of vegetables done in 1950 and in 1999, and found substantial decreases in six of 13 nutrients measured, including 6% of protein and 38% of riboflavin. Reductions in calcium, phosphorus, iron and ascorbic acid were also found. The study, conducted at the Biochemical Institute, University of Texas at Austin, concluded in summary: *"We suggest that any real declines are generally most easily explained by changes in cultivated varieties between 1950 and 1999, in which there may be trade-offs between yield and nutrient content."*

The debate surrounding genetically modified food during the 1990s peaked in 1999 in terms of media coverage and risk perception, and continues today - for example, *"Germany has thrown its weight behind a growing European mutiny over genetically modified crops by banning the planting of a widely grown pest-resistant corn variety."* The debate encompasses the ecological impact of genetically modified plants, the safety of genetically modified food and concepts used for safety evaluation like substantial equivalence. Such concerns are not new to plant breeding. Most countries have regulatory processes in place to help ensure that new crop varieties entering the marketplace are both safe and meet farmers' needs. Examples include variety registration, seed schemes, regulatory authorizations for GM plants, etc.

Plant breeders' rights is also a major and controversial issue. Today, production of new varieties is dominated by commercial plant breeders, who seek to protect their work and collect royalties through national and international agreements based in intellectual property rights. The range of related issues is complex. In the simplest terms, critics of the increasingly restrictive regulations argue that, through a combination of technical and economic pressures, commercial breeders are reducing biodiversity and significantly constraining individuals (such as farmers) from developing and trading seed on a regional level. Efforts to strengthen breeders' rights, for example, by lengthening periods of variety protection, are ongoing.

When new plant breeds or cultivars are bred, they must be maintained and propagated. Some plants are propagated by asexual means while others are propagated by seeds. Seed propagated cultivars require specific control over seed source and production procedures to maintain the integrity of the plant breeds results. Isolation is necessary to prevent cross contamination with related plants or the mixing of seeds after harvesting. Isolation is normally accomplished by planting distance but in certain crops, plants are enclosed in greenhouses or cages (most commonly used when producing F1 hybrids.)

Role of Plant Breeding in Organic Agriculture

Critics of organic agriculture claim it is too low-yielding to be a viable alternative to conventional agriculture. However, part of that poor performance may be the result of growing poorly adapted varieties. It is estimated that over 95% of organic agriculture is based on conventionally adapted varieties, even though the production environments found in organic vs. conventional farming systems are vastly different due to their distinctive management practices. Most notably, organic farmers have fewer inputs available than conventional growers to control their production environments. Breeding varieties specifically adapted to the unique conditions of organic agriculture is critical for this sector to realize its full potential. This requires selection for traits such as:

- Water use efficiency

- Nutrient use efficiency (particularly nitrogen and phosphorus)

- Weed competitiveness

- Tolerance of mechanical weed control

- Pest/disease resistance

- Early maturity (as a mechanism for avoidance of particular stresses)

- Abiotic stress tolerance (i.e. drought, salinity, etc...)

Currently, few breeding programs are directed at organic agriculture and until recently those that did address this sector have generally relied on indirect selection (i.e. selection in conventional environments for traits considered important for organic agriculture). However, because the difference between organic and conventional environments is large, a given genotype may perform very differently in each environment due to an interaction between genes and the environment. If this interaction is severe enough, an important trait required for the organic environment may not be revealed in the conventional environment, which can result in the selection of poorly adapted individuals. To ensure the most adapted varieties are identified, advocates of organic breeding now promote the use of direct selection (i.e. selection in the target environment) for many agronomic traits.

There are many classical and modern breeding techniques that can be utilized for crop improvement in organic agriculture despite the ban on genetically modified organisms. For instance, controlled crosses between individuals allow desirable genetic variation to be recombined and transferred to seed progeny via natural processes. Marker assisted selection can also be employed as a diagnostics tool to facilitate selection of progeny who possess the desired trait(s), greatly speeding up the breeding process. This technique has proven particularly useful for the introgression of resistance genes into new backgrounds, as well as the efficient selection of many resistance genes pyramided into a single individual. Unfortunately, molecular markers are not currently available for many important traits, especially complex ones controlled by many genes.

Addressing Global Food Security Through Plant Breeding

For future agriculture to thrive there are necessary changes which must be made in accordance to arising global issues. These issues are arable land, harsh cropping conditions and food security which involves, being able to provide the world population with food containing sufficient nutrients. These crops need to be able to mature in several environments allowing for worldwide access, this is involves issues such as drought tolerance. These global issues are achievable through the process of plant breeding, as it offers the ability to select specific genes allowing the crop to perform at a level which yields the desired results.

Minimal Land Degradation

Land degradation is a major issue, as it can negatively impact the capability of the land to be productive. Poor agricultural management has a huge impact on the degradation of soil worldwide and it is Africa and Asia that are most affected. Through education and development of modified plants, these statistics can be reduced and agricultural land can become more productive. Plant breeding allows for an increase in yield with out the extra strain on the land. The genetically modified, Bt white maize, was introduced to South Africa and was surveyed in 33 large commercial farms and 368 small landholders properties and in both cases a higher yield was recorded.

Increased Yield Without Expansion

With an increasing population, the production of food needs to increase with it. It is estimated that a 70% increase in food production is needed by 2050 in order to meet the Declaration of the World Summit on Food Security. But with the natural degradation of agricultural land, simply planting more crops is no longer a viable option. Therefore, new varieties of plants need to be developed through plant breeding that generates an increase of yield without relying on an increase in land area. An example of this can be seen in Asia, where food production per capita has increased twofold. This has been achieved through not only the use of fertilisers, but through the use of better crops that have been specifically designed for the area.

Breeding for Increased Nutritional Value

Plant breeding can contribute to global food security as it is a cost-effective tool for increasing nutritional value of forage and crops. Improvements in nutritional value for forage crops from the use of analytical chemistry and rumen fermentation technology have been recorded since 1960; this science and technology gave breeders the ability to screen thousands of samples within a small amount of time, meaning breeders could identify a high performing hybrid quicker. The main area genetic increases were made was in vitro dry matter digestibility (IVDMD) resulting in 0.7-2.5% increase, at just 1% increase in IVDMD a single Bos Taurus also known as beef cattle reported 3.2% increase in daily gains. This improvement indicates plant breeding is an essential tool in gearing future agriculture to perform at a more advanced level.

Breeding for Tolerance

Plant breeding of hybrid crops has become extremely popular worldwide in an effort to combat the harsh environment. With long periods of drought and lack of water or nitrogen stress tolerance has become a significant part of agriculture. Plant breeders have focused on identifying crops which will ensure crops perform under these conditions; a way to achieve this is finding strains of the crop that is resistance to drought conditions with low nitrogen. It is evident from this that plant breeding is vital for future agriculture to survive as it enables farmers to produce stress resistant crops hence improving food security.

Participatory Plant Breeding

The development of agricultural science, with phenomenon like the Green Revolution arising, have left millions of farmers in developing countries, most of whom operate small farms under unstable and difficult growing conditions, in a precarious situation. The adoption of new plant varieties by this group has been hampered by the constraints of poverty and the international policies promoting an industrialized model of agriculture. Their response has been the creation of a novel and promising set of research methods collectively known as participatory plant breeding. Participatory means that farmers are more involved in the breeding process and breeding goals are defined by farmers instead of international seed companies with their large-scale breeding programs. Farmers' groups and NGOs, for example, may wish to affirm local people's rights over genetic resources, produce seeds themselves, build farmers' technical expertise, or develop new products for niche markets, like organically grown food.

Plant Propagation

Plant propagation is the process of creating new plants from a variety of sources: seeds, cuttings, bulbs and other plant parts. Plant propagation can also refer to the artificial or natural dispersal of plants.

Gentian seedlings in a plant nursery

Sexual Propagation (Seed)

One way to propagate an avocado seed

Seeds and spores can be used for reproduction (through e.g. sowing). Seeds are typically produced from sexual reproduction within a species, because genetic recombination has occurred. A plant grown from seeds may have different characteristics from its parents. Some species produce seeds that require special conditions to germinate, such as cold treatment. The seeds of many Australian plants and plants from southern Africa and the American west require smoke or fire to germinate. Some plant species, including many trees do not produce seeds until they reach maturity, which may take many years. Seeds can be difficult to acquire and some plants do not produce seed at all. Some plants (like certain F1/F2 hybrids and GMO plants) may produce seed, but not fertile seed. In certain cases (like with GMO's), this is done to prevent the accidental spreading of these plants (which are generally non-native crops), for example by birds and other animals.

Asexual Propagation

Softwood stemcuttings rooting in a controlled environment

Plants have a number of mechanisms for asexual or vegetative reproduction. Some of these have been taken advantage of by horticulturists and gardeners to multiply or clone plants rapidly. People also use methods that plants do not use, such as tissue culture and grafting. Plants are produced using material from a single parent and as such there is no exchange of genetic material, therefore vegetative propagation methods almost always produce plants that are identical to the parent. Vegetative reproduction uses plants parts such as roots, stems and leaves. In some plants seeds can be produced without fertilization and the seeds contain only the genetic material of the parent plant. Therefore, propagation via asexual seeds or apomixis is asexual reproduction but not vegetative propagation.

Techniques for vegetative propagation include:

- Air or ground layering
- Division
- Grafting and bud grafting, widely used in fruit tree propagation
- Micropropagation
- Stolons or runners
- Storage organs such as bulbs, corms, tubers and rhizomes
- Striking or cuttings
- Twin-scaling
- Offsets

Heated Propagator

A heated propagator is a horticultural device to maintain a warm and damp environment for seeds and cuttings to grow in.

This can be in the form of a clear enclosed bin sitting over a hotpad, or even a portable heater pointed at the bin. The key is to keep the moisture in the clear bin, while keeping lighting over the top of it, usually.

Seed Propagation Mat

An electric seed-propagation mat is a heated rubber mat covered by a metal cage which is used in gardening. The mats are made so that planters containing seedlings can be placed on top of the metal cage without the risk of starting a fire. In extreme cold, gardeners place a loose plastic cover over the planters/mats which creates a sort of miniature greenhouse. The constant and predictable heat allows people to garden in the winter months when the weather is generally too cold for seedlings to survive naturally. When combined with a lighting system, many plants can be grown indoors using these mats.

Plant Morphology

Plant morphology or phytomorphology is the study of the physical form and external structure of plants. This is usually considered distinct from plant anatomy, which is the study of the internal structure of plants, especially at the microscopic level. Plant morphology is useful in the visual identification of plants.

Inflorescences emerging from protective coverings

Scope

Plant morphology "represents a study of the development, form, and structure of plants, and, by implication, an attempt to interpret these on the basis of similarity of plan and origin." There are four major areas of investigation in plant morphology, and each overlaps with another field of the biological sciences.

Looking up into the branch structure of a *Pinus sylvestris* tree

Asclepias syriaca showing complex morphology of the flowers.

First of all, morphology is comparative, meaning that the morphologist examines structures in many different plants of the same or different species, then draws comparisons and formulates ideas about similarities. When structures in different species are believed to exist and develop as a result of common, inherited genetic pathways, those structures are termed homologous. For example, the leaves of pine, oak, and cabbage all look very different, but share certain basic structures and arrangement of parts. The homology of leaves is an easy conclusion to make. The plant morphologist goes further, and discovers that the spines of cactus also share the same basic structure and development as leaves in other plants, and therefore cactus spines are homologous to leaves as well. This aspect of plant morphology overlaps with the study of plant evolution and paleobotany.

Secondly, plant morphology observes both the vegetative (somatic) structures of plants, as well as the reproductive structures. The vegetative structures of vascular plants includes the study of the shoot system, composed of stems and leaves, as well as the root system. The reproductive structures are more varied, and are usually specific to a particular group of plants, such as flowers and seeds, fern sori, and moss capsules. The detailed study of reproductive structures in plants led to the discovery of the alternation of generations found in all plants and most algae. This area of plant morphology overlaps with the study of biodiversity and plant systematics.

Thirdly, plant morphology studies plant structure at a range of scales. At the smallest scales are ultrastructure, the general structural features of cells visible only with the aid of an electron microscope, and cytology, the study of cells using optical microscopy. At this scale, plant morphology overlaps with plant anatomy as a field of study. At the largest scale is the study of plant growth habit, the overall architecture of a plant. The pattern of branching in a tree will vary from species to species, as will the appearance of a plant as a tree, herb, or grass.

Fourthly, plant morphology examines the pattern of development, the process by which structures originate and mature as a plant grows. While animals produce all the body parts they will ever have from early in their life, plants constantly produce new tissues and structures throughout their life. A living plant always has embryonic tissues. The way in which new structures mature as they are produced may be affected by the point in the plant's life when they begin to develop, as well as by the environment to which the structures are exposed. A morphologist studies this process, the causes, and its result. This area of plant morphology overlaps with plant physiology and ecology.

A Comparative Science

A plant morphologist makes comparisons between structures in many different plants of the same or different species. Making such comparisons between similar structures in different plants tackles the question of *why* the structures are similar. It is quite likely that similar underlying causes of genetics, physiology, or response to the environment have led to this similarity in appearance. The result of scientific investigation into these causes can lead to one of two insights into the underlying biology:

1. Homology - the structure is similar between the two species because of shared ancestry and common genetics.

2. Convergence - the structure is similar between the two species because of independent adaptation to common environmental pressures.

Understanding which characteristics and structures belong to each type is an important part of understanding plant evolution. The evolutionary biologist relies on the plant morphologist to interpret structures, and in turn provides phylogenies of plant relationships that may lead to new morphological insights.

Homology

When structures in different species are believed to exist and develop as a result of common, inherited genetic pathways, those structures are termed *homologous*. For example, the leaves of pine, oak, and cabbage all look very different, but share certain basic structures and arrangement of parts. The homology of leaves is an easy conclusion to make. The plant morphologist goes further, and discovers that the spines of cactus also share the same basic structure and development as leaves in other plants, and therefore cactus spines are homologous to leaves as well.

Convergence

When structures in different species are believed to exist and develop as a result of common adaptive responses to environmental pressure, those structures are termed *convergent*. For example, the fronds of *Bryopsis plumosa* and stems of *Asparagus setaceus* both have the same feathery branching appearance, even though one is an alga and one is a flowering plant. The similarity in overall structure occurs independently as a result of convergence. The growth form of many cacti and species of *Euphorbia* is very similar, even though they belong to widely distant families. The similarity results from common solutions to the problem of surviving in a hot, dry environment.

Astrophytum asterias, a cactus.   *Euphorbia obesa*, a spurge

Vegetative and Reproductive Characteristics

Plant morphology treats both the vegetative structures of plants, as well as the reproductive structures.

The vegetative (somatic) structures of vascular plants include two major organ systems: (1) a shoot system, composed of stems and leaves, and (2) a root system. These two systems are common to nearly all vascular plants, and provide a unifying theme for the study of plant morphology.

By contrast, the reproductive structures are varied, and are usually specific to a particular group of plants. Structures such as flowers and fruits are only found in the angiosperms; sori are only found in ferns; and seed cones are only found in conifers and other gymnosperms. Reproductive characters are therefore regarded as more useful for the classification of plants than vegetative characters.

Use in Identification

Plant biologists use morphological characters of plants which can be compared, measured, counted and described to assess the differences or similarities in plant taxa and use these characters for plant identification, classification and descriptions.

When characters are used in descriptions or for identification they are called diagnostic or key characters which can be either qualitative and quantitative.

1. Quantitative characters are morphological features that can be counted or measured for example a plant species has flower petals 10–12 mm wide.

2. Qualitative characters are morphological features such as leaf shape, flower color or pubescence.

Both kinds of characters can be very useful for the identification of plants.

Alternation of Generations

The detailed study of reproductive structures in plants led to the discovery of the alternation of generations, found in all plants and most algae, by the German botanist Wilhelm Hofmeister. This discovery is one of the most important made in all of plant morphology, since it provides a common basis for understanding the life cycle of all plants.

Pigmentation in Plants

The primary function of pigments in plants is photosynthesis, which uses the green pigment chlorophyll along with several red and yellow pigments that help to capture as much light energy as possible. Pigments are also an important factor in attracting insects to flowers to encourage pollination.

Plant pigments include a variety of different kinds of molecule, including porphyrins, carotenoids, anthocyanins and betalains. All biological pigments selectively absorb certain wavelengths of light while reflecting others. The light that is absorbed may be used by the plant to power chemical reactions, while the reflected wavelengths of light determine the color the pigment will appear to the eye.

Morphology in Development

Plant development is the process by which structures originate and mature as a plant grows. It is a subject studies in plant anatomy and plant physiology as well as plant morphology.

The process of development in plants is fundamentally different from that seen in vertebrate animals. When an animal embryo begins to develop, it will very early produce all of the body parts that it will ever have in its life. When the animal is born (or hatches from its egg), it has all its body parts and from that point will only grow larger and more mature. By contrast, plants constantly produce new tissues and structures throughout their life from meristems located at the tips of organs, or between mature tissues. Thus, a living plant always has embryonic tissues.

The properties of organization seen in a plant are emergent properties which are more than the sum of the individual parts. "The assembly of these tissues and functions into an integrated multicellular organism yields not only the characteristics of the separate parts and processes but also quite a new set of characteristics which would not have been predictable on the basis of examination of the separate parts." In other words, knowing everything about the molecules in a plant are not enough to predict characteristics of the cells; and knowing all the properties of the cells will not predict all the properties of a plant's structure.

Growth

A vascular plant begins from a single celled zygote, formed by fertilisation of an egg cell by a sperm cell. From that point, it begins to divide to form a plant embryo through the process of embryogenesis. As this happens, the resulting cells will organize so that one end becomes the first root, while the other end forms the tip of the shoot. In seed plants, the embryo will develop one or more "seed leaves" (cotyledons). By the end of embryogenesis, the young plant will have all the parts necessary to begin in its life.

Once the embryo germinates from its seed or parent plant, it begins to produce additional organs (leaves, stems, and roots) through the process of organogenesis. New roots grow from root meristems located at the tip of the root, and new stems and leaves grow from shoot meristems located at the tip of the shoot. Branching occurs when small clumps of cells left behind by the meristem, and which have not yet undergone cellular differentiation to form a specialized tissue, begin to grow as the tip of a new root or shoot. Growth from any such meristem at the tip of a root or shoot is termed primary growth and results in the lengthening of that root or shoot. Secondary growth results in widening of a root or shoot from divisions of cells in a cambium.

In addition to growth by cell division, a plant may grow through cell elongation. This occurs when individual cells or groups of cells grow longer. Not all plant cells will grow to the same length. When cells on one side of a stem grow longer and faster than cells on the other side, the stem will bend to the side of the slower growing cells as a result. This directional growth can occur via a plant's response to a particular stimulus, such as light (phototropism), gravity (gravitropism), water, (hydrotropism), and physical contact (thigmotropism).

Plant growth and development are mediated by specific plant hormones and plant growth regulators (PGRs) (Ross et al. 1983). Endogenous hormone levels are influenced by plant age, cold hardiness, dormancy, and other metabolic conditions; photoperiod, drought, temperature, and other external environmental conditions; and exogenous sources of PGRs, e.g., externally applied and of rhizospheric origin.

Morphological Variation

Plants exhibit natural variation in their form and structure. While all organisms vary from individual to individual, plants exhibit an additional type of variation. Within a single individual, parts are repeated which may differ in form and structure from other similar parts. This variation is most easily seen in the leaves of a plant, though other organs such as stems and flowers may show similar variation. There are three primary causes of this variation: positional effects, environmental effects, and juvenility.

Evolution of Plant Morphology

Transcription factors and transcriptional regulatory networks play key roles in plant morphogenesis and their evolution. During plant landing, many novel transcription factor families emerged and are preferentially wired into the networks of multicellular development, reproduction, and organ development, contributing to more complex morphogenesis of land plants.

Positional Effects

Variation in leaves from the giant ragweed illustrating positional effects. The lobed leaves come from the base of the plant, while the unlobed leaves come from the top of the plant.

Although plants produce numerous copies of the same organ during their lives, not all copies of a particular organ will be identical. There is variation among the parts of a mature plant resulting from the relative position where the organ is produced. For example, along a new branch the leaves may vary in a consistent pattern along the branch. The form of leaves produced near the base of the branch will differ from leaves produced at the tip of the plant, and this difference is consistent from branch to branch on a given plant and in a given species. This difference persists after the leaves at both ends of the branch have matured, and is not the result of some leaves being younger than others.

Environmental Effects

The way in which new structures mature as they are produced may be affected by the point in the plants life when they begin to develop, as well as by the environment to which the structures are exposed. This can be seen in aquatic plants and emergent plants.

Temperature

Temperature has a multiplicity of effects on plants depending on a variety of factors, including the size and condition of the plant and the temperature and duration of exposure. The smaller and more succulent the plant, the greater the susceptibility to damage or death from temperatures that are too high or too low. Temperature affects the rate of biochemical and physiological processes, rates generally (within limits) increasing with temperature. However, the Van't Hoff relationship for monomolecular reactions (which states that the velocity of a reaction is doubled or trebled by a temperature increase of 10 °C) does not strictly hold for biological processes, especially at low and high temperatures.

When water freezes in plants, the consequences for the plant depend very much on whether the freezing occurs intracellularly (within cells) or outside cells in intercellular (extracellular) spaces. Intracellular freezing usually kills the cell regardless of the hardiness of the plant and its tissues. Intracellular freezing seldom occurs in nature, but moderate rates of decrease in temperature, e.g., 1 °C to 6 °C/hour, cause intercellular ice to form, and this "extraorgan ice" may or may not be lethal, depending on the hardiness of the tissue.

At freezing temperatures, water in the intercellular spaces of plant tissues freezes first, though the water may remain unfrozen until temperatures fall below 7 °C. After the initial formation of ice intercellularly, the cells shrink as water is lost to the segregated ice. The cells undergo freeze-drying, the dehydration being the basic cause of freezing injury.

The rate of cooling has been shown to influence the frost resistance of tissues, but the actual rate of freezing will depend not only on the cooling rate, but also on the degree of supercooling and the properties of the tissue. Sakai (1979a) demonstrated ice segregation in shoot primordia of Alaskan white and black spruces when cooled slowly to 30 °C to -40 °C. These freeze-dehydrated buds survived immersion in liquid nitrogen when slowly rewarmed. Floral primordia responded similarly. Extraorgan freezing in the primordia accounts for the ability of the hardiest of the boreal conifers to survive winters in regions when air temperatures often fall to -50 °C or lower. The hardiness of the winter buds of such conifers is enhanced by the smallness of the buds, by the evolution of faster translocation of water, and an ability to tolerate intensive freeze dehydration. In boreal species of

Picea and *Pinus*, the frost resistance of 1-year-old seedlings is on a par with mature plants, given similar states of dormancy.

Juvenility

The organs and tissues produced by a young plant, such as a seedling, are often different from those that are produced by the same plant when it is older. This phenomenon is known as juvenility or heteroblasty. For example, young trees will produce longer, leaner branches that grow upwards more than the branches they will produce as a fully grown tree. In addition, leaves produced during early growth tend to be larger, thinner, and more irregular than leaves on the adult plant. Specimens of juvenile plants may look so completely different from adult plants of the same species that egg-laying insects do not recognize the plant as food for their young. Differences are seen in rootability and flowering and can be seen in the same mature tree. Juvenile cuttings taken from the base of a tree will form roots much more readily than cuttings originating from the mid to upper crown. Flowering close to the base of a tree is absent or less profuse than flowering in the higher branches especially when a young tree first reaches flowering age.

Juvenility in a seedling of European beech. There is a marked difference in shape between the first dark green "seed leaves" and the lighter second pair of leaves.

The transition from early to late growth forms is referred to as 'vegetative phase change', but there is some disagreement about terminology.

Some Recent Developments

Rolf Sattler has revised fundamental concepts of comparative morphology such as the concept of homology. He emphasized that homology should also include partial homology and quantitative homology. This leads to a continuum morphology that demonstrates a continuum between the morphological categories of root, shoot, stem (caulome), leaf (phyllome), and hair (trichome). How intermediates between the categories are best described has been discussed by Bruce K. Kirchoff et al.

Honoring Agnes Arber, author of the partial-shoot theory of the leaf, Rutishauser and Isler called the continuum approach Fuzzy Arberian Morphology (FAM). "Fuzzy" refers to fuzzy logic, "Arberian" to Agnes Arber. Rutishauser and Isler emphasized that this approach is not only supported by many morphological data but also by evidence from molecular genetics. More recent evidence from molecular genetics provides further support for continuum morphology. James (2009) concluded that "it is now widely accepted that... radiality [characteristic of most shoots] and dorsiventrality [characteristic of leaves] are but extremes of a continuous spectrum. In fact, it is simply the timing of the KNOX gene expression!." Eckardt and Baum (2010) concluded that "it is now generally accepted that compound leaves express both leaf and shoot properties.".

Process morphology (dynamic morphology) describes and analyzes the dynamic continuum of plant form. According to this approach, structures do not *have* process(es), they *are* process(es). Thus, the structure/process dichotomy is overcome by "an enlargement of our concept of 'structure' so as to include and recognize that in the living organism it is not merely a question of spatial structure with an 'activity' as something over or against it, but that the concrete organism is a spatio-*temporal* structure and that this spatio-temporal structure is the activity itself."

For Jeune, Barabé and Lacroix, classical morphology (that is, mainstream morphology, based on a qualitative homology concept implying mutually exclusive categories) and continuum morphology are sub-classes of the more encompassing process morphology (dynamic morphology).

Plant Ecology

Plant ecology is a subdiscipline of ecology which studies the distribution and abundance of plants, the effects of environmental factors upon the abundance of plants, and the interactions among and between plants and other organisms. Examples of these are the distribution of temperate deciduous forests in North America, the effects of drought or flooding upon plant survival, and competition among desert plants for water, or effects of herds of grazing animals upon the composition of grasslands.

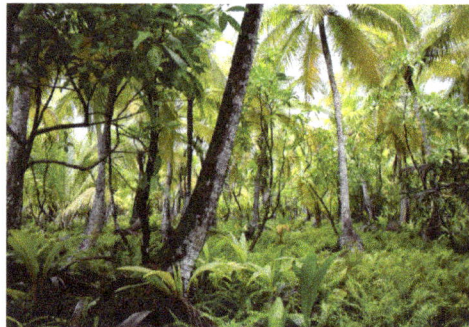

A tropical plant community on Diego Garcia

A global overview of the Earth's major vegetation types is provided by O.W. Archibold. He recognizes 11 major vegetation types: tropical forests, tropical savannas, arid regions (deserts), Mediterranean ecosystems, temperate forest ecosystems, temperate grasslands, coniferous forests, tundra (both polar and high mountain), terrestrial wetlands, freshwater ecosystems and coastal/marine systems. This breadth of topics shows the complexity of plant ecology, since it includes plants from floating single-celled algae up to large canopy forming trees.

One feature that defines plants is photosynthesis. One of the most important aspects of plant ecology is the role plants have played in creating the oxygenated atmosphere of earth, an event that occurred some 2 billion years ago. It can be dated by the deposition of banded iron formations, distinctive sedimentary rocks with large amounts of iron oxide. At the same time, plants began removing carbon dioxide from the atmosphere, thereby initiating the process of controlling Earth's climate. A long term trend of the Earth has been toward increasing oxygen and decreasing carbon dioxide, and many other events in the Earths history, like the first movement of life onto land, are likely tied to this sequence of events.

One of the early classic books on plant ecology was written by J.E. Weaver and F.E. Clements. It talks broadly about plant communities, and particularly the importance of forces like competition and processes like succession. Although some of the terminology is dated, this important book can still often be obtained in used book stores.

Plant ecology can also be divided by levels of organization including plant ecophysiology, plant population ecology, community ecology, ecosystem ecology, landscape ecology and biosphere ecology.

The study of plants and vegetation is complicated by their form. First, most plants are rooted in the soil, which makes it difficult to observe and measure nutrient uptake and species interactions. Second, plants often reproduce vegetatively, that is asexually, in a way that makes it difficult to distinguish individual plants. Indeed, the very concept of an individual is doubtful, since even a tree may be regarded as a large collection of linked meristems. Hence, plant ecology and animal ecology have different styles of approach to problems that involve processes like reproduction, dispersal and mutualism. Some plant ecologists have placed considerable emphasis upon trying to treat plant populations as if they were animal populations, focusing on population ecology. Many other ecologists believe that while it is useful to draw upon population ecology to solve certain scientific problems, plants demand that ecologists work with multiple perspectives, appropriate to the problem, the scale and the situation.

History

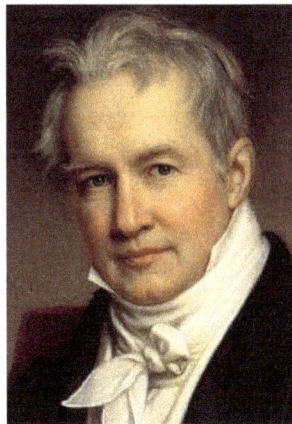

Alexander von Humboldt's work connecting plant distributions with environmental factors played an important role in the genesis of the discipline of plant ecology.

Plant ecology has its origin in the application of plant physiology to the questions raised by plant geographers. Carl Ludwig Willdenow was one of the first to note that similar climates produced similar types of vegetation, even when they were located in different parts of the world. Willdenow's student, Alexander von Humboldt, used physiognomy to describe vegetation types and observed that the distribution vegetation types was based on environmental factors. Later plant geographers who built upon Humboldt's work included Joakim Frederik Schouw, A.P. de Candolle, August Grisebach and Anton Kerner von Marilaun. Schouw's work, published in 1822, linked plant distributions to environmental factors (especially temperature) and established the practice of naming plant associations by adding the suffix *-etum* to the name of the dominant species. Working from herbarium collections, De Candolle searched for general rules of plant distribution and settled on using temperature as well. Grisebach's two-volume work, *Die Vegetation der Erde nach Ihrer Klimatischen Anordnung*, published in 1872, saw plant geography reach its "ultimate form" as a descriptive field.

Starting in the 1870s, Swiss botanist Simon Schwendener, together with his students and colleagues, established the link between plant morphology and physiological adaptations, laying the groundwork for the first ecology textbooks, Eugenius Warming's *Plantesamfund* (published in 1895) and Andreas Schimper's 1898 *Pflanzengeographie auf Physiologischer Grundlage*. Warming successfully incorporated plant morphology, physiology taxonomy and biogeography into plant geography to create the field of plant ecology. Although more morphological than physiological, Schimper's has been considered the beginning of plant physiological ecology. Plant ecology was initially built around static ideas of plant distribution; incorporating the concept of succession added an element to change through time to the field. Henry Chandler Cowles' studies of plant succession on the Lake Michigan sand dunes (published in 1899) and Frederic Clements' 1916 monograph on the subject established it as a key element of plant ecology.

Plant ecology developed within the wider discipline of ecology over the twentieth century. Inspired by Warming's *Plantesamfund*, Arthur Tansley set out to map British plant communities. In 1904 he teamed up with William Gardner Smith and others involved in vegetation mapping to establish the Central Committee for the Survey and Study of British Vegetation, later shortened to British Vegetation Committee. In 1913, the British Vegetation Committee organised the British Ecological Society (BES), the first professional society of ecologists. This was followed in 1917 by the establishment of the Ecological Society of America (ESA); plant ecologists formed the largest subgroup among the inaugural members of the ESA.

Cowles' students played an important role in the development of the field of plant ecology during the first half of the twentieth century, among them William S. Cooper, E. Lucy Braun and Edgar Transeau.

Graphical timeline of plant ecologists

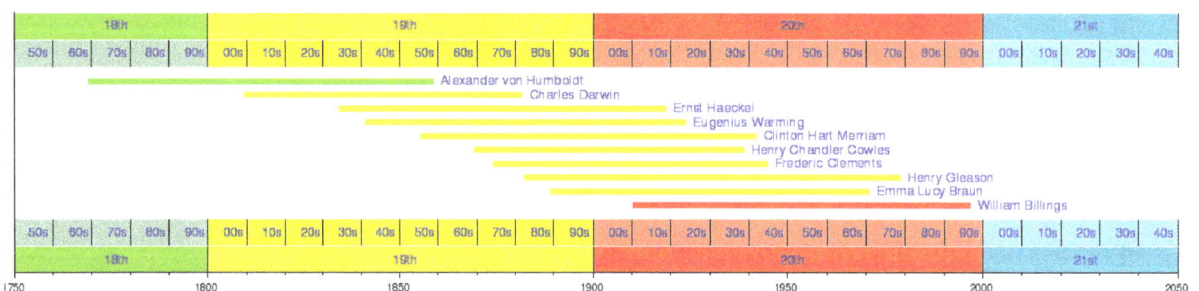

Distribution

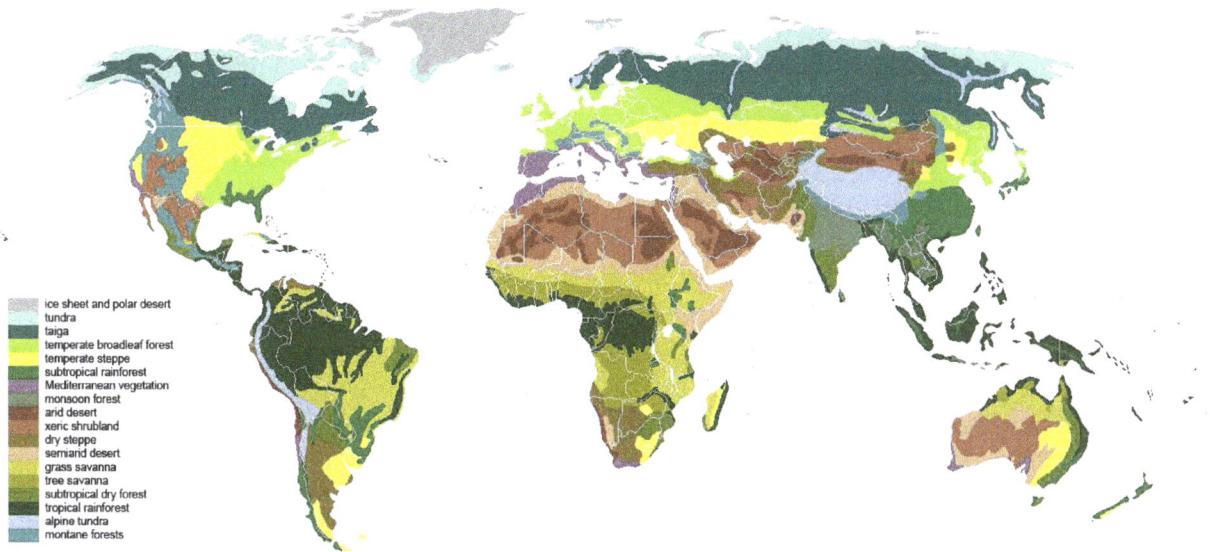

World biomes are based upon the type of dominant plant.

Plant distributions is governed by a combination of historical factors, ecophysiology and biotic interactions. The set of species that can be present at a given site is limited by historical contingency. In order to show up, a species must either have evolved in an area or dispersed there (either naturally or through human agency), and must not have gone locally extinct. The set of species present locally is further limited to those that possess the physiological adaptations to survive the environmental conditions that exist. This group is further shaped through interactions with other species.

Plant communities are broadly distributed into biomes based on the form of the dominant plant species. For example, grasslands are dominated by grasses, while forests are dominated by trees. Biomes are determined by regional climates, mostly temperature and precipitation, and follow general latitudinal trends. Within biomes, there may be many ecological communities, which are impacted not only by climate and a variety of smaller-scale features, including soils, hydrology, and disturbance regime. Biomes also change with elevation, high elevations often resembling those found at higher latitudes.

Biological Interactions

Competition

Plants, like most life forms, require relatively few basic elements: carbon, hydrogen, oxygen, nitrogen, phosphorus and sulphur; hence they are known as CHNOPS life forms. There are also lesser elements needed as well, frequently termed micronutrients, such as magnesium and sodium. When plants grow in close proximity, they may deplete supplies of these elements and have a negative impact upon neighbours. Competition for resources vary from complete symmetric (all individuals receive the same amount of resources, irrespective of their size) to perfectly size symmetric (all individuals exploit the same amount of resource per unit biomass) to absolutely size-asymmetric (the largest individuals exploit all the available resource). The degree of size asymmetry has major effects on the structure and diversity of ecological communities.In many cases (perhaps most) the negative effects upon neighbours arise from size asymmetric competition for light. In other cases, there may be competition below ground for water, nitrogen, or phosphorus. To detect and measure competition, experiments are necessary; these experiments require removing neighbours, and measuring responses in the remaining plants. Many such studies are required before useful generalizations can be drawn.

Overall, it appears that light is the most important resource for which plants compete, and the increase in plant height over evolutionary time likely reflects selection for taller plants to better intercept light. Many plant communities are therefore organized into hierarchies based upon the relative competitive abilities for light. In some systems, particularly infertile or arid systems, below ground competition may be more significant. Along natural gradients of soil fertility, it is likely that the ratio of above ground to below ground competition changes, with higher above ground competition in the more fertile soils. Plants that are relatively weak competitors may escape in time (by surviving as buried seeds) or in space (by dispersing to a new location away from strong competitors.)

In principle, it is possible to examine competition at the level of the limiting resources if a detailed knowledge of the physiological processes of the competing plants is available. However, in most terrestrial ecological studies, there is only little information on the uptake and dynamics of the resources that limit the growth of different plant species, and, instead, competition is inferred from observed negative effects of neighbouring plants without knowing precisely which resources the plants were competing for. In certain situations, plants may compete for a single growth-limiting resource, perhaps for light in agricultural systems with sufficient water and nutrients, or in dense stands of marsh vegetation, but in many natural ecosystems plants may be colimited by several resources, e.g. light, phosphorus and nitrogen at the same time.

Therefore, there are many details that remain to be uncovered, particularly the kinds of competition that arise in natural plant communities, the specific resource(s), the relative importance of different resources, and the role of other factors like stress or disturbance in regulating the importance of competition.

Mutualism

Mutualism is defined as an interaction "between two species or individuals that is beneficial to both". Probably the most widespread example in plants is the mutual beneficial relationship between plants and fungi, known as mycorrhizae. The plant is assisted with nutrient uptake, while the fungus receives carbohydrates. Some the earliest known fossil plants even have fossil mycorrhizae on their rhizomes.

The flowering plants are a group that have evolved by using two major mutualisms. First, flowers are pollinated by insects. This relationship seems to have its origins in beetles feeding on primitive flowers, eating pollen and also acting (unwittingly) as pollinators. Second, fruits are eaten by animals, and the animals then disperse the seeds. Thus, the flowering plants actually have three major types of mutualism, since most higher plants also have mycorrhizae.

Plants may also have beneficial effects upon one another, but this is less common. Examples might include "nurse plants" whose shade allows young cacti to establish. Most examples of mutualism, however, are largely beneficial to only one of the partners, and may not really be true mutualism. The term used for these more one-sided relationships, which are mostly beneficial to one participant, is facilitation. Facilitation among neighboring plants may act by reducing the negative impacts of a stressful environment. In general, facilitation is more likely to occur in physically stressful environments than in favorable environments, where competition may be the most important interaction among species.

Commensalism is similar to facilitation, in that one plant is mostly exploiting another. A familiar example is the ephiphytes which grow on branches of tropical trees, or even mosses which grow on trees in deciduous forests.

It is important to keep track of the benefits received by each species to determine the appropriate term. Although people are often fascinated by unusual examples, it is important to remember that in plants, the main mutualisms are mycorrhizae, pollination, and seed dispersal.

Herbivory

An important ecological function of plants is that they produce organic compounds for herbivores in the bottom of the food web. A large number of plant traits, from thorns to chemical defenses, can be related to the intensity of herbivory. Large herbivores can also have many effects on vegetation. These include removing selected species, creating gaps for regeneration of new individuals, recycling nutrients, and dispersing seeds. Certain ecosystem types, such as grasslands, may be dominated by the effects of large herbivores, although fire is also an equally important factor in this biome. In few cases, herbivores are capable of nearly removing all the vegetation at a site (for example, geese in the Hudson Bay Lowlands of Canada, and nutria in the marshes of Louisiana) but normally heribovres have a more selective impact, particularly when large predators control the abundance of herbivores. The usual method of studying the effects of herbivores is to build exclosures, where they cannot feed, and compare the plant communities in the exclosures to those outside over many years. Often such long term experiments show that herbivores have a significant effect upon the species that make up the plant community.

Other Topics

Abundance

The ecological success of a plant species in a specific environment may be quantified by its abundance, and depending on the life form of the plant different measures of abundance may be relevant, e.g. density, biomass, or plant cover.

The change in the abundance of a plant species may be due to both abiotic factors, e.g. climate change, or biotic factors, e.g. herbivory or interspecific competition.

Colonisation and Local Extinction

Whether a plant species is present at a local area depends on the processes of colonisation and local extinction. The probability of colonisation decreases with distance to neighboring habitats where the species is present and increases with plant abundance and fecundity in neighboring habitats and the dispersal distance of the species. The probability of local extinction decreases with abundance (both living plants and seeds in the soil seed bank).

Plant Physiology

Plant physiology is a subdiscipline of botany concerned with the functioning, or physiology, of plants. Closely related fields include plant morphology (structure of plants), plant ecology (interactions with the environment), phytochemistry (biochemistry of plants), cell biology, genetics, biophysics and molecular biology.

A germination rate experiment

Fundamental processes such as photosynthesis, respiration, plant nutrition, plant hormone functions, tropisms, nastic movements, photoperiodism, photomorphogenesis, circadian rhythms, environmental stress physiology, seed germination, dormancy and stomata function and transpiration, both parts of plant water relations, are studied by plant physiologists.

Aims

The field of plant physiology includes the study of all the internal activities of plants—those chemical and physical processes associated with life as they occur in plants. This includes study at many levels of scale of size and time. At the smallest scale are molecular interactions of photosynthesis and internal diffusion of water, minerals, and nutrients. At the largest scale are the processes of plant development, seasonality, dormancy, and reproductive control. Major subdisciplines of plant physiology include phytochemistry (the study of the biochemistry of plants) and phytopathology (the study of disease in plants). The scope of plant physiology as a discipline may be divided into several major areas of research.

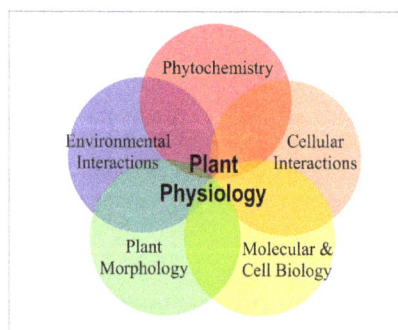

Five key areas of study within plant physiology.

First, the study of phytochemistry (plant chemistry) is included within the domain of plant physiology. To function and survive, plants produce a wide array of chemical compounds not found in other organisms. Photosynthesis requires a large array of pigments, enzymes, and other compounds to function. Because they cannot move, plants must also defend themselves chemically from herbivores, pathogens and competition from other plants. They do this by producing toxins and foul-tasting or smelling chemicals. Other compounds defend plants against disease, permit survival during drought, and prepare plants for dormancy, while other compounds are used to attract pollinators or herbivores to spread ripe seeds.

Secondly, plant physiology includes the study of biological and chemical processes of individual plant cells. Plant cells have a number of features that distinguish them from cells of animals, and which lead to major differences in the way that plant life behaves and responds differently from animal life. For example, plant cells have a cell wall which restricts the shape of plant cells and thereby limits the flexibility and mobility of plants. Plant cells also contain chlorophyll, a chemical compound that interacts with light in a way that enables plants to manufacture their own nutrients rather than consuming other living things as animals do.

Thirdly, plant physiology deals with interactions between cells, tissues, and organs within a plant. Different cells and tissues are physically and chemically specialized to perform different functions. Roots and rhizoids function to anchor the plant and acquire minerals in the soil. Leaves catch light in order to manufacture nutrients. For both of these organs to remain living, minerals that the roots acquire must be transported to the leaves, and the nutrients manufactured in the leaves must be transported to the roots. Plants have developed a number of ways to achieve this transport, such as vascular tissue, and the functioning of the various modes of transport is studied by plant physiologists.

Fourthly, plant physiologists study the ways that plants control or regulate internal functions. Like animals, plants produce chemicals called hormones which are produced in one part of the plant to signal cells in another part of the plant to respond. Many flowering plants bloom at the appropriate time because of light-sensitive compounds that respond to the length of the night, a phenomenon known as photoperiodism. The ripening of fruit and loss of leaves in the winter are controlled in part by the production of the gas ethylene by the plant.

Finally, plant physiology includes the study of plant response to environmental conditions and their variation, a field known as environmental physiology. Stress from water loss, changes in air chemistry, or crowding by other plants can lead to changes in the way a plant functions. These changes may be affected by genetic, chemical, and physical factors.

Biochemistry of Plants

The chemical elements of which plants are constructed—principally carbon, oxygen, hydrogen, nitrogen, phosphorus, sulfur, etc.—are the same as for all other life forms animals, fungi, bacteria and even viruses. Only the details of the molecules into which they are assembled differs.

Despite this underlying similarity, plants produce a vast array of chemical compounds with unique properties which they use to cope with their environment. Pigments are used by plants to absorb or detect light, and are extracted by humans for use in dyes. Other plant products may be used for the manufacture of commercially important rubber or biofuel. Perhaps the most celebrated compounds from plants are those with pharmacological activity, such as salicylic acid from which aspirin is made, morphine, and digoxin. Drug companies spend billions of dollars each year researching plant compounds for potential medicinal benefits.

Latex being collected from a tapped rubber tree.

Constituent Elements

Plants require some nutrients, such as carbon and nitrogen, in large quantities to survive. Such nutrients are termed macronutrients, where the prefix *macro-* (large) refers to the quantity needed, not the size of the nutrient particles themselves. Other nutrients, called micronutrients, are required only in trace amounts for plants to remain healthy. Such micronutrients are usually absorbed as ions dissolved in water taken from the soil, though carnivorous plants acquire some of their micronutrients from captured prey.

The following tables list element nutrients essential to plants. Uses within plants are generalized.

Macronutrients – necessary in large quantities		
Element	Form of uptake	Notes
Nitrogen	NO_3^-, NH_4^+	Nucleic acids, proteins, hormones, etc.
Oxygen	O_2 H_2O	Cellulose, starch, other organic compounds
Carbon	CO_2	Cellulose, starch, other organic compounds
Hydrogen	H_2O	Cellulose, starch, other organic compounds
Potassium	K^+	Cofactor in protein synthesis, water balance, etc.
Calcium	Ca^{2+}	Membrane synthesis and stabilization
Magnesium	Mg^{2+}	Element essential for chlorophyll
Phosphorus	$H_2PO_4^-$	Nucleic acids, phospholipids, ATP
Sulfur	SO_4^{2-}	Constituent of proteins

Micronutrients – necessary in small quantities		
Element	Form of uptake	Notes
Chlorine	Cl^-	Photosystem II and stomata function
Iron	Fe^{2+}, Fe^{3+}	Chlorophyll formation
Boron	HBO_3	Crosslinking pectin
Manganese	Mn^{2+}	Activity of some enzymes
Zinc	Zn^{2+}	Involved in the synthesis of enzymes and chlorophyll
Copper	Cu^+	Enzymes for lignin synthesis
Molybdenum	MoO_4^{2-}	Nitrogen fixation, reduction of nitrates
Nickel	Ni^{2+}	Enzymatic cofactor in the metabolism of nitrogen compounds

Pigments

Among the most important molecules for plant function are the pigments. Plant pigments include a variety of different kinds of molecules, including porphyrins, carotenoids, and anthocyanins. All biological pigments selectively absorb certain wavelengths of light while reflecting others. The light that is absorbed may be used by the plant to power chemical reactions, while the reflected wavelengths of light determine the color the pigment appears to the eye.

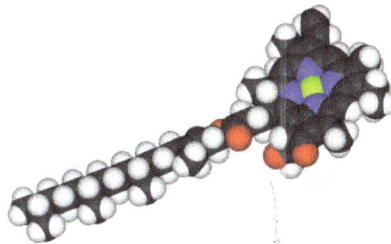

Space-filling model of the chlorophyll molecule.

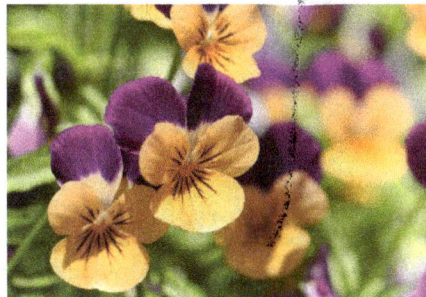

Anthocyanin gives these pansies their dark purple pigmentation.

Chlorophyll is the primary pigment in plants; it is a porphyrin that absorbs red and blue wavelengths of light while reflecting green. It is the presence and relative abundance of chlorophyll that gives plants their green color. All land plants and green algae possess two forms of this pigment: chlorophyll a and chlorophyll b. Kelps, diatoms, and other photosynthetic heterokonts contain chlorophyll c instead of b, red algae possess chlorophyll a and " d". All chlorophylls serve as the primary means plants use to intercept light to fuel photosynthesis.

Carotenoids are red, orange, or yellow tetraterpenoids. They function as accessory pigments in plants, helping to fuel photosynthesis by gathering wavelengths of light not readily absorbed by chlorophyll. The most familiar carotenoids are carotene (an orange pigment found in carrots), lutein (a yellow pigment found in fruits and vegetables), and lycopene (the red pigment responsible for the color of tomatoes). Carotenoids have been shown to act as antioxidants and to promote healthy eyesight in humans.

Anthocyanins (literally "flower blue") are water-soluble flavonoid pigments that appear red to blue, according to pH. They occur in all tissues of higher plants, providing color in leaves, stems, roots, flowers, and fruits, though not always in sufficient quantities to be noticeable. Anthocyanins are most visible in the petals of flowers, where they may make up as much as 30% of the dry weight of the tissue. They are also responsible for the purple color seen on the underside of tropical shade plants such as *Tradescantia zebrina*. In these plants, the anthocyanin catches light that has passed through the leaf and reflects it back towards regions bearing chlorophyll, in order to maximize the use of available light

Betalains are red or yellow pigments. Like anthocyanins they are water-soluble, but unlike anthocyanins they are indole-derived compounds synthesized from tyrosine. This class of pigments is found only in the Caryophyllales (including cactus and amaranth), and never co-occur in plants with anthocyanins. Betalains are responsible for the deep red color of beets, and are used commercially as food-coloring agents. Plant physiologists are uncertain of the function that betalains have in plants which possess them, but there is some preliminary evidence that they may have fungicidal properties.

Signals and Regulators

A mutation that stops *Arabidopsis thaliana* responding to auxin causes abnormal growth (right)

Plants produce hormones and other growth regulators which act to signal a physiological response in their tissues. They also produce compounds such as phytochrome that are sensitive to light and which serve to trigger growth or development in response to environmental signals.

Plant Hormones

Plant hormones, known as plant growth regulators (PGRs) or phytohormones, are chemicals that regulate a plant's growth. According to a standard animal definition, hormones are signal molecules produced at specific locations, that occur in very low concentrations, and cause altered processes in target cells at other locations. Unlike animals, plants lack specific hormone-producing tissues or organs. Plant hormones are often not transported to other parts of the plant and production is not limited to specific locations.

Plant hormones are chemicals that in small amounts promote and influence the growth, development and differentiation of cells and tissues. Hormones are vital to plant growth; affecting processes in plants from flowering to seed development, dormancy, and germination. They regulate which tissues grow upwards and which grow downwards, leaf formation and stem growth, fruit development and ripening, as well as leaf abscission and even plant death.

The most important plant hormones are abscissic acid (ABA), auxins, ethylene, gibberellins, and cytokinins, though there are many other substances that serve to regulate plant physiology.

Photomorphogenesis

While most people know that light is important for photosynthesis in plants, few realize that plant sensitivity to light plays a role in the control of plant structural development (morphogenesis). The use of light to control structural development is called photomorphogenesis, and is dependent upon the presence of specialized photoreceptors, which are chemical pigments capable of absorbing specific wavelengths of light.

Plants use four kinds of photoreceptors: phytochrome, cryptochrome, a UV-B photoreceptor, and protochlorophyllide *a*. The first two of these, phytochrome and cryptochrome, are photoreceptor proteins, complex molecular structures formed by joining a protein with a light-sensitive pigment. Cryptochrome is also known as the UV-A photoreceptor, because it absorbs ultraviolet light in the long wave "A" region. The UV-B receptor is one or more compounds not yet identified with certainty, though some evidence suggests carotene or riboflavin as candidates. Protochlorophyllide *a*, as its name suggests, is a chemical precursor of chlorophyll.

The most studied of the photoreceptors in plants is phytochrome. It is sensitive to light in the red and far-red region of the visible spectrum. Many flowering plants use it to regulate the time of flowering based on the length of day and night (photoperiodism) and to set circadian rhythms. It also regulates other responses including the germination of seeds, elongation of seedlings, the size, shape and number of leaves, the synthesis of chlorophyll, and the straightening of the epicotyl or hypocotyl hook of dicot seedlings.

Photoperiodism

Many flowering plants use the pigment phytochrome to sense seasonal changes in day length, which they take as signals to flower. This sensitivity to day length is termed photoperiodism. Broadly speaking, flowering plants can be classified as long day plants, short day plants, or day neutral plants, depending on their particular response to changes in day length. Long day plants require a certain minimum length of daylight to starts flowering, so these plants flower in the spring or summer. Conversely, short day plants flower when the length of daylight falls below a certain critical level. Day neutral plants do not initiate flowering based on photoperiodism, though some may use temperature sensitivity (vernalization) instead.

Although a short day plant cannot flower during the long days of summer, it is not actually the period of light exposure that limits flowering. Rather, a short day plant requires a minimal length of uninterrupted darkness in each 24-hour period (a short daylength) before floral development can begin. It has been determined experimentally that a short day plant (long night) does not flower if a flash of phytochrome activating light is used on the plant during the night.

The poinsettia is a short-day plant, requiring two months of long nights prior to blooming.

Plants make use of the phytochrome system to sense day length or photoperiod. This fact is utilized by florists and greenhouse gardeners to control and even induce flowering out of season, such as the *Poinsettia*.

Environmental Physiology

Paradoxically, the subdiscipline of environmental physiology is on the one hand a recent field of study in plant ecology and on the other hand one of the oldest. Environmental physiology is the preferred name of the subdiscipline among plant physiologists, but it goes by a number of other names in the applied sciences. It is roughly synonymous with ecophysiology, crop ecology, horticulture and agronomy. The particular name applied to the subdiscipline is specific to the viewpoint and goals of research. Whatever name is applied, it deals with the ways in which plants respond to their environment and so overlaps with the field of ecology.

Phototropism in *Arabidopsis thaliana* is regulated by blue to UV light.

Environmental physiologists examine plant response to physical factors such as radiation (including light and ultraviolet radiation), temperature, fire, and wind. Of particular importance are water relations (which can be measured with the Pressure bomb) and the stress of drought or inundation, exchange of gases with the atmosphere, as well as the cycling of nutrients such as nitrogen and carbon.

Environmental physiologists also examine plant response to biological factors. This includes not only negative interactions, such as competition, herbivory, disease and parasitism, but also positive interactions, such as mutualism and pollination.

Tropisms and Nastic Movements

Plants may respond both to directional and non-directional stimuli. A response to a directional stimulus, such as gravity or sunlight, is called a tropism. A response to a nondirectional stimulus, such as temperature or humidity, is a nastic movement.

Tropisms in plants are the result of differential cell growth, in which the cells on one side of the plant elongates more than those on the other side, causing the part to bend toward the side with less growth. Among the common tropisms seen in plants is phototropism, the bending of the plant toward a source of light. Phototropism allows the plant to maximize light exposure in plants which require additional light for photosynthesis, or to minimize it in plants subjected to intense light and heat. Geotropism allows the roots of a plant to determine the direction of gravity and grow downwards. Tropisms generally result from an interaction between the environment and production of one or more plant hormones.

Nastic movements results from differential cell growth (e.g. epinasty and hiponasty), or from changes in turgor pressure within plant tissues (e.g., nyctinasty), which may occur rapidly. A familiar example is thigmonasty (response to touch) in the Venus fly trap, a carnivorous plant. The traps consist of modified leaf blades which bear sensitive trigger hairs. When the hairs are touched by an insect or other animal, the leaf folds shut. This mechanism allows the plant to trap and digest small insects for additional nutrients. Although the trap is rapidly shut by changes in internal cell pressures, the leaf must grow slowly to reset for a second opportunity to trap insects.

Plant Disease

Powdery mildew on crop leaves

Economically, one of the most important areas of research in environmental physiology is that of phytopathology, the study of diseases in plants and the manner in which plants resist or cope with infection. Plant are susceptible to the same kinds of disease organisms as animals, including viruses, bacteria, and fungi, as well as physical invasion by insects and roundworms.

Because the biology of plants differs with animals, their symptoms and responses are quite different. In some cases, a plant can simply shed infected leaves or flowers to prevent the spread of disease, in a process called abscission. Most animals do not have this option as a means of controlling disease. Plant diseases organisms themselves also differ from those causing disease in animals because plants cannot usually spread infection through casual physical contact. Plant pathogens tend to spread via spores or are carried by animal vectors.

One of the most important advances in the control of plant disease was the discovery of Bordeaux mixture in the nineteenth century. The mixture is the first known fungicide and is a combination of copper sulfate and lime. Application of the mixture served to inhibit the growth of downy mildew that threatened to seriously damage the French wine industry.

History

Early History

Sir Francis Bacon published one of the first plant physiology experiments in 1627 in the book, *Sylva Sylvarum*. Bacon grew several terrestrial plants, including a rose, in water and concluded that soil was only needed to keep the plant upright. Jan Baptist van Helmont published what is considered the first quantitative experiment in plant physiology in 1648. He grew a willow tree for five years in a pot containing 200 pounds of oven-dry soil. The soil lost just two ounces of dry weight and van Helmont concluded that plants get all their weight from water, not soil. In 1699, John Woodward published experiments on growth of spearmint in different sources of water. He found that plants grew much better in water with soil added than in distilled water.

Jan Baptist van Helmont.

Stephen Hales is considered the Father of Plant Physiology for the many experiments in the 1727 book; though Julius von Sachs unified the pieces of plant physiology and put them together as a discipline. His *Lehrbuch der Botanik* was the plant physiology bible of its time.

Researchers discovered in the 1800s that plants absorb essential mineral nutrients as inorganic ions in water. In natural conditions, soil acts as a mineral nutrient reservoir but the soil itself is not essential to plant growth. When the mineral nutrients in the soil are dissolved in water, plant roots absorb nutrients readily, soil is no longer required for the plant to thrive. This observation is the basis for hydroponics, the growing of plants in a water solution rather than soil, which has become a standard technique in biological research, teaching lab exercises, crop production and as a hobby.

Economic Applications

Food Production

In horticulture and agriculture along with food science, plant physiology is an important topic relating to fruits, vegetables, and other consumable parts of plants. Topics studied include: *climatic requirements, fruit drop, nutrition, ripening, fruit set.* The production of food crops also hinges on the study of plant physiology covering such topics as optimal planting and harvesting times and post harvest storage of plant products for human consumption and the production of secondary products like drugs and cosmetics.

Plant Pathology

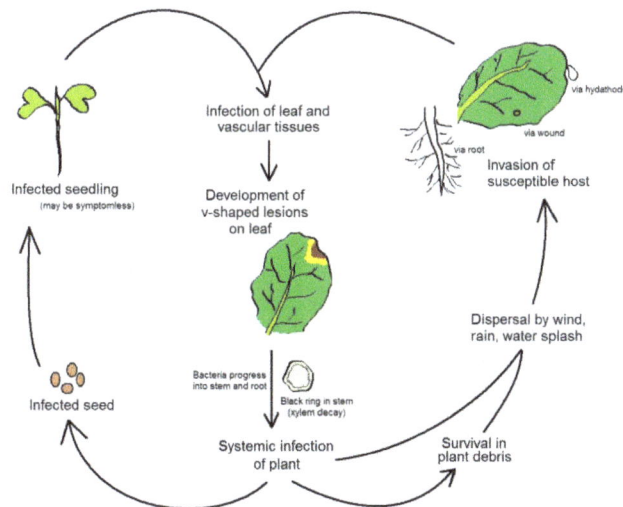

Life cycle of the black rot pathogen, *Xanthomonas campestris* pathovar *campes*

Plant pathology (also phytopathology) is the scientific study of diseases in plants caused by pathogens (infectious organisms) and environmental conditions (physiological factors). Organisms that cause infectious disease include fungi, oomycetes, bacteria, viruses, viroids, virus-like organisms, phytoplasmas, protozoa, nematodes and parasitic plants. Not included are ectoparasites like insects, mites, vertebrate, or other pests that affect plant health by consumption of plant tissues. Plant pathology also involves the study of pathogen identification, disease etiology, disease cycles, economic impact, plant disease epidemiology, plant disease resistance, how plant diseases affect humans and animals, pathosystem genetics, and management of plant diseases.

Overview

Control of plant diseases is crucial to the reliable production of food, and it provides significant reductions in agricultural use of land, water, fuel and other inputs. Plants in both natural and cultivated populations carry inherent disease resistance, but there are numerous examples of devastating plant disease impacts, as well as recurrent severe plant diseases. However, disease control is reasonably successful for most crops. Disease control is achieved by use of plants that have been bred for good resistance to many diseases, and by plant cultivation approaches such as crop rotation, use of pathogen-free seed, appropriate planting date and plant density, control of field moisture, and pesticide use. Across large regions and many crop species, it is estimated that diseases typically reduce plant yields by 10% every year in more developed settings, but yield loss to diseases often exceeds 20% in less developed settings. Continuing advances in the science of plant pathology are needed to improve disease control, and to keep up with changes in disease pressure caused by the ongoing evolution and movement of plant pathogens and by changes in agricultural practices. Plant diseases cause major economic losses for farmers worldwide. The Food and Agriculture Organization estimates indeed that pests and diseases are responsible for about 25% of crop loss. To solve this issue, new methods are needed to detect diseases and pests early, such as novel sensors that detect plant odours and spectroscopy and biophotonics that are able to diagnos-tic plant health and metabolism.

Plant Pathogens

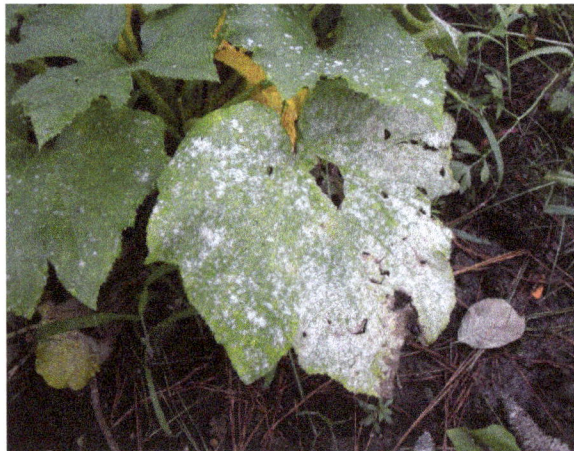

Powdery mildew, a biotrophic fungus

Fungi

Most phytopathogenic fungi belong to the Ascomycetes and the Basidiomycetes.

The fungi reproduce both sexually and asexually via the production of spores and other structures. Spores may be spread long distances by air or water, or they may be soilborne. Many soil inhabiting fungi are capable of living saprotrophically, carrying out the part of their life cycle in the soil. These are known as facultative saprotrophs.

Fungal diseases may be controlled through the use of fungicides and other agriculture practices. However, new races of fungi often evolve that are resistant to various fungicides.

Rice blast, caused by a necrotrophic fungus

Biotrophic fungal pathogens colonize living plant tissue and obtain nutrients from living host cells. Necrotrophic fungal pathogens infect and kill host tissue and extract nutrients from the dead host cells.

Significant fungal plant pathogens include:

Ascomycetes

- *Fusarium* spp. (causal agents of Fusarium wilt disease)

- *Thielaviopsis* spp. (causal agents of: canker rot, black root rot, *Thielaviopsis* root rot)

- *Verticillium* spp.

- *Magnaporthe grisea* (causal agent of rice blast)

- *Sclerotinia sclerotiorum* (causal agent of cottony rot)

Basidiomycetes

- *Ustilago* spp. (causal agents of smut)

- *Rhizoctonia* spp.

- *Phakospora pachyrhizi* (causal agent of soybean rust)

- *Puccinia* spp. (causal agents of severe rusts of virtually all cereal grains and cultivated grasses)

- *Armillaria* spp. (the so-called honey fungus species, which are virulent pathogens of trees and produce edible mushrooms)

Fungus-like Organisms

Oomycetes

The oomycetes are not true fungi but are fungus-like organisms. They include some of the most destructive plant pathogens including the genus *Phytophthora*, which includes the causal agents of potato late blight and sudden oak death. Particular species of oomycetes are responsible for root rot.

Despite not being closely related to the fungi, the oomycetes have developed very similar infection strategies. Oomycetes are capable of using effector proteins to turn off a plant's defenses in its infection process. Plant pathologists commonly group them with fungal pathogens.

Significant oomycete plant pathogens

- *Pythium* spp.

- *Phytophthora* spp., including the causal agent of the Great Irish Famine (1845–1849)

Phytomyxea

Some slime molds in Phytomyxea cause important diseases, including club root in cabbage and its relatives and powdery scab in potatoes. These are caused by species of *Plasmodiophora* and *Spongospora*, respectively.

Bacteria

Most bacteria that are associated with plants are actually saprotrophic and do no harm to the plant itself. However, a small number, around 100 known species, are able to cause disease. Bacterial diseases are much more prevalent in subtropical and tropical regions of the world.

Crown gall disease caused by Agrobacterium

Most plant pathogenic bacteria are rod-shaped (bacilli). In order to be able to colonize the plant they have specific pathogenicity factors. Five main types of bacterial pathogenicity factors are known: uses of cell wall–degrading enzymes, toxins, effector proteins, phytohormones and exopolysaccharides.

Pathogens such as *Erwinia* species use cell wall–degrading enzymes to cause soft rot. *Agrobacterium* species change the level of auxins to cause tumours with phytohormones. Exopolysaccharides are produced by bacteria and block xylem vessels, often leading to the death of the plant.

Bacteria control the production of pathogenicity factors via quorum sensing.

Vitis vinifera with "Ca. Phytoplasma vitis" infection

Significant bacterial plant pathogens:

- Burkholderia

- Proteobacteria

  - *Xanthomonas* spp.

  - *Pseudomonas* spp.

- Pseudomonas syringae pv. tomato causes tomato plants to produce less fruit, and it "continues to adapt to the tomato by minimizing its recognition by the tomato immune system."

Phytoplasmas ('Mycoplasma-like Organisms') and Spiroplasmas

Phytoplasma and *Spiroplasma* are a genre of bacteria that lack cell walls and are related to the mycoplasmas, which are human pathogens. Together they are referred to as the mollicutes. They also tend to have smaller genomes than most other bacteria. They are normally transmitted by sap-sucking insects, being transferred into the plants phloem where it reproduces.

Tobacco mosaic virus

Viruses, Viroids and Virus-like Organisms

There are many types of plant virus, and some are even asymptomatic. Under normal circumstances, plant viruses cause only a loss of crop yield. Therefore, it is not economically viable to try to control them, the exception being when they infect perennial species, such as fruit trees.

Most plant viruses have small, single-stranded RNA genomes. However some plant viruses also have double stranded RNA or single or double stranded DNA genomes. These genomes may encode only three or four proteins: a replicase, a coat protein, a movement protein, in order to allow cell to cell movement through plasmodesmata, and sometimes a protein that allows transmission by a vector. Plant viruses can have several more proteins and employ many different molecular translation methods.

Plant viruses are generally transmitted from plant to plant by a vector, but mechanical and seed transmission also occur. Vector transmission is often by an insect (for example, aphids), but some fungi, nematodes, and protozoa have been shown to be viral vectors. In many cases, the insect and virus are specific for virus transmission such as the beet leafhopper that transmits the curly top virus causing disease in several crop plants.

Nematodes

Nematodes are small, multicellular wormlike animals. Many live freely in the soil, but there are some species that parasitize plant roots. They are a problem in tropical and subtropical regions of the world, where they may infect crops. Potato cyst nematodes (*Globodera pallida* and *G. rosto-chiensis*) are widely distributed in Europe and North and South America and cause $300 million worth of damage in Europe every year. Root knot nematodes have quite a large host range, whereas cyst nematodes tend to be able to infect only a few species. Nematodes are able to cause radical changes in root cells in order to facilitate their lifestyle.

Root-knot nematode galls

Protozoa and Algae

There are a few examples of plant diseases caused by protozoa (e.g., *Phytomonas*, a kinetoplastid). They are transmitted as zoospores that are very durable, and may be able to survive in a resting state in the soil for many years. They have also been shown to transmit plant viruses.

When the motile zoospores come into contact with a root hair they produce a plasmodium and invade the roots.

Some colourless parasitic algae (e.g., *Cephaleuros*) also cause plant diseases.

Parasitic Plants

Parasitic plants such as mistletoe and dodder are included in the study of phytopathology. Dodder, for example, is used as a conduit either for the transmission of viruses or virus-like agents from a host plant to a plant that is not typically a host or for an agent that is not graft-transmissible.

Common Pathogenic Infection Methods

- Cell wall-degrading enzymes: These are used to break down the plant cell wall in order to release the nutrients inside.

- Toxins: These can be non-host-specific, which damage all plants, or host-specific, which cause damage only on a host plant.

- Effector proteins: These can be secreted into the extracellular environment or directly into the host cell, often via the Type three secretion system. Some effectors are known to suppress host defense processes. This can include: reducing the plants internal signaling mechanisms or reduction of phytochemicals production. Bacteria, fungus and oomycetes are known for this function.

Physiological Plant Disorders

Significant abiotic disorders can be caused by:

Natural

Drought

Frost damage and breakage by snow and hail

Flooding and poor drainage

Nutrient deficiency

Salt deposition and other soluble mineral excesses (e.g., gypsum)

Wind (windburn and breakage by hurricanes and tornadoes)

Lightning and wildfire (also often man-made)

Man-made (arguably not abiotic, but usually regarded as such)

Soil compaction

Pollution of air, soil, or both

Salt from winter road salt application or irrigation

Herbicide over-application

Poor education and training of people working with plants (e.g. lawnmower damage to trees)

Vandalism

Orchid leaves with viral infections

Epidemiology

the study of phytopathology epidemics

Disease Resistance

Management

Quarantine

A diseased patch of vegetation or individual plants can be isolated from other, healthy growth. Specimens may be destroyed or relocated into a greenhouse for treatment or study. Another option is to avoid the introduction of harmful nonnative organisms by controlling all human traffic and activity (e.g., AQIS), although legislation and enforcement are crucial in order to ensure lasting effectiveness.

Cultural

Farming in some societies is kept on a small scale, tended by peoples whose culture includes farming traditions going back to ancient times. (An example of such traditions would be lifelong training in techniques of plot terracing, weather anticipation and response, fertilization, grafting, seed care, and dedicated gardening.) Plants that are intently monitored often benefit from not only active external protection but also a greater overall vigor. While primitive in the sense of being the most labor-intensive solution by far, where practical or necessary it is more than adequate.

Plant Resistance

Sophisticated agricultural developments now allow growers to choose from among systematically cross-bred species to ensure the greatest hardiness in their crops, as suited for a particular region's pathological profile. Breeding practices have been perfected over centuries, but with the advent of genetic manipulation even finer control of a crop's immunity traits is possible. The engineering of food plants may be less rewarding, however, as higher output is frequently offset by popular suspicion and negative opinion about this "tampering" with nature.

Chemical

Many natural and synthetic compounds can be employed to combat the above threats. This method works by directly eliminating disease-causing or-ganisms or curbing their spread; however, it has been shown to have too broad an effect, typically, to be good for the local ecosystem. From an economic standpoint, all but the sim-plest natural additives may disqualify a product from "organic" status, potentially reducing the value of the yield.

Biological

Crop rotation may be an effective means to prevent a parasitic population from becoming well-established, as an organism affecting leaves would be starved when the leafy crop is replaced by a tuberous type, etc. Other means to undermine parasites without attacking them directly may exist.

Integrated

The use of two or more of these methods in combination offers a higher chance of effectiveness.

References

- Evert, Ray Franklin and Esau, Katherine (2006) Esau's Plant anatomy: meristems, cells, and tissues of the plant body - their structure, function and development Wiley, Hoboken, New Jersey, page xv, ISBN 0-471-73843-3

- Raven, P. H., R. F. Evert, & S. E. Eichhorn. Biology of Plants, 7th ed., page 9. (New York: W. H. Freeman, 2005). ISBN 0-7167-1007-2.

- Harold C. Bold, C. J. Alexopoulos, and T. Delevoryas. Morphology of Plants and Fungi, 5th ed., page 3. (New York: Harper-Collins, 1987). ISBN 0-06-040839-1.

- Michael A Dirr; Charles W Heuser, jr. (2006). "2". The Reference Manual of Woody Plant Propagation (Second ed.). Varsity Press Inc. pp. 26, 28, 29. ISBN 0942375092.

- Schulze, Ernst-Detlef; et al. (2005). Plant Ecology – (Section 1.10.1: Herbivory). Springer. Retrieved April 24, 2012. ISBN 3-540-20833-X

- Fosket, Donald E. (1994). Plant Growth and Development: A Molecular Approach. San Diego: Academic Press. pp. 498–509. ISBN 0-12-262430-0.

- Kingsley Rowland Stern; Shelley Jansky (1991). Introductory Plant Biology. WCB/McGraw-Hill. p. 309. ISBN 978-0-697-09948-8.

- Jackson RW (editor). (2009). Plant Pathogenic Bacteria: Genomics and Molecular Biology. Caister Academic Press. ISBN 978-1-904455-37-0.

- "Plasmopara viticola, the Cause of Downy Mildew of Grapes". The Origin of Plant Pathology and The Potato Famine, and Other Stories of Plant Diseases. Retrieved 4 February 2015.

- "Fusarium oxysporum : The End of the Banana Industry?». The Origin of Plant Pathology and The Potato Famine, and Other Stories of Plant Diseases. Retrieved 4 February 2015.

- Bänziger (2000). "Breeding for drought and nitrogen stress tolerance in maize: from theory to practice". From Theory to Practice: 7–9. Retrieved 7-11-2013.

- Oldeman, l (1994). "The global extent of soil degradation" (PDF). Soil resilience and sustainable land use. 32 (5967): 818–822. Retrieved 7-11-2013.

- Nicole Davis (September 9, 2009). "Genome of Irish potato famine pathogen decoded". Haas et al. Broad Institute of MIT and Harvard. Retrieved 24 July 2012.

- "Scientists discover how deadly fungal microbes enter host cells". (VBI) at Virginia Tech affiliates. Physorg. July 22, 2010. Retrieved July 31, 2012.

Zoology: The Study of Animals

Zoology is the branch of biology that studies animals, their structures, evolution and habits. Zoology includes both living animals and the extinct ones as well. Animal science, embryology, ethology and behavioral ecology are the topics explained in the following section. The text helps the reader in developing an in-depth understanding of the subject.

Zoology

Zoology or animal biology is the branch of biology that studies the animal kingdom, including the structure, embryology, evolution, classification, habits, and distribution of all animals, both living and extinct, and how they interact with their ecosystems.

History

Ancient History to Darwin

Conrad Gesner (1516–1565). His *Historiae animalium* is considered the beginning of modern zoology.

The history of zoology traces the study of the animal kingdom from ancient to modern times. Although the concept of *zoology* as a single coherent field arose much later, the zoological sciences emerged from natural history reaching back to the works of Aristotle and Galen in the ancient Greco-Roman world. This ancient work was further developed in the Middle Ages by Muslim physicians and scholars such as Albertus Magnus. During the Renaissance and early modern period, zoological thought was revolutionized in Europe by a renewed interest in empiricism and the discovery of many novel organisms. Prominent in this movement were Vesalius and William Harvey, who used experimentation and careful observation in physiology, and naturalists such as Carl Linnaeus and Buffon who began to classify the diversity of life and the fossil record, as well as the development and behavior of organisms. Microscopy revealed the previously unknown world of microorganisms, laying the groundwork for cell theory. The growing importance of natural theology, partly a response to the rise of mechanical philosophy, encouraged the growth of natural history (although it entrenched the argument from design).

Over the 18th and 19th centuries, zoology became an increasingly professional scientific discipline. Explorer-naturalists such as Alexander von Humboldt investigated the interaction between organisms and their environment, and the ways this relationship depends on geography, laying the foundations for biogeography, ecology and ethology. Naturalists began to reject essentialism and consider the importance of extinction and the mutability of species. Cell theory provided a new perspective on the fundamental basis of life.

Post-darwin

These developments, as well as the results from embryology and paleontology, were synthesized in Charles Darwin's theory of evolution by natural selection. In 1859, Darwin placed the theory of organic evolution on a new footing, by his discovery of a process by which organic evolution can occur, and provided observational evidence that it had done so.

Darwin gave new direction to morphology and physiology, by uniting them in a common biological theory: the theory of organic evolution. The result was a reconstruction of the classification of animals upon a genealogical basis, fresh investigation of the development of animals, and early attempts to determine their genetic relationships. The end of the 19th century saw the fall of spontaneous generation and the rise of the germ theory of disease, though the mechanism of inheritance remained a mystery. In the early 20th century, the rediscovery of Mendel's work led to the rapid development of genetics by Thomas Hunt Morgan and his students, and by the 1930s the combination of population genetics and natural selection in the "neo-Darwinian synthesis".

Research

Structural

Cell biology studies the structural and physiological properties of cells, including their behavior, interactions, and environment. This is done on both the microscopic and molecular levels, for single-celled organisms such as bacteria as well as the specialized cells in multicellular organisms such as humans. Understanding the structure and function of cells is fundamental to all of the biological sciences. The similarities and differences between cell types are particularly relevant to molecular biology.

Anatomy considers the forms of macroscopic structures such as organs and organ systems. It focuses on how organs and organ systems work together in the bodies of humans and animals, in addition to how they work independently. Anatomy and cell biology are two studies that are closely related, and can be categorized under "structural" studies.

Physiological

Physiology studies the mechanical, physical, and biochemical processes of living organisms by attempting to understand how all of the structures function as a whole. The theme of "structure to function" is central to biology. Physiological studies have traditionally been divided into plant physiology and animal physiology, but some principles of physiology are universal, no matter what particular organism is being studied. For example, what is learned about the physiology of yeast cells can also apply to human cells. The field of animal physiology extends the tools and methods of human physiology to non-human species. Physiology studies how for example nervous, immune, endocrine, respiratory, and circulatory systems, function and interact.

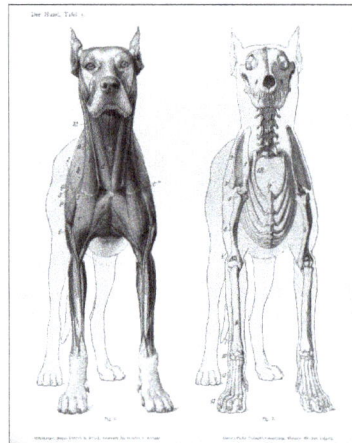

Animal anatomical engraving from *Handbuch der Anatomie der Tiere für Künstler*.

Evolutionary

Evolutionary research is concerned with the origin and descent of species, as well as their change over time, and includes scientists from many taxonomically oriented disciplines. For example, it generally involves scientists who have special training in particular organisms such as mammalogy, ornithology, herpetology, or entomology, but use those organisms as systems to answer general questions about evolution.

Evolutionary biology is partly based on paleontology, which uses the fossil record to answer questions about the mode and tempo of evolution, and partly on the developments in areas such as population genetics and evolutionary theory. Following the development of DNA fingerprinting techniques in the late 20th century, the application of these techniques in zoology has increased the understanding of animal populations. In the 1980s, developmental biology re-entered evolutionary biology from its initial exclusion from the modern synthesis through the study of evolutionary developmental biology. Related fields often considered part of evolutionary biology are phylogenetics, systematics, and taxonomy.

Classification

Scientific classification in zoology, is a method by which zoologists group and categorize organisms by biological type, such as genus or species. Biological classification is a form of scientific taxonomy. Modern biological classification has its root in the work of Carl Linnaeus, who grouped species according to shared physical characteristics. These groupings have since been revised to improve consistency with the Darwinian principle of common descent. Molecular phylogenetics, which uses DNA sequences as data, has driven many recent revisions and is likely to continue to do so. Biological classification belongs to the science of zoological systematics.

Linnaeus's table of the animal kingdom from the first edition of *Systema Naturae* (1735).

Many scientists now consider the five-kingdom system outdated. Modern alternative classification systems generally start with the three-domain system: Archaea (originally Archaebacteria); Bacteria (originally Eubacteria); Eukaryota (including protists, fungi, plants, and animals) These domains reflect whether the cells have nuclei or not, as well as differences in the chemical composition of the cell exteriors.

Further, each kingdom is broken down recursively until each species is separately classified. The order is: Domain; kingdom; phylum; class; order; family; genus; species. The scientific name of an organism is generated from its genus and species. For example, humans are listed as *Homo sapiens*. *Homo* is the genus, and *sapiens* the specific epithet, both of them combined make up the species name. When writing the scientific name of an organism, it is proper to capitalize the first letter in the genus and put all of the specific epithet in lowercase. Additionally, the entire term may be italicized or underlined.

The dominant classification system is called the Linnaean taxonomy. It includes ranks and binomial nomenclature. The classification, taxonomy, and nomenclature of zoological organisms is administered by the International Code of Zoological Nomenclature, and International Code of Nomenclature of Bacteria for animals and bacteria, respectively. The classification of viruses, viroids, prions, and all other sub-viral agents that demonstrate biological characteristics is conducted by the International Code of Virus classification and nomenclature. However, several other viral classification systems do exist.

A merging draft, BioCode, was published in 1997 in an attempt to standardize nomenclature in these areas, but has yet to be formally adopted. The BioCode draft has received little attention since 1997; its originally planned implementation date of January 1, 2000, has passed unnoticed. However, a 2004 paper concerning the cyanobacteria does advocate a future adoption of a BioCode and interim steps consisting of reducing the differences between the codes. The International Code of Virus Classification and Nomenclature (ICVCN) remains outside the BioCode.

Ethology

Ethology is the scientific and objective study of animal behavior under natural conditions, as opposed to behaviourism, which focuses on behavioral response studies in a laboratory setting. Ethologists have been particularly concerned with the evolution of behavior and the understanding of behavior in terms of the theory of natural selection. In one sense, the first modern ethologist was Charles Darwin, whose book, *The Expression of the Emotions in Man and Animals,* influenced many future ethologists.

Kelp gull chicks peck at red spot on mother's beak to stimulate the regurgitating reflex.

Biogeography

Biogeography studies the spatial distribution of organisms on the Earth, focusing on topics like plate tectonics, climate change, dispersal and migration, and cladistics. The creation of this study is widely accredited to Alfred Russel Wallace, a British biologist who had some of his work jointly published with Charles Darwin.

Branches of Zoology

Although the study of animal life is ancient, its scientific incarnation is relatively modern. This mirrors the transition from natural history to biology at the start of the 19th century. Since Hunter and Cuvier, comparative anatomical study has been associated with morphography, shaping the modern areas of zoological investigation: anatomy, physiology, histology, embryology, teratology and ethology. Modern zoology first arose in German and British universities. In Britain, Thomas Henry Huxley was a prominent figure. His ideas were centered on the morphology of animals. Many consider him the greatest comparative anatomist of the latter half of the 19th century. Similar to Hunter, his courses were composed of lectures and laboratory practical classes in contrast to the previous format of lectures only.

Gradually zoology expanded beyond Huxley's comparative anatomy to include the following sub-disciplines:

- Zoography, also known as *descriptive zoology*, describes animals and their habitats

- Comparative anatomy studies the structure of animals

- Animal physiology

- Behavioral ecology

- Ethology studies animal behavior

- Invertebrate zoology

- Vertebrate zoology

- Soil zoology

- Comparative zoology

- The various taxonomically oriented disciplines such as mammalogy, herpetology, ornithology and entomology identify and classify species and study the structures and mechanisms specific to those groups.

Related Fields:

- Evolutionary biology: Development of both animals and plants is considered in the articles on evolution, population genetics, heredity, variation, Mendelism, reproduction.

- Molecular biology studies the common genetic and developmental mechanisms of animals and plants

- Palaeontology

- Systematics, cladistics, phylogenetics, phylogeography, biogeography and taxonomy classify and group species via common descent and regional associations.

Animal Science

Animal Science (also Animal Bioscience) is described as "studying the biology of animals that are under the control of humankind". Historically, the degree was called animal husbandry and the animals studied were livestock species, like cattle, sheep, pigs, poultry, and horses. Today, courses available now look at a far broader area to include companion animals like dogs and cats, and many exotic species. Degrees in Animal Science are offered at a number of colleges and universities. In the United States, the universities offering such a program were Land Grant Universities and include University of Nebraska–Lincoln, Cornell University, UC Davis, Michigan State University, Purdue University, The Ohio State University, The Pennsylvania State University, Iowa State University and the University of Minnesota. Typically, the Animal Science curriculum not only provides a strong science background, but also hands-on experience working with animals on campus-based farms.

Education

Professional education in animal science prepares students for career opportunities in areas such as animal breeding, food and fiber production, nutrition, animal agribusiness, animal behavior and welfare, and biotechnology. Courses in a typical Animal Science program may include genetics, microbiology, animal behavior, nutrition, physiology, and reproduction. Courses in support areas, such as genetics, soils, agricultural economics and marketing, legal aspects, and the environment also are offered. All of these courses are essential to entering an animal science profession.

Bachelor Degree

At many universities, a Bachelor of Science (BS) degree in Animal Science allows emphasis in certain areas. Typical areas are species-specific or career-specific. Species-specific areas of emphasis prepare students for a career in dairy management, beef management, swine management, sheep or small ruminant management, poultry production, or the horse industry. Other career-specific areas of study include pre-veterinary medicine studies, livestock business and marketing, animal welfare and behavior, animal nutrition science, animal reproduction science, or genetics. Youth programs are also an important part of animal science programs.

Pre-veterinary Emphasis

Many schools that offer a degree option in Animal Science also offer a pre-veterinary emphasis such as the University of Nebraska-Lincoln and the University of Minnesota, for example. This option provides an in-depth knowledge base of the biological and physical sciences including nutrition, reproduction, physiology, and genetics. This can prepare students for graduate studies in animal science, veterinary school, and pharmaceutical or animal science industries.

Graduate Studies

In a Master of Science degree option, students take required courses in areas that support their main interest. These courses are above courses normally required for a Bachelor of Science degree in the Animal Science major. For example, in a Ph.D. degree program students take courses related to their major that are more in depth than those for the Master of Science degree, with an emphasis on research or teaching.

Graduate studies in animal sciences are considered preparation for upper level positions in production, management, education, research, or agriservices. Professional study in veterinary medicine, law, and business administration are among the most commonly chosen programs by graduates. Other areas of study include growth biology, physiology, nutrition, and production systems.

Embryology

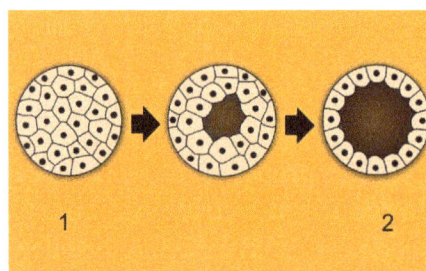

1 - morula, 2 - blastula

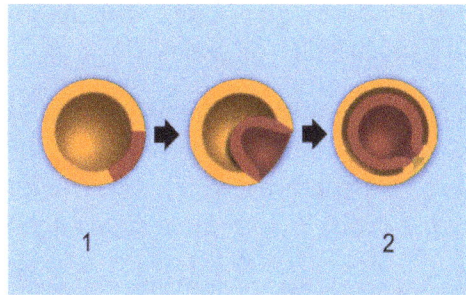

1 - blastula, 2 - gastrula with blastopore; orange - ectoderm, red - endoderm.

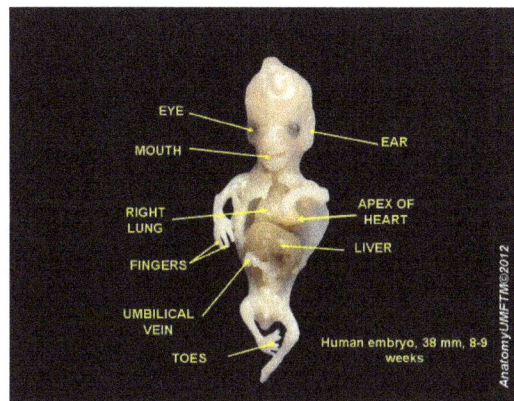

Dissection of human embryo, 38 mm - 8 weeks

Embryology is the branch of biology that studies the prenatal development of gametes (sex cells), fertilization, and development of embryos and fetuses. Additionally, embryology is the study of congenital disorders that occur before birth.

Embryonic Development of Animals

After cleavage, the dividing cells, or morula, becomes a hollow ball, or blastula, which develops a hole or pore at one end.

Bilaterals

In bilateral animals, the blastula develops in one of two ways that divides the whole animal kingdom into two halves. If in the blastula the first pore (blastopore) becomes the mouth of the animal, it is a protostome; if the first pore becomes the anus then it is a deuterostome. The protostomes include most invertebrate animals, such as insects, worms and molluscs, while the deuterostomes include the vertebrates. In due course, the blastula changes into a more differentiated structure called the gastrula.

The gastrula with its blastopore soon develops three distinct layers of cells (the germ layers) from which all the bodily organs and tissues then develop:

- The innermost layer, or endoderm, gives a rise to the digestive organs, the gills, lungs or swim bladder if present, and kidneys or nephrites.

- The middle layer, or mesoderm, gives rise to the muscles, skeleton if any, and blood system.

- • The outer layer of cells, or ectoderm, gives rise to the nervous system, including the brain, and skin or carapace and hair, bristles, or scales.

Embryos in many species often appear similar to one another in early developmental stages. The reason for this similarity is because species have a shared evolutionary history. These similarities among species are called homologous structures, which are structures that have the same or similar function and mechanism, having evolved from a common ancestor.

Drosophila Melanogaster (Fruit Fly)

Drosophila melanogaster, a fruit fly, is a model organism in biology on which much research into embryology has been done. Before fertilization, the female gamete produces an abundance of mRNA - transcribed from the genes that encode bicoid protein and nanos protein. These mRNA molecules are stored to be used later in what will become a developing embryo. The male and female *Drosophila* gametes exhibit anisogamy (differences in morphology and sub-cellular biochemistry). The female gamete is larger than the male gamete because it harbors more cytoplasm and, within the cytoplasm, the female gamete contains an abundance of the mRNA previously mentioned. At fertilization, the male and female gametes fuse (plasmogamy) and then the nucleus of the male gamete fuses with the nucleus of the female gamete (karyogamy). Note that before the gametes' nuclei fuse, they are known as pronuclei. A series of nuclear divisions will occur without cytokinesis (division of the cell) in the zygote to form a multi-nucleated cell (a cell containing multiple nuclei) known as a syncytium. All the nuclei in the syncytium are identical, just as all the nuclei in every somatic cell of any multicellular organism are identical *in terms of the DNA sequence of the genome*. Before the nuclei can differentiate in transcriptional activity, the embryo (syncytium) must be divided into segments. In each segment, a unique set of regulatory proteins will cause specific genes in the nuclei to be transcribed. The resulting combination of proteins will transform clusters of cells into early embryo tissues that will each develop into multiple fetal and adult tissues later in development (note: this happens after each nucleus becomes wrapped with its own cell membrane).

Drosophila melanogaster

Drosophila melanogaster larvae contained in lab apparatus to be used for experiments in genetics and embryology.

Outlined below is the process that leads to cell and tissue differentiation.

Maternal-effect genes - subject to Maternal (cytoplasmic) inheritance.

- Egg-polarity genes establish the Anteroposterior axis.

Zygotic-effect genes - subject to Mendelian (classical) inheritance.

- Segmentation genes establish 14 segments of the embryo using the anteroposterior axis as a guide.

1. Gap genes establish 3 broad segments of the embryo.

2. Pair-rule genes define 7 segments of the embryo within the confines of the second broad segment that was defined by the gap genes.

3. Segment-polarity genes define another 7 segments by dividing each of the pre-existing 7 segments into anterior and posterior halves.

- Homeotic (homeobox) genes use the 14 segments as pinpoints for specific types of cell differentiation and the histological developments that correspond to each cell type.

Humans

Humans are bilaterals and deuterostomes.

In humans, the term embryo refers to the ball of dividing cells from the moment the zygote implants itself in the uterus wall until the end of the eighth week after conception. Beyond the eighth week after conception (tenth week of pregnancy), the developing human is then called a fetus.

History

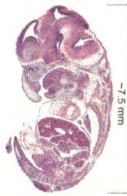

Histological film 10-day mouse embryo

Human embryo at six weeks gestational age

As recently as the 18th century, the prevailing notion in western human embryology was preformation: the idea that semen contains an embryo – a preformed, miniature infant, or *homunculus* – that simply becomes larger during development. The competing explanation of embryonic development was *epigenesis*, originally proposed 2,000 years earlier by Aristotle. Much early embryology came from the work of the Italian anatomists Aldrovandi, Aranzio, Leonardo da Vinci, Marcello Malpighi, Gabriele Falloppio, Girolamo Cardano, Emilio Parisano, Fortunio Liceti, Stefano Lorenzini, Spallanzani, Enrico Sertoli, and Mauro Rusconi. According to epigenesis, the form of an animal emerges gradually from a relatively formless egg. As microscopy improved during the 19th century, biologists could see that embryos took shape in a series of progressive steps, and epigenesis displaced preformation as the favoured explanation among embryologists.

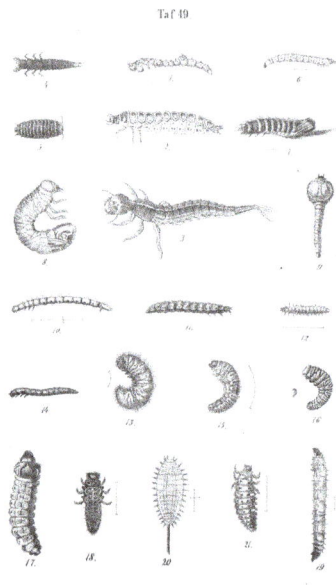

Beetle larvae

After 1827

Karl Ernst von Baer and Heinz Christian Pander proposed the germ layer theory of development; von Baer discovered the mammalian ovum in 1827. Modern embryological pioneers include Charles Darwin, Ernst Haeckel, J.B.S. Haldane, and Joseph Needham. Other important contributors include William Harvey, Kaspar Friedrich Wolff, Heinz Christian Pander, August Weismann, Gavin de Beer, Ernest Everett Just, and Edward B. Lewis.

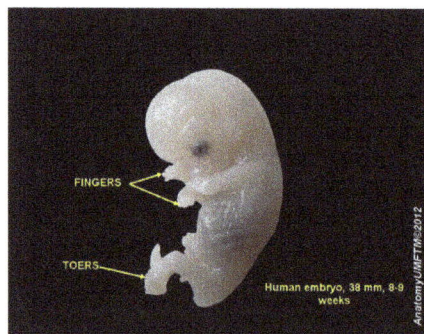

8–9-week human embryo

After 1950

After the 1950s, with the DNA helical structure being unravelled and the increasing knowledge in the field of molecular biology, developmental biology emerged as a field of study which attempts to correlate the genes with morphological change, and so tries to determine which genes are responsible for each morphological change that takes place in an embryo, and how these genes are regulated.

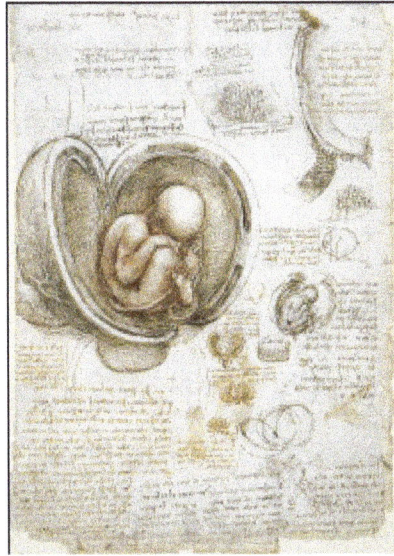

A study of embryos by Leonardo da Vinci

Vertebrate and Invertebrate Embryology

Many principles of embryology apply to invertebrates as well as to vertebrates. Therefore, the study of invertebrate embryology has advanced the study of vertebrate embryology. However, there are many differences as well. For example, numerous invertebrate species release a larva before development is complete; at the end of the larval period, an animal for the first time comes to resemble an adult similar to its parent or parents. Although invertebrate embryology is similar in some ways for different invertebrate animals, there are also countless variations. For instance, while spiders proceed directly from egg to adult form, many insects develop through at least one larval stage.

Modern Embryology Research

Currently, embryology has become an important research area for studying the genetic control of the development process (e.g. morphogens), its link to cell signalling, its importance for the study of certain diseases and mutations, and in links to stem cell research.

Ethology

Ethology is the scientific and objective study of animal behaviour, usually with a focus on behaviour under natural conditions, and viewing behaviour as an evolutionarily adaptive trait. Behaviourism is a term that also describes the scientific and objective study of animal behaviour, usually referring to measured responses to stimuli or trained behavioural responses in a laboratory context, without a particular emphasis on evolutionary adaptivity. Many naturalists have studied aspects of animal behaviour throughout history. Ethology has its scientific roots in the work of Charles Darwin and of American and German ornithologists of the late 19th and early 20th century, including Charles O. Whitman, Oskar Heinroth, and Wallace Craig. The modern discipline of ethology is generally considered to have begun during the 1930s with the work of Dutch biologist

Nikolaas Tinbergen and by Austrian biologists Konrad Lorenz and Karl von Frisch, joint awardees of the 1973 Nobel Prize in Physiology or Medicine. Ethology is a combination of laboratory and field science, with a strong relation to some other disciplines such as neuroanatomy, ecology, and evolutionary biology. Ethologists are typically interested in a behavioural process rather than in a particular animal group, and often study one type of behaviour, such as aggression, in a number of unrelated animals.

A range of animal behaviours

Ethology is a rapidly growing field. Since the dawn of the 21st century, many aspects of animal communication, emotions, culture, learning and sexuality that the scientific community long thought it understood have been re-examined, and new conclusions reached. New fields, such as neuroethology, have developed.

Understanding ethology or animal behaviour can be important in animal training. Considering the natural behaviours of different species or breeds enables the trainer to select the individuals best suited to perform the required task. It also enables the trainer to encourage the performance of naturally occurring behaviours and also the discontinuance of undesirable behaviours.

Etymology

The term was first popularized by American myrmecologist (a person who studies ants) William Morton Wheeler in 1902. An earlier, slightly different sense of the term was proposed by John Stuart Mill in his 1843 *System of Logic*. He recommended the development of a new science, "ethology", the purpose of which would be explanation of individual and national differences in character, on the basis of associationistic psychology. This use of the word was never adopted.

Relationship with Comparative Psychology

Comparative psychology also studies animal behaviour, but, as opposed to ethology, is construed as a sub-topic of psychology rather than as one of biology. Historically, where comparative psychology has included research on animal behaviour in the context of what is known about human psychology, ethology involves research on animal behaviour in the context of what is known about animal anatomy, physiology, neurobiology, and phylogenetic history. Furthermore, early comparative psychologists concentrated on the study of learning and tended to research behaviour in artificial situations, whereas early ethologists concentrated on behaviour in natural situations, tending to describe it as instinctive.

The two approaches are complementary rather than competitive, but they do result in different perspectives, and occasionally conflicts of opinion about matters of substance. In addition, for most of the twentieth century, comparative psychology developed most strongly in North America, while ethology was stronger in Europe. From a practical standpoint, early comparative psychologists concentrated on gaining extensive knowledge of the behaviour of very few species. Ethologists were more interested in understanding behaviour across a wide range of species to facilitate principled comparisons across taxonomic groups. Ethologists have made much more use of such cross-species comparisons than comparative psychologists have.

History

Scala Naturae and Lamarck's Theories

Until the 19th century, the most common theory among scientists was still the concept of *scala naturae*, proposed by Aristotle. According to this theory, living beings were classified on an ideal pyramid that represented non-living things (such as minerals and sediment) and the simplest animals on the lower levels, with complexity increasing progressively towards the top, occupied by human beings. In the Western world of the time, people believed animal species were eternal and immutable, created with a specific purpose, as this seemed the only possible explanation for the incredible variety of living beings and their surprising adaptation to their habitats.

Jean-Baptiste Lamarck (1744–1829)

Jean-Baptiste Lamarck (1744 - 1829) was the first biologist to describe a complex theory of evolution. His theory substantially comprised two statements: first, that animal organs and behaviour can change according to the way they are used; and second, that those characteristics can transmit from one generation to the next (the example of the giraffe whose neck becomes longer while trying to reach the upper leaves of a tree is well-known). The second statement is that every living organism, humans included, tends to reach a greater level of perfection. When Charles Darwin went to the Galapagos Islands, he was well aware of Lamarck's theories and was influenced by them.

Theory of Evolution by Natural Selection and The Beginnings of Ethology

Because ethology is considered a topic of biology, ethologists have been concerned particularly with the evolution of behaviour and the understanding of behaviour in terms of the theory of natural selection. In one sense, the first modern ethologist was Charles Darwin, whose book *The Expression of the Emotions in Man and Animals* influenced many ethologists. He pursued his interest in behaviour by encouraging his protégé George Romanes, who investigated animal learning and intelligence using an anthropomorphic method, anecdotal cognitivism, that did not gain scientific support.

Charles Darwin (1809–1882)

Other early ethologists, such as Charles O. Whitman, Oskar Heinroth, Wallace Craig and Julian Huxley, instead concentrated on behaviours that can be called instinctive, or natural, in that they occur in all members of a species under specified circumstances. Their beginning for studying the behaviour of a new species was to construct an ethogram (a description of the main types of behaviour with their frequencies of occurrence). This provided an objective, cumulative data-base of behaviour, which subsequent researchers could check and supplement.

Social Ethology and Recent Developments

In 1970, the English ethologist John H. Crook published an important paper in which he distinguished comparative ethology from social ethology, and argued that much of the ethology that had existed so far was really comparative ethology—examining animals as individuals—whereas, in the future, ethologists would need to concentrate on the behaviour of social groups of animals and the social structure within them.

Also in 1970, Robert Ardrey's book *The Social Contract: A Personal Inquiry into the Evolutionary Sources of Order and Disorder* was published. The book and study investigated animal behaviour and then compared human behaviour to it as a similar phenomenon.

E. O. Wilson's book *Sociobiology: The New Synthesis* appeared in 1975, and since that time, the study of behaviour has been much more concerned with social aspects. It has also been driven by the stronger, but more sophisticated, Darwinism associated with Wilson, Robert Trivers, and William Hamilton. The related development of behavioural ecology has also helped transform ethology. Furthermore, a substantial reapprochement with comparative psychology has occurred, so the modern scientific study of behaviour offers a more or less seamless spectrum of approaches: from animal cognition to more traditional comparative psychology, ethology, sociobiology, and behavioural ecology.

Growth of The Field

Due to the work of Lorenz and Tinbergen, ethology developed strongly in continental Europe during the years prior to World War II. After the war, Tinbergen moved to the University of Oxford, and ethology became stronger in the UK, with the additional influence of William Thorpe, Robert Hinde, and Patrick Bateson at the Sub-department of Animal Behaviour of the University of Cambridge, located in the village of Madingley. In this period, too, ethology began to develop strongly in North America.

Lorenz, Tinbergen, and von Frisch were jointly awarded the Nobel Prize in Physiology or Medicine in 1973 for their work of developing ethology.

Ethology is now a well-recognized scientific discipline, and has a number of journals covering developments in the subject, such as *Animal Behaviour, Animal Welfare, Applied Animal Behaviour Science, Behaviour, Behavioral Ecology* and *Journal of Ethology*. In 1972, the International Society for Human Ethology was founded to promote exchange of knowledge and opinions concerning human behaviour gained by applying ethological principles and methods and published their journal, *The Human Ethology Bulletin*. In 2008, in a paper published in the journal *Behaviour*, ethologist Peter Verbeek introduced the term "Peace Ethology" as a sub-discipline of Human Ethology that is concerned with issues of human conflict, conflict resolution, reconciliation, war, peacemaking, and peacekeeping behaviour.

Today, along with ethologists, many biologists, zoologists, primatologists, anthropologists, veterinarians, and physicians study ethology and other related fields such as animal psychology, the study of animal social groups, animal cognition and animal welfare science.

Instinct

Kelp gull chicks peck at red spot on mother's beak to stimulate regurgitating reflex

The Merriam-Webster dictionary defines instinct as "A largely inheritable and unalterable tendency of an organism to make a complex and specific response to environmental stimuli without involving reason".

Fixed Action Patterns

An important development, associated with the name of Konrad Lorenz though probably due more to his teacher, Oskar Heinroth, was the identification of fixed action patterns (FAPs). Lorenz popularized FAPs as instinctive responses that would occur reliably in the presence of identifiable stimuli called sign stimuli or "releasing stimuli". FAPs are now considered to be instinctive behavioural sequences that are relatively invariant within the species and almost inevitably run to completion.

One example of a releaser is the beak movements of many bird species performed by newly hatched chicks, which stimulates the mother to regurgitate food for her offspring. Other examples are the classic studies by Tinbergen on the egg-retrieval behaviour and the effects of a "supernormal stimulus" on the behaviour of graylag geese.

One investigation of this kind was the study of the waggle dance ("dance language") in bee communication by Karl von Frisch. Lorenz subsequently developed a theory of the evolution of animal communication based on his observations of fixed action patterns and the circumstances in which they are expressed.

Learning

Habituation

Habituation is a simple form of learning and occurs in many animal taxa. It is the process whereby an animal ceases responding to a stimulus. Often, the response is an innate behaviour. Essentially, the animal learns not to respond to irrelevant stimuli. For example, prairie dogs (*Cynomys ludovicianus*) give alarm calls when predators approach, causing all individuals in the group to quickly scramble down burrows. When prairie dog towns are located near trails used by humans, giving alarm calls every time a person walks by is expensive in terms of time and energy. Habituation to humans is therefore an important adaptation in this context.

Associative Learning

Associative learning in animal behaviour is any learning process in which a new response becomes associated with a particular stimulus. The first studies of associative learning were made by Russian physiologist Ivan Pavlov. Examples of associative learning include when a goldfish swims to the water surface when a human is going to feed it, or the excitement of a dog whenever it sees a leash as a prelude for a walk.

Imprinting

Being able to discriminate the members of one's own species is also of fundamental importance for reproductive success. Such discrimination can be based on a number of factors. However, this important type of learning only takes place in a very limited period of time. This kind of learning is called imprinting, and was a second important finding of Lorenz. Lorenz observed that the young of birds such as geese and chickens followed their mothers spontaneously from almost the first day after they were hatched, and he discovered that this response could be imitated by an arbitrary stimulus if the eggs were incubated artificially and the stimulus were presented during a critical period that continued for a few days after hatching.

Example of imprinting in a moose

Cultural Learning

Observational Learning

Imitation

Imitation is an advanced behaviour whereby an animal observes and exactly replicates the behaviour of another. The National Institutes of Health reported that capuchin monkeys preferred the company of researchers who imitated them to that of researchers who did not. The monkeys not only spent more time with their imitators but also preferred to engage in a simple task with them even when provided with the option of performing the same task with a non-imitator. Imitation has been observed in recent research on chimpanzees; not only did these chimps copy the actions of another individual, when given a choice, the chimps preferred to imitate the actions of the higher-ranking elder chimpanzee as opposed to the lower-ranking young chimpanzee.

Stimulus and Local Enhancement

There are various ways animals can learn using observational learning but without the process of imitation. One of these is *stimulus enhancement* in which individuals become interested in an object as the result of observing others interacting with the object. Increased interest in an object can result in object manipulation which allows for new object-related behaviours by trial-and-error learning. Haggerty (1909) devised an experiment in which a monkey climbed up the side of a cage, placed its arm into a wooden chute, and pulled a rope in the chute to release food. Another monkey was provided an opportunity to obtain the food after watching a monkey go through this process on four separate occasions. The monkey performed a different method and finally succeeded after trial-and-error. Another example familiar to some cat and dog owners is the ability of their animals to open doors. The action of humans operating the handle to open the door results in the animals becoming interested in the handle and then by trial-and-error, they learn to operate the handle and open the door.

In local enhancement, a demonstrator attracts an observer's attention to a particular location. Local enhancement has been observed to transmit foraging information among birds, rats and pigs. The stingless bee (*Trigona corvina*) uses local enhancement to locate other members of their colony and food resources.

Social Transmission

A well-documented example of social transmission of a behaviour occurred in a group of macaques on Hachijojima Island, Japan. The macaques lived in the inland forest until the 1960s, when a group of researchers started giving them potatoes on the beach: soon, they started venturing onto the beach, picking the potatoes from the sand, and cleaning and eating them. About one year later, an individual was observed bringing a potato to the sea, putting it into the water with one hand, and cleaning it with the other. This behaviour was soon expressed by the individuals living in contact with her; when they gave birth, this behaviour was also expressed by their young - a form of social transmission.

Teaching

Teaching is a highly specialized aspect of learning in which the "teacher" (demonstrator) adjusts their behaviour to increase the probability of the "pupil" (observer) achieving the desired end-result of the behaviour. For example, killer whales are known to intentionally beach themselves to catch pinniped prey. Mother killer whales teach their young to catch pinnipeds by pushing them onto the shore and encouraging them to attack the prey. Because the mother killer whale is altering her behaviour to help her offspring learn to catch prey, this is evidence of teaching. Teaching is not limited to mammals. Many insects, for example, have been observed demonstrating various forms of teaching to obtain food. Ants, for example, will guide each other to food sources through a process called "tandem running," in which an ant will guide a companion ant to a source of food. It has been suggested that the pupil ant is able to learn this route to obtain food in the future or teach the route to other ants.This behaviour of teaching is also exemplified by crows. Specifically New Caledonian crows. The adults (whether individual or in families) teach their young adolescent offspring how to construct and utilize tools. For example; *Pandanus* branches are used to extract insects and other larvae from holes within trees.

Mating and The Fight for Supremacy

Individual reproduction is the most important phase in the proliferation of individuals or genes within a species: for this reason, there exist complex mating rituals, which can be very complex even if they are often regarded as FAPs. The stickleback's complex mating ritual, studied by Tinbergen, is regarded as a notable example of a FAP.

Often in social life, animals fight for the right to reproduce, as well as social supremacy. A common example of fighting for social and sexual supremacy is the so-called pecking order among poultry. Every time a group of poultry cohabitate for a certain time length, they establish a pecking order. In these groups, one chicken dominates the others and can peck without being pecked. A second chicken can peck all the others except the first, and so on. Higher level chickens are easily distinguished by their well-cured aspect, as opposed to lower level chickens. While the pecking order is establishing, frequent and violent fights can happen, but once established, it is broken only when other individuals enter the group, in which case the pecking order re-establishes from scratch.

Living in Groups

Several animal species, including humans, tend to live in groups. Group size is a major aspect of their social environment. Social life is probably a complex and effective survival strategy. It may be regarded as a sort of symbiosis among individuals of the same species: a society is composed of a group of individuals belonging to the same species living within well-defined rules on food management, role assignments and reciprocal dependence.

When biologists interested in evolution theory first started examining social behaviour, some apparently unanswerable questions arose, such as how the birth of sterile castes, like in bees, could be explained through an evolving mechanism that emphasizes the reproductive success of as many individuals as possible, or why, amongst animals living in small groups like squirrels, an individual would risk its own life to save the rest of the group. These behaviours may be examples of altruism. Of course, not all behaviours are altruistic, as indicated by the table below. For example, revengeful behaviour was at one point claimed to have been observed exclusively in *Homo sapiens*. However, other species have been reported to be vengeful, including reports of vengeful camels and chimpanzees.

Classification of social behaviours		
Type of behaviour	**Effect on the donor**	**Effect on the receiver**
Egoistic	Increases fitness	Decreases fitness
Cooperative	Increases fitness	Increases fitness
Altruistic	Decreases fitness	Increases fitness
Revengeful	Decreases fitness	Decreases fitness

Altruistic behaviour has been explained by the gene-centred view of evolution.

Benefits and Costs of Group Living

One advantage of group living can be decreased predation. If the number of predator attacks stays the same despite increasing prey group size, each prey may have a reduced risk of predator attacks through the dilution effect. Additionally, a predator that is confused by a mass of individuals can find it more difficult to single out one target. For this reason, the zebra's stripes offer not only camouflage in a habitat of tall grasses, but also the advantage of blending into a herd of other zebras. In groups, prey can also actively reduce their predation risk through more effective defense tactics, or through earlier detection of predators through increased vigilance.

Another advantage of group living can be an increased ability to forage for food. Group members may exchange information about food sources between one another, facilitating the process of resource location. Honeybees are a notable example of this, using the waggle dance to communicate the location of flowers to the rest of their hive. Predators also receive benefits from hunting in groups, through using better strategies and being able to take down larger prey.

Some disadvantages accompany living in groups. Living in close proximity to other animals can facilitate the transmission of parasites and disease, and groups that are too large may also experience greater competition for resources and mates.

Group Size

Theoretically, social animals should have optimal group sizes that maximize the benefits and minimize the costs of group living. However, in nature, most groups are stable at slightly larger than optimal sizes. Because it generally benefits an individual to join an optimally-sized group, despite slightly decreasing the advantage for all members, groups may continue to increase in size until it is more advantageous to remain alone than to join an overly full group.

Tinbergen's Four Questions for Ethologists

Niko Tinbergen argued that ethology always needed to include four kinds of explanation in any instance of behaviour:

- Function – How does the behaviour affect the animal's chances of survival and reproduction? Why does the animal respond that way instead of some other way?

- Causation – What are the stimuli that elicit the response, and how has it been modified by recent learning?

- Development – How does the behaviour change with age, and what early experiences are necessary for the animal to display the behaviour?

- Evolutionary history – How does the behaviour compare with similar behaviour in related species, and how might it have begun through the process of phylogeny?

These explanations are complementary rather than mutually exclusive—all instances of behaviour require an explanation at each of these four levels. For example, the function of eating is to acquire nutrients (which ultimately aids survival and reproduction), but the immediate cause of eating is hunger (causation). Hunger and eating are evolutionarily ancient and are found in many species (evolutionary history), and develop early within an organism's lifespan (development). It is easy to confuse such questions—for example, to argue that people eat because they're hungry and not to acquire nutrients—without realizing that the reason people experience hunger is because it causes them to acquire nutrients.

Behavioral Ecology

Behavioral ecology, also spelled behavioural ecology, is the study of the evolutionary basis for animal behavior due to ecological pressures. Behavioral ecology emerged from ethology after Niko Tinbergen outlined four questions to address when studying animal behavior which are the proximate causes, ontogeny, survival value, and phylogeny of behavior.

If an organism has a trait which provides them with a selective advantage (i.e. has an adaptive significance) in its environment, then natural selection can potentially favor it. Adaptive significance therefore refers to the beneficial qualities (such as in terms of increased survival and reproduction), any given modified trait conveys. For example, genetic differences between individuals may lead to behavioral differences, some of which in turn may drive differences in reproductive success, and ultimately over generations, the increased dominance of individuals with those favoured traits, i.e. evolution.

Individuals are always in competition with others for limited resources, including food, territories, and mates. Conflict will occur between predators and prey, between rivals for mates, between siblings, mates, and even between parents and their offspring.

Competing for Resources

The value of a social behavior depends in part on the social behavior of an animal's neighbors. For example, the more likely a rival male is to back down from a threat, the more value a male gets out of making the threat. The more likely, however, that a rival will attack if threatened, the less useful it is to threaten other males. When a population exhibits a number of interacting social behaviors such as this, it can evolve a stable pattern of behaviors known as an evolutionarily stable strategy (or ESS). This term, derived from economic game theory, became prominent after John Maynard Smith (1982) recognized the possible application of the concept of a Nash equilibrium to model the evolution of behavioral strategies.

Evolutionarily Stable Strategy

In short, evolutionary game theory asserts that only strategies that, when common in the population, cannot be "invaded" by any alternative (mutant) strategy will be an ESS, and thus maintained in the population. In other words, at equilibrium every player should play the best strategic response to each other. When the game is two player and symmetric each player should play the strategy which is the best response to itself.

Therefore, the ESS is considered to be the evolutionary end point subsequent to the interactions. As the fitness conveyed by a strategy is influenced by what other individuals are doing (the relative frequency of each strategy in the population), behavior can be governed not only by optimality but the frequencies of strategies adopted by others and are therefore frequency dependent (frequency dependence).

Behavioral evolution is therefore influenced by both the physical environment and interactions between other individuals.

An example of how changes in geography can make a strategy susceptible to alternative strategies is the parasitization of the African honey bee, *A. m. scutellata*.

Resource Defense

The term economic defendability was first introduced by Jerram Brown in 1964. Economic defendability states that defense of a resource have costs, such as energy expenditure or risk of injury, as well as benefits of priority access to the resource. Territorial behavior arises when benefits are greater than the costs.

Studies of the golden-winged sunbird have validated the concept of economic defendability. Comparing the energetic costs a sunbird expends in a day to the extra nectar gained by defending a territory, researchers showed that birds only became territorial when they were making a net energetic profit. When resources are at low density, the gains from excluding others may not be sufficient to pay for the cost of territorial defense. In contrast, when resource availability is high, there may be so many intruders that the defender would have no time to make use of the resources made available by defense.

Sometimes the economics of resource competition favors shared defense. An example is the feeding territories of the white wagtail. The white wagtails feed on insects washed up by the river onto the bank, which acts as a renewing food supply. If any intruders harvested their territory then the prey would quickly become depleted, but sometimes territory owners tolerate a second bird, known as a satellite. The two sharers would then move out of phase with one another, resulting in decreased feeding rate but also increased defense, illustrating advantages of group living.

Ideal Free Distribution

One of the major models used to predict the distribution of competing individuals amongst resource patches is the ideal free distribution model. Within this model, resource patches can be of variable quality, and there is no limit to the number of individuals that can occupy and extract resources from a particular patch. Competition within a particular patch means that the benefit each individual receives from exploiting a patch decreases logarithmically with increasing number of competitors sharing that resource patch. The model predicts that individuals will initially flock to higher-quality patches until the costs of crowding bring the benefits of exploiting them in line with the benefits of being the only individual on the lesser-quality resource patch. After this point has been reached, individuals will alternate between exploiting the higher-quality patches and the lower-quality patches in such a way that the average benefit for all individuals in both patches is the same. This model is *ideal* in that individuals have complete information about the quality of a resource patch and the number of individuals currently exploiting it, and *free* in that individuals are freely able to choose which resource patch to exploit.

An experiment by Manfred Malinski in 1979 demonstrated that feeding behavior in three-spined sticklebacks follows an ideal free distribution. Six fish were placed in a tank, and food items were dropped into opposite ends of the tank at different rates. The rate of food deposition at one end was set at twice that of the other end, and the fish distributed themselves with four individuals at the faster-depositing end and two individuals at the slower-depositing end. In this way, the average feeding rate was the same for all of the fish in the tank.

Mating Strategies and Tactics

As with any competition of resources, species across the animal kingdom may also engage in competitions for mating. If one considers mates or potentials mates as a resource, these sexual partners can be randomly distributed amongst resource pools within a given environment. Following the ideal free distribution model, suitors will distribute themselves amongst the potential mates in an effort to maximize their chances or the number of potential mates that they can consummate. For all competitors, males of a species in most cases, there will be variations in both the strategies and tactics used to obtain matings. Strategies generally refer to the genetically determined behaviors which can be described as conditional. Tactics refer to the subset of behaviors within a given genetic strategy. Thus it is not difficult for a great many variations in mating strategies to exist in a given environment or species.

An experiment conducted by Anthony Arak, where playback of synthetic calls from male natterjack toads was used to manipulate behavior of the males in a chorus, the difference between strategies and tactics is clear. While small and immature, male natterjack toads adopted a satellite tactic to parasitize larger males. Though large males on average still retained greater reproductive success, smaller males were able to intercept matings. When the large males of the chorus were removed, smaller males adopted a calling behavior, no longer competing against the loud calls of larger males. When smaller males got larger and their calls more competitive, then they started calling and competing directly for mates.

Sexual Selection

Mate Choice by Resources

In many sexually reproducing species, such as mammals, birds and amphibians, the females are responsible for bearing the offspring for a certain period of time, during which the males are free to mate with other available females and therefore can father many more offspring, thus continue to pass on their genes. The fundamental difference between male and female reproduction mechanisms determines the different strategies each sex employs to maximize their reproductive success. For males, their reproductive success is limited by access to females, while females are limited by their access to resources. In this sense, females can be way choosier than males because they have to bet on the resources provided by the males to ensure reproductive success.

Resources usually include nest sites, food and protection. In some cases, the males provide all of them (e.g. sedge warblers). The females dwell in their chosen males' territories for access to these resources. The males gain ownership to the territories through male-male competition that often involves physical aggression. Only the largest and strongest males manage to defend the best quality nest sites. Females choose males by inspecting the quality of different territories or by looking at some male traits that can indicate the quality of resources. Sometimes, males leave after mating. The only resource that a male provides is a nuptial gift, such as protection or food. The female can evaluate the quality of the protection or food provided by the male so as to decide whether to mate or not or how long she is willing to copulate.

Mate Choice by Genes

When males' only contribution to offspring is their sperm, females are particularly choosy. With this high level of female choice, sexual ornaments are seen in males, where the ornaments reflect the male's social status. Two hypotheses have been proposed to conceptualize the genetic benefits from female mate choice.

First, the good genes hypothesis suggests that female choice is for higher genetic quality and that this preference is favored because it increases fitness of the offspring. This includes Zahavi's handicap hypothesis and Hamilton and Zuk's host and parasite arms race. Zahavi's handicap hypothesis was proposed within the context of looking at elaborate male sexual displays. He suggested that females favor ornamented traits because they are handicaps and are indicators of the male's genetic quality. Since these ornamented traits are hazards, the male's survival must be indicative of his high genetic quality in other areas. In this way, the degree that a male expresses his sexual display indicates to the female his genetic quality. Zuk and Hamilton proposed a hypothesis after observing disease as a powerful selective pressure on a rabbit population. They suggested that sexual displays were indicators of resistance of disease on a genetic level.

Such 'choosiness' from the female individuals can be seen in wasp species too, especially among *Polistes dominula* wasps. The females tend to prefer males with smaller, more elliptically shaped spots than those with larger and more irregularly shaped spots. Those males would have reproductive superiority over males with irregular spots.

Fisher's hypothesis of runaway sexual selection suggests that female preference is genetically correlated with male traits and that the preference co-evolves with the evolution of that trait, thus the preference is under indirect selection. Fisher suggests that female preference began because the trait indicated the male's quality. The female preference spread, so that the females' offspring now benefited from the higher quality from specific trait but also greater attractiveness to mates. Eventually, the trait will only represent attractiveness to mates and no longer represent increased survival.

An example of mate choice by genes is seen in the cichlid fish *Tropheus moorii* where males provide no parental care. An experiment found that a female *T. moorii* is more likely to choose a mate with the same color morph as her own. In another experiment, females have been shown to share preferences for the same males when given two to choose from, meaning some males get to reproduce more often than others.

Sensory Bias

The sensory bias hypothesis states that the preference for a trait evolves in a non-mating context and is then exploited by one sex in order to obtain more mating opportunities. The competitive sex evolves traits that exploit a pre-existing bias that the choosy sex already possesses. This mechanism is thought to explain remarkable trait differences in closely related species because it produces a divergence in signaling systems which leads to reproductive isolation.

Sensory bias has been demonstrated in guppies, freshwater fish from Trinidad and Tobago. In this mating system, female guppies prefer to mate with males with more orange body coloration. However, outside of a mating context, both sexes prefer animate orange objects which suggests that preference originally evolved in another context, like foraging. Orange fruits are a rare treat that fall into streams where the guppies live. The ability to find these fruits quickly is an adaptive quality that has evolved outside of a mating context. Sometime after the affinity for orange objects arose, male guppies exploited this preference by incorporating large orange spots to attract females.

Another example of sensory exploitation is in the water mite *Neumania papillator*, an ambush predator which hunts copepods (small crustaceans) passing by in the water column. When hunting, *N. papillator* adopts a characteristic stance termed the 'net stance' - their first four legs are held out into the water column, with their four hind legs resting on aquatic vegetation; this allows them to detect vibrational stimuli produced by swimming prey and use this to orient towards and clutch at prey. During courtship, males actively search for females - if a male finds a female, he slowly circles around the female whilst trembling his first and second leg near her. Male leg trembling causes females (who were in the 'net stance') to orient towards often clutch the male. This did not damage the male or deter further courtship; the male then deposited spermatophores and began to vigorously fan and jerk his fourth pair of legs over the spermatophore, generating a current of water that passed over the spermatophores and towards the female. Sperm packet uptake by the female would sometimes follow. Heather Proctor hypothesised that the vibrations trembling male legs made were done to mimic the vibrations that females detect from swimming prey - this would trigger the female prey-detection responses causing females to orient and then clutch at males, mediating courtship. If this was true and males were exploiting female predation responses, then hungry females should be more receptive to male trembling – Proctor found that unfed captive females did orient and clutch at males significantly more than fed captive females did, consistent with the sensory exploitation hypothesis.

Other examples for the sensory bias mechanism include traits in auklets, wolf spiders, and manakins. Further experimental work is required to reach a fuller understanding of the prevalence and mechanisms of sensory bias.

Sexual Conflict

Sexual conflict, in some form or another, may very well be inherent in the ways most animals reproduce. Females invest more in offspring prior to mating, due to the differences in gametes in species that exhibit anisogamy, and often invest more in offspring after mating. This unequal investment leads, on one hand, to intense competition between males for mates and, on the other hand, to females choosing among males for better access to resources and good genes. Because of differences in mating goals, males and females may have very different preferred outcomes to mating.

Sexual conflict occurs whenever the preferred outcome of mating is different for the male and female. This difference, in theory, should lead to each sex evolving adaptations that bias the outcome of reproduction towards its own interests. This sexual competition leads to sexually antagonistic coevolution between males and females, resulting in what has been described as an evolutionary arms race between males and females.

Conflict Over Mating

Males' reproductive successes are often limited by access to mates, whereas females' reproductive successes are more often limited by access to resources. Thus, for a given sexual encounter, it will benefit the male to mate but the female to be choosy and resist. For example, male small tortoiseshell butterfly will compete in order to gain the best territory to mate. Another example of this conflict can be found in the Eastern carpenter bee, *Xylocopa virginica*. Males of this species are limited in reproduction primarily by access to mates, so they will claim a territory and wait for a female to pass through. Big males are, therefore, more successful in mating because they claim territories near the female nesting sites that are more sought after. Smaller males, on the other hand, will monopolize less competitive sites in foraging areas so that they may mate with reduced conflict.

Male scorpionfly

Extreme manifestations of this conflict are seen throughout nature. For example, the male *Panorpa* scorpionflies attempt to force copulation. Male scorpionflies usually acquire mates by presenting them with edible nuptial gifts in the forms of salivary secretions or dead insects. However, some males attempt to force copulation by grabbing females with a specialized abdominal organ without offering a gift. Forced copulation is costly to the female as she does not receive the food from the male and has to search for food herself (costing time and energy), while it is beneficial for the male as he does not need to find a nuptial gift.

In other cases, however, it pays for the female to gain more matings and her social mate to prevent these so as to guard paternity. For example, in many socially monogamous birds, males will follow females closely during her fertile period and attempt to chase away any other males so as to prevent extra-pair matings. The female may attempt to sneak off to achieve these extra matings. In species where males are incapable of constant guarding, the social male will often frequently copulate with the female so as to swamp rival males' sperm.

Female red junglefowl in Thailand

Sexual conflict after mating has also been shown to occur in both males and females. Males employ a diverse array of tactics to increase their success in sperm competition. These can include removing other male's sperm from females, displacing other male's sperm by flushing out prior inseminations with large amounts of their own sperm, creating copulatory plugs in females' reproductive tracts to prevent future matings with other males, spraying females with anti-aphrodisiacs to discourage other males from mating with the female, and producing sterile parasperm to protect fertile eusperm in the female's reproductive tract. Furthermore, males may control the strategic allocation of sperm, producing more sperm when females are more promiscuous. All these methods are meant to ensure that females will be more likely to produce offspring belonging to the males who uses the method.

Females also control the outcomes of matings, and there exists the possibility that females choose sperm (cryptic female choice). A dramatic example of this is the feral fowl *Gallus gallus*. In this species, females prefer to copulate with dominant males, but subordinate males can force matings. In these cases, the female is able to eject the subordinate male's sperm using cloacal contractions.

Parental Care and Family Conflicts

Parental care is the investment a parent will put into their offspring, which includes protecting and feeding the young, preparing burrows or nests, and providing eggs with yolk. There is great variation in parental care in the animal kingdom. In some species, the parents may not care for their offspring at all, while in others the parents exhibit single-parental or even bi-parental care. As with other topics in behavioral ecology, interactions within a family will involve conflicts. These conflicts can be broken down into three general types: sexual (male-female) conflict, parent-offspring conflict, and sibling conflict.

Types of Parental Care

There are many different patterns of parental care in the animal kingdom. The patterns can be explained by physiological constraints or ecological conditions, such as mating opportunities. In invertebrates, there is no parental care in most species because it is more favorable for parents to produce a large number of eggs whose fate is left to chance than to protect a few individual young. For example, female *L. figueresi* die after stocking their larvae's cells with pollen and nectar and before their larvae hatch. In birds, biparental care is the most common, because reproductive success directly depends on the parents' ability to feed their chicks. Two parents can feed twice as many young, so it is more favorable for birds to have both parents delivering food. In mammals, female-only care is the most common. This is most likely because females are internally fertilized and so are holding the young inside for a prolonged period of gestation, which provides males with the opportunity to desert. Females also feed the young through lactation after birth, so males are not required for feeding. Male parental care is only observed in species where they contribute to feeding or carrying of the young, such as in marmosets. In fish there is no parental care in 79% of bony fish. In fish with parental care, it usually limited to selecting, preparing, and defending a nest, as seen in sockeye salmon, for example. Also, parental care in fish, if any, is primarily done by males, as seen in gobies and redlip blennies. The cichlid fish *V. moorii* exhibits biparental care. In species with internal fertilization, the female is usually the one to take care of the young. In cases where fertilization is external the male becomes the main caretaker.

Familial Conflict

Familial conflict is a result of trade-offs as a function of lifetime parental investment. Parental investment was defined by Robert Trivers in 1972 as "any investment by the parent in an individual offspring that increases the offspring's chance of surviving at the cost of the parent's ability to invest in other offspring". Parental investment includes behaviors like guarding and feeding. Each parent has a limited amount of parental investment over the course of their lifetime. Investment trade-offs in offspring quality and quantity within a brood and trade offs between current and future broods leads to conflict over how much parental investment to provide and to whom parents should invest in. There are three major types of familial conflict: sexual, parent-offspring, and sibling-sibling conflict.

Sexual Conflict

Great tit

There is conflict among parents as to who should provide the care as well as how much care to provide. Each parent must decide whether or not to stay and care for their offspring, or to desert their offspring. This decision is best modeled by game theoretic approaches to evolutionarily stable strategies (ESS) where the best strategy for one parent depends on the strategy adopted by the other parent. Recent research has found response matching in parents who determine how much care to invest in their offspring. Studies found that parent great tits will match their partner's increased care-giving efforts with increased provisioning rates of their own. This cued parental response is a type of behavioral negotiation between parents that leads to stabilized compensation.

Parent-offspring Conflict

According to Robert Trivers's theory on relatedness, each offspring is related to itself by 1, but is only 0.5 related to their parents and siblings. Genetically, offspring are predisposed to behave in their own self-interest while parents are predisposed to behave equally to all their offspring, including both current and future ones. Offspring will selfishly attempt to take more than their fair shares of parental investment while parents will attempt to spread out their parental investment equally amongst their present young and future young. There are many examples of parent-offspring conflict in nature. One manifestation of this is asynchronous hatching in birds. A behavioral ecology hypothesis is known as Lack's brood reduction hypothesis (named after David Lack). Lack's hypothesis posits an evolutionary and ecological explanation as to why birds lay a series of eggs with an asynchronous delay leading to nestlings of mixed age and weights. According to Lack, this brood behavior is an ecological insurance that allows the larger birds to survive in poor years and all birds to survive when food is plentiful. We also see sex-ratio conflict between the queen and her workers in social hymenoptera. Because of haplodiploidy, the workers (offspring) prefer a 3:1 female to male sex allocation while the queen prefers a 1:1 sex ratio. Both the queen and the workers will attempt to bias the sex ratio in their favor. In some species, the workers gain control of the sex ratio, while in other species, like *B. terrestris*, the queen has a considerable amount of control over the colony sex ratio. Lastly, there has been recent evidence regarding genomic imprinting that is a result of parent-offspring conflict. Paternal genes in offspring will demand more maternal resources than maternal genes in the same offspring and vice versa. This has been show in imprinted genes like insulin-like growth factor-II.

Blackbird chicks in a nest

Parent-offspring Conflict Resolution

Parents need an honest signal from their offspring indicating their level of hunger or need, so that the parents can distribute resources accordingly. Offspring want to get more than their fair share of resources, so they will want to exaggerate their signals to wheedle more investment from their parents. However, this conflict is resolved by the cost of excessive begging. Not only does excessive begging attract predators, but it also retards chick growth if begging goes unrewarded. Thus, the cost of increased begging will enforce offspring honesty.

Another resolution for parent-offspring conflict is that parental provisioning and offspring demand have actually coevolved, so that there is no obvious underlying conflict. Cross-fostering experiments in great tits (*Parus major*) have shown that offspring beg more when their biological mothers are more generous. Therefore, it seems that the willingness to invest in offspring is co-adapted to offspring demand.

Sibling-sibling Conflict

The lifetime parental investment is the fixed amount of parental resources available for all of a parent's young, and an offspring will want as much of it as possible. Siblings in a brood will often compete for parental resources by trying to gain more than their fair share of what their parents have to offer. There are numerous examples in nature where sibling rivalry is escalated to such an extreme that one sibling will try to kill off his other broodmates in order to maximize his own parental investment. In the Galapagos fur seal, the second pup of a female is usually born when the first pup is still suckling. This competition for the mother's milk is especially fierce during periods of food shortage such as an El Niño year, and this usually results in the older pup directly attacking and killing the younger one.

Galápagos fur seals

In some bird species, sibling rivalry is also abetted by the asynchronous hatching of eggs. In the blue-footed booby, for example, the first egg in a nest is hatched four days before the second one, resulting in the elder chick having a four-day head start in growth. When the elder chick falls 20-25% below its expected weight threshold, it will attack its younger sibling and drive it from the nest.

Sibling relatedness in a brood also influences the level of sibling-sibling conflict. In a study on passerine birds, it was found that chicks begged more loudly in species with higher levels of extra-pair paternity.

Brood Parasitism

Some animals deceive other species into providing all parental care. These brood parasites selfishly exploit their hosts' parents and host offspring. The common cuckoo is a well known example of a brood parasite. Female cuckoos lay a single egg in the nest of the host species and when the cuckoo chick hatches, it ejects all the host eggs and young. Other examples of brood parasites include honeyguides, cowbirds, and the large blue butterfly. Brood parasite offspring have many strategies to induce their host parents to invest parental care. Studies show that the common cuckoo uses vocal mimicry to reproduce the sound of multiple hungry host young to solicit more food. Other cuckoos will use visual deception with their wings to exaggerate the begging display. False gapes from brood parasite offspring cause host parents to collect more food. Another example of a brood parasite is *Phengaris* butterflies such as *Phengaris rebeli* and *Phengaris arion*, which differ from the cuckoo in that the butterflies do not oviposit directly in the nest of the host, an ant species *Myrmica schencki*. Rather, the butterfly larvae release chemicals that deceive the ants into believing that they are ant larvae, causing the ants to bring the butterfly larvae back to their own nests to feed them. Other examples of brood parasites are *Polistes sulcifer*, a paper wasp that has lost the ability to build its own nests so females lay their eggs in the nest of a host species, *Polistes dominula*, and rely on the host workers to take care of their brood, as well as *Bombus bohemicus*, a bumblebee that relies on host workers of various other *Bombus* species. Similarly, in *Eulaema meriana*, some Leucospidae wasps exploit the brood cells and nest for shelter and food from the bees. *Vespula austriaca* is another wasp in which the females force the host workers to feed and take care of the brood. In particular, *Bombus hyperboreus*, an Arctic bee species, is also classified as a brood parasite in that it attacks and enslaves other species within their subgenus, *Alpinobombus* to propagate their population.

Adult reed warbler feeding a common cuckoo chick

Mating Systems

Various types of mating systems include monogamy, polygyny, polyandry, promiscuity, and polygamy. Each is differentiated by the sexual behavior between mates, such as which males mate with certain females. An influential paper by Stephen Emlen and Lewis Oring (1977) argued that two main factors of animal behavior influence the diversity of mating systems: the relative accessibility that each sex has to mates, and the parental desertion by either sex.

Mating Systems with No Male Parental Care

In a system that does not have male parental care, resource Dispersion, predation, and the effects of social living primarily influence female dispersion, which in turn influences male dispersion. Since males' primary concern is female acquisition, the males will either indirectly or directly compete for the females. In direct competition, the males are directly focused on the females. Blue-headed wrasse demonstrate the behavior in which females follow resources—such as good nest sites—and males follow the females. Conversely, species with males that exemplify indirectly competitive behavior tend towards the males' anticipation of the resources desired by females and their subsequent effort to control or acquire these resources, which helps them to achieve success with females. Grey-sided voles demonstrate indirect male competition for females. The males were experimentally observed to home in on the sites with the best food in anticipation of females settling in these areas. Males of *Euglossa imperialis*, a non-social bee species, also demonstrate indirect competitive behavior by forming aggregations of territories, which can be considered leks, to defend fragrant-rich primary territories. The purpose of these aggregations is largely only facultative, since the more suitable fragrant-rich sites there are, the more habitable territories there are to inhabit, giving females of this species a large selection of males with whom to potentially mate. Leks and choruses have also been deemed another behavior among the phenomena of male competition for females. Due to the resource-poor nature of the territories that lekking males often defend, it is difficult to categorize them as indirect competitors. Additionally, it is difficult to classify them as direct competitors seeing as they put a great deal of effort into their defense of their territories before females arrive, and upon female arrival they put for the great mating displays to attract the females to their individual sites. These observations make it difficult to determine whether female or resource dispersion primarily influences male aggregation, especially in lieu of the apparent difficulty that males may have defending resources and females in such densely populated areas. Because the reason for male aggregation into leks is unclear, five hypothesis have been proposed. These postulates propose the following as reasons for male lekking: hotspot, predation reduction, increased female attraction, hotshot males, facilitation of female choice. With all of the mating behaviors discussed, the primary factors influencing differences within and between species are ecology, social conflicts, and life history differences.

In some other instances, neither direct nor indirect competition is seen. Instead, in species like the Edith's checkerspot butterfly, males' efforts are directed at acquisition of females and they exhibit indiscriminate mate location behavior, where, given the low cost of mistakes, they will blindly attempt to mate both correctly with females and incorrectly with other objects.

Mating Systems with Male Parental Care

Monogamy

Monogamy is the mating system in 90% of birds, possibly because each male and female will have a greater number of offspring if they share in raising a brood. In obligate monogamy, males feed females on the nest, or share in incubation and chick-feeding. In some species, males and females form lifelong pair bonds. Monogamy may also arise from limited opportunities for polygamy, due to strong competition among males for mates, females suffering from loss of male help, and female-female aggression.

Polygyny

In birds, polygyny occurs when males indirectly monopolize females by controlling resources. In species where males normally do not contribute much to parental care, females suffer relatively little or not at all. In other species, however, females suffer through the loss of male contribution, and the cost of having to share resources that the male controls, such as nest sites or food. In some cases, a polygynous male may control a high-quality territory so for the female, the benefits of polygyny may outweigh the costs.

Polyandry Threshold

There also seems to be a "polyandry threshold" where males may do better by agreeing to share a female instead of attempting to be in a monogamous mating system. Situations that may lead to cooperation among males include when food is scarce, and when there is intense competition for territories or females. For example, male lions sometimes form coalitions to gain control of a pride of females. In some populations of Galapagos hawks, groups of males would cooperate to defend one breeding territory. The males would share matings with the female and share paternity with the offspring.

Female Desertion and Sex Role Reversal

In birds, desertion often happens when food is abundant, so the remaining partner is better able to raise the young unaided. Desertion also occurs if there is a great chance of a parent to gain another mate, which depends on environmental and populational factors. Some birds, such as the phalaropes, have reversed sex roles where the female is larger and more brightly colored, and compete for males to incubate their clutches. In jacanas, the female is larger than the male and her territory could overlap the multiple territories of up to four males.

Social Behaviors

Animals cooperate with each other in order to increase their own fitness. These altruistic, and sometimes spiteful behaviors can be explained by Hamilton's rule, which states that $rB-C > 0$ where r= relatedness, B= benefits, and C= costs.

Kin Selection

Kin selection refers to evolutionary strategies where an individual acts to favor the reproductive success of relatives, or kin, even if the action incurs some cost to the organism's own survival and ability to procreate. John Maynard Smith coined the term in 1964, although the concept was referred to by Charles Darwin who cited that helping relatives would be favored by group selection. Mathematical descriptions of kin selection were initially offered by R. A. Fisher in 1930 and J. B. S. Haldane in 1932. and 1955. W. D. Hamilton popularized the concept later, including the mathematical treatment by George Price in 1963 and 1964.

Kin selection predicts that individuals will harbor personal costs in favor of one or multiple individuals because this can maximize their genetic contribution to future generations. For example, an organism may be inclined to expend great time and energy in parental investment to rear offspring since this future generation may be better suited for propagating genes that are highly shared between the parent and offspring. Ultimately, the initial actor performs apparent altruistic actions for kin in order to enhance its own reproductive fitness. In particular, organisms are hypothesized to act in favor of kin depending on their genetic relatedness. So, individuals will be inclined to act altruistically for siblings, grandparents, cousins and other relatives, but to differing degrees.

Inclusive Fitness

Inclusive fitness describes the component of reproductive success in both a focal individual and their relatives. Importantly, the measure embodies the sum of direct and indirect fitness and the change in their reproductive success based on the actor's behavior. That is, the effect an individual's behaviors have on: being personally better-suited to reproduce offspring, and aiding descendent and non-descendent relatives in their reproductive efforts. Natural selection is predicted to push individuals to behave in ways that maximize their inclusive fitness. Studying inclusive fitness is often done using predictions from Hamilton's rule.

Kin Recognition

Genetic Cues

One possible method of kin selection is based on genetic cues that can be recognized phenotypically. Genetic recognition has been exemplified in a species that is usually not thought of as a social creature: amoebae. Social amoebae form fruiting bodies when starved for food. These amoebae preferentially formed slugs and fruiting bodies with members of their own lineage, which is clonially related. The genetic cue comes from variable lag genes, which are involved in signaling and adhesion between cells.

Kin can also be recognized a genetically determined odor, as studied in the primitively social sweat bee, *Lasioglossum zephyrum*. These bees can even recognize relatives they have never met and roughly determine relatedness. The Brazilian stingless bee *Schwarziana quadripunctata* uses a distinct combination of chemical hydrocarbons to recognize and locate kin. Each chemical odor, emitted from the organism's epicuticles, is unique and varies according to age, sex, location, and hierarchical position. Similarly, individuals of the stingless bee species *Trigona fulviventris* can distinguish kin from non-kin through recognition of a number of compounds, including hydrocarbons and fatty acids that are present in their wax and floral oils from plants used to construct their nests. In the species, *Osmia rufa,* kin selection has also been associated with mating selection. Females, specifically, will select males for mating with whom they are genetically more related to.

Environmental Cues

There are two simple rules that animals follow to determine who is kin. These rules can be exploited, but exist because they are generally successful.

The first rule is 'treat anyone in my home as kin.' This rule is readily seen in the reed warbler, a bird species that will only focus on chicks in their own nest. Interestingly, if its own kin is placed outside of the nest, a parent bird will ignore that chick. This rule can sometimes lead to odd results, especially if there is a parasitic bird that lays eggs in the reed warbler nest. For example, an adult cuckoo may sneak its egg into the nest. Once the cuckoo hatches, the reed warbler parent will feed the invading bird like its own child. Even with the risk for exploitation, the rule generally proves successful.

The second rule, named by Konrad Lorenz as 'imprinting,' states that those who you grow up with are kin. Several species exhibit this behavior, including, but not limited to the Belding's ground squirrel. Experimentation with these squirrels showed that regardless of true genetic relatedness, those that were reared together rarely fought. Further research suggests that there is partially some genetic recognition going on as well, as siblings that were raised apart were less aggressive toward one another compared to non-relatives reared apart.

Cooperation

Cooperation is broadly defined as behavior that provides a benefit to another individual that specifically evolved for that benefit. This excludes behavior that has not been expressly selected for to provide a benefit for another individual, because there are many commensal and parasitic relationships where the behavior one individual (which has evolved to benefit that individual and no others) is taken advantage of by other organisms. For cooperative behavior to be stable, it must provide a benefit to both the actor and recipient, although the benefit to the actor can take many different forms.

Within Species

Within species cooperation occurs among members of the same species. Examples of intraspecific cooperation include cooperative breeding (such as in weeper capuchins) and cooperative foraging (such as in wolves). There are also forms of cooperative defense mechanisms, such as the "fighting swarm" behavior used by the stingless bee *Tetragonula carbonaria*. Much of this behavior occurs due to kin selection. Kin selection allows cooperative behavior to evolve where the actor receives no direct benefits from the cooperation.

Cooperation (without kin selection) must evolve to provide benefits to both the actor and recipient of the behavior. This includes reciprocity, where the recipient of the cooperative behavior repays the actor at a later time. This may occur in vampire bats but it is uncommon in non-human animals. Cooperation can occur willingly between individuals when both benefit directly as well. Cooperative breeding, where one individual cares for the offspring of another, occurs in several species, including wedge-capped capuchin monkeys.

Cooperative behavior may also be enforced, where there failure to cooperate results in negative consequences. One of the best examples of this is worker policing, which occurs in social insect colonies.

Between Species

Cooperation can occur between members of different species. For interspecific cooperation to be evolutionarily stable, it must benefit individuals in both species. Examples include pistol shrimp and goby fish, nitrogen fixing microbes and legumes, ants and aphids. In ants and aphids, aphids secrete a sugary liquid called honeydew, which that ants eat. The ants provide protection to the aphids against predators, and, in some instances, raise the aphid eggs and larvae inside the ant colony. This behavior is analogous to human domestication. The genus of goby fish, *Elacatinus* also demonstrate cooperation by removing and feeding on ectoparasites of their clients. The species of wasp *Polybia rejecta* and ants *Azteca chartifex* show a cooperative behavior protecting one another's nests from predators.

Spite

Hamilton's rule can also predict spiteful behaviors between non-relatives. A spiteful behavior is one that is harmful to both the actor and to the recipient. Spiteful behavior will be favored if the actor is less related to the recipient than to the average member of the population making r negative and if rB-C is still greater than zero. Spite can also be thought of as a type of altruism because harming a non-relative, by taking his resources for example, could also benefit a relative, by allowing him access to those resources. Furthermore, certain spiteful behaviors may provide harmful short term consequences to the actor but also give long term reproductive benefits. Many behaviors that are commonly thought of as spiteful are actually better explained as being selfish, that is benefiting the actor and harming the recipient, and true spiteful behaviors are rare in the animal kingdom.

An example of spite is the sterile soldiers of the polyembryonic parasitoid wasp. A female wasp will lay a male and a female egg in a caterpillar. The eggs will divide asexually creating many genetically identical male and female larvae. Sterile soldier wasps will also develop and attack the relatively unrelated brother larvae so that the genetically identical sisters will have more access to food.

Another example is bacteria that release bacteriocins. The bacteria that releases the bacteriocin may have to die in order to do so; however most of the harm will be done to unrelated individuals who will be killed by the bacteriocin. This is because the ability to produce and release the bacteriocin is linked with immunity to it. Therefore, close relatives to the releasing cell will be less likely to die than non-relatives.

Altruism and Conflict in Social Insects

Many insect species of the order Hymenoptera (bees, ants, wasps) are eusocial. Within the nests or hives of social insects, individuals engage in specialized tasks to ensure the survival of the colony. Dramatic examples of these specializations include changes in body morphology or unique behaviors, such as the engorged bodies of the honeypot ant *Myrmecocystus mexicanus* or the waggle dance of honey bees and a wasp species, *Vespula vulgaris*.

Honeypot ant

In many, but not all social insects, reproduction is monopolized by the queen of the colony. Due to the effects of a haplodiploid mating system, in which unfertilized eggs become male drones and fertilized eggs become worker females, average relatedness values between sister workers can be higher than those seen in humans or other eutherian mammals. This has led to the suggestion that kin selection may be a driving force in the evolution of eusociality, as individuals could provide cooperative care that establishes a favorable benefit to cost ratio (rB-c > 0). However, not all social insects follow this rule. In the social wasp *Polistes dominula*, 35% of the nest mates are unrelated. In many other species, unrelated individuals only help the queen when no other options are present. In this case, subordinates work for unrelated queens even when other options may be present. No other social insect submits to unrelated queens in this way. This seemingly unfavorable behavior parallels some vertebrate systems. It is thought that this unrelated assistance is evidence of altruism in *P. dominula*.

Naked mole-rats

Cooperation in social organisms has numerous ecological factors that can determine the benefits and costs associated with this form of organization. One suggested benefit is a type of "life insurance" for individuals who participate in the care of the young. In this instance, individuals may have a greater likelihood of transmitting genes to the next generation when helping in a group compared to individual reproduction. Another suggested benefit is the possibility of "fortress defense", where soldier castes will threaten or attack intruders, thus protecting related individuals inside the territory. Such behaviors are seen in the snapping shrimp *Synalpheus regalis* and gall-forming aphid *Pemphigus spyrothecae*. A third ecological factor that is posited to promote eusociality is the distribution of resources: when food is sparse and concentrated in patches, eusociality is favored. Evidence supporting this third factor comes from studies of naked mole-rats and Damaraland mole-rats, which have communities containing a single pair of reproductive individuals.

Conflicts in Social Insects

Although eusociality has been shown to offer many benefits to the colony, there is also potential for conflict. Examples include the sex-ratio conflict and worker policing seen in certain species of social Hymenoptera such as *Dolichovespula media*, *Dolichovespula sylvestris*, *Dolichovespula norwegica* and *Vespula vulgaris*. The queen and the worker wasps either indirectly kill the laying-workers' offspring by neglecting them or directly condemn them by cannibalizing and scavenging.

The sex-ratio conflict arises from a relatedness asymmetry, which is caused by the haplodiploidy nature of Hymenoptera. For instance, workers are most related to each other because they share half of the genes from the queen and inherit all of the father's genes. Their total relatedness to each other would be 0.5+ (0.5 x 0.5) = 0.75. Thus, sisters are three-fourths related to each other. On the other hand, males arise from unfertilized larva, meaning they only inherit half of the queen's genes and none from the father. As a result, a female is related to her brother by 0.25, because 50% of her genes that come from her father have no chance of being shared with a brother. Her relatedness to her brother would therefore be 0.5 x 0.5=0.25.

According to Trivers and Hare's population-level sex-investment ratio theory, the ratio of relatedness between sexes determines the sex investment ratios. As a result, it has been observed that there is a tug-of-war between the queen and the workers, where the queen would prefer a 1:1 female to male ratio because she is equally related to her sons and daughters (r=0.5 in each case). However, the workers would prefer a 3:1 female to male ratio because they are 0.75 related to each other and only 0.25 related to their brothers. Allozyme data of a colony may indicate who wins this conflict.

Conflict can also arise between workers in colonies of social insects. In some species, worker females retain their ability to mate and lay eggs. The colony's queen will be related to her sons by half of her genes and a quarter to the sons of her worker daughters. Workers, however, are related to their sons by half of their genes and to their brothers by a quarter. Thus, the queen and her worker daughters would compete for reproduction to maximize their own reproductive fitness. Worker reproduction is limited by other workers who are more related to the queen than their sisters, a situation occurring in many polyandrous hymenopteran species. Workers police the egg-laying females by engaging in oophagy or directed acts of aggression.

The Monogamy Hypothesis

The monogamy hypothesis states that the presence of monogamy in insects is crucial for eusociality to occur. This is thought to be true because of Hamilton's rule that states that rB-C>0. By having a monogamous mating system, all of the offspring have high relatedness to each other. This means that it is equally beneficial to help out a sibling, as it is to help out an offspring. If there were many fathers the relatedness of the colony would be lowered.

This monogamous mating system has been observed in insects such as termites, ants, bees and wasps. In termites the queen commits to a single male when founding a nest. In ants, bees and wasps the queens have a functional equivalent to lifetime monogamy. The male can even die before the founding of the colony. The queen can store and use the sperm from a single male throughout their lifetime, sometimes up to 30 years.

In an experiment looking at the mating of 267 hymenopteran species, the results were mapped onto a phylogeny. It was found that monogamy was the ancestral state in all the independent transitions to eusociality. This indicates that monogamy is the ancestral, likely to be crucial state for the development of eusociality. In species where queens mated with multiple mates, it was found that these were developed from lineages where sterile castes already evolved, so the multiple mating was secondary. In these cases, multiple mating is likely to be advantageous for reasons other than those important at the origin of eusociality. Most likely reasons are that a diverse worker pool attained by multiple mating by the queen increases disease resistance and may facilitate a division of labor among workers.

Communication and Signalling

Communication is varied at all scales of life, from interactions between microscopic organisms to those of large groups of people. Nevertheless, the signals used in communication abide by a fundamental property: they must be a quality of the receiver that can transfer information to a receiver that is capable of interpreting the signal and modifying its behavior accordingly. Signals are distinct from cues in that evolution has selected for signalling between both parties, whereas cues are merely informative to the observer and may not have originally been used for the intended purpose. The natural world is replete with examples of signals, from the luminescent flashes of light from fireflies, to chemical signaling in red harvester ants to prominent mating displays of birds such as the Guianan cock-of-the-rock, which gather in leks, the pheromones released by the corn earworm moth, the dancing patterns of the blue-footed booby, or the alarm sound *Synoeca cyanea* make by rubbing their mandibles against their nest.The nature of communication poses evolutionary concerns, such as the potential for deceit or manipulation on the part of the sender. In this situation, the receiver must be able to anticipate the interests of the sender and act appropriately to a given signal. Should any side gain advantage in the short term, evolution would select against the signal or the response. The conflict of interests between the sender and the receiver results in an evolutionarily stable state only if both sides can derive an overall benefit.

Although the potential benefits of deceit could be great in terms of mating success, there are several possibilities for how dishonesty is controlled, which include indices, handicaps, and common interests. Indices are reliable indicators of a desirable quality, such as overall health, fertility, or fighting ability of the organism. Handicaps, as the term suggests, place a restrictive cost on the organisms that own them, and thus lower quality competitors experience a greater relative cost compared to their higher quality counterparts. In the common interest situation, it is beneficial to both sender and receiver to communicate honestly such that the benefit of the interaction is maximized.

Signals are often honest, but there are exceptions. Prime examples of dishonest signals include the luminescent lure of the anglerfish, which is used to attract prey, or the mimicry of non-poisonous butterfly species, like the Batesian mimic *Papilio polyxenes* of the poisonous model *Battus philenor*. Although evolution should normally favor selection against the dishonest signal, in these cases it appears that the receiver would benefit more on average by accepting the signal.

References

- Paul S. Agutter & Denys N. Wheatley (2008). Thinking about Life: The History and Philosophy of Biology and Other Sciences. Springer. p. 43. ISBN 1-4020-8865-5.
- Saint Albertus Magnus (1999). On Animals: A Medieval Summa Zoologica. Johns Hopkins University Press. ISBN 0-8018-4823-7.
- Lois N. Magner (2002). A History of the Life Sciences, Revised and Expanded. CRC Press. pp. 133–144. ISBN 0-8247-0824-5.
- Vassiliki Betty Smocovitis (1996). Unifying Biology: The Evolutionary Synthesis and Evolutionary Biology. Princeton University Press. ISBN 0-691-03343-9.
- Heather Silyn-Roberts (2000). Writing for Science and Engineering: Papers, Presentation. Oxford: Butterworth-Heinemann. p. 198. ISBN 0-7506-4636-5.
- Lois N. Magner (2005). History of the Life Sciences. New York. Basel: Marcel Dekker. p. 166. ISBN 9780824743604.

- McGreevy, Paul; Robert Boakes (2011). Carrots and Sticks: Principles of Animal Training. Darlington Press. pp. xi–23. ISBN 978-1-921364-15-0. Retrieved 9 September 2016.

- Bourg, Julian (2007). From Revolution to Ethics: May 1968 and Contemporary French Thought. McGill-Queen's Press - MQUP. p. 155. ISBN 978-0-7735-7621-6.

- Bateson, P. (1991). The Development and Integration of Behaviour: Essays in Honour of Robert Hinde. Cambridge University Press. p. 479. ISBN 978-0-521-40709-0.

- Buchmann, Stephen (2006). Letters from the Hive: An Intimate History of Bees, Honey, and Humankind. Random House of Canada. p. 105. ISBN 978-0-553-38266-2.

- Keil, Frank C.; Robert Andrew Wilson (2001). The MIT encyclopedia of the cognitive sciences. MIT Press. p. 184. ISBN 978-0-262-73144-7.

- Mercer, Jean (2006). Understanding attachment: parenting, child care, and emotional development. Greenwood Publishing Group. p. 19. ISBN 978-0-275-98217-1.

- Hoppitt, W.; Laland, K.N. (2013). Social Learning: An Introduction to Mechanisms, Methods, and Models. Princeton University Press. ISBN 978-1-4008-4650-4.

- Cummings, Mark; Carolyn Zahn-Waxler; Ronald Iannotti (1991). Altruism and aggression: biological and social origins. Cambridge University Press. p. 7. ISBN 978-0-521-42367-0.

- Nicholas B. Davies; John R. Krebs; Stuart A. West (2012). An Introduction to Behavioral Ecology. West Sussex, UK: Wiley-Blackwell. pp. 193–202. ISBN 978-1-4051-1416-5.

- Davies, Nicholas B.; Krebs, John R.; West, Stuart A. (2012). An Introduction to Behavioral Ecology. West Sussex, UK: Wiley-Blackwell. pp. 307–333. ISBN 978-1-4051-1416-5.

- Wiley, R. H. (1981). "Social structure and individual ontogenies: problems of description, mechanism, and evolution" (PDF). Perspectives in ethology. 4: 105–133. Retrieved 21 December 2012.

- Davies, Nicholas B., Krebs, John R., and West, Stuart A. An Introduction to Behavioral Ecology, 4th ed. Oxford: Wiley-Blackwell, 2012.

- Ratnieks, Francis L.W.; P. Kirk Visscher (December 1989). "Worker policing in the honeybee". Nature. 342 (6251): 796–797. Bibcode:1989Natur.342..796R. doi:10.1038/342796a0. Retrieved 26 November 2012.

Permissions

Index

www.ingramcontent.com/pod-product-compliance
Lightning Source LLC
Chambersburg PA
CBHW061320190326
41458CB00011B/3847